Climate Change and European Emissions Trading

NEW HORIZONS IN ENVIRONMENTAL LAW

Series Editors: Kurt Deketelaere, *Professor of Law and Director, Institute of Environmental and Energy Law, University of Leuven, Belgium* and Zen Makuch, *Department of Environmental Science and Technology, Imperial College, London, UK*

Environmental law is an increasingly important area of legal research. Given the increasingly interdependent web of global society and the significant steps being made towards environmental democracy in decision-making processes, there are few people that are untouched by environmental lawmaking processes.

At the same time, environmental law is at a crossroads. The command and control methodology that evolved in the 1960s and 1970s for air, land and water protection may have reached the limit of its environmental protection achievements. New life needs to be injected into our environmental protection regimes. This series seeks to press forward the boundaries of environmental law through innovative research into environmental protection standards, procedures, alternative instruments and case law. Adopting a wide interpretation of environmental law, it includes contributions from both leading and emerging European and international scholars.

Titles in the series include:

Whaling Diplomacy
Defining Issues in International Environmental Law
Alexander Gillespie

EU Climate Change Policy
The Challenge of New Regulatory Initiatives
Edited by Marjan Peeters and Kurt Deketelaere

Environmental Law in Development
Lessons from the Indonesian Experience
Edited by Michael Faure and Nicole Niessen

Finding Solutions for Environmental Conflicts
Power and Negotiation
Edward Christie

China and International Environmental Liability
Legal Remedies for Transboundary Pollution
Edited by Michael Faure and Şong Ying

Climate Change and European Emissions Trading
Lessons for Theory and Practice
Edited by Michael Faure and Marjan Peeters

Climate Change and European Emissions Trading

Lessons for Theory and Practice

Edited by

Michael Faure
Professor of Comparative and International Environmental Law, Maastricht University and Professor of Comparative Private Law and Economics, Erasmus University Rotterdam, The Netherlands

and

Marjan Peeters
Professor of Environmental Policy and Law, Maastricht University, The Netherlands

NEW HORIZONS IN ENVIRONMENTAL LAW

Edward Elgar
Cheltenham, UK • Northampton, MA, USA

Published by
Edward Elgar Publishing Limited
The Lypiatts
15 Lansdown Road
Cheltenham
Glos GL50 2JA
UK

Edward Elgar Publishing, Inc.
William Pratt House
9 Dewey Court
Northampton
Massachusetts 01060
USA

A catalogue record for this book
is available from the British Library

Library of Congress Control Number: 2008935950

PEFC
PEFC/16-33-111
CATG-PEFC-052
www.pefc.org

ISBN 978 1 84720 898 9

Typeset by Cambrian Typesetters, Camberley, Surrey
Printed and bound in Great Britain by MPG Books Ltd, Bodmin, Cornwall

Contents

PART III ALTERNATIVES AND NEW DEVELOPMENTS

PART IV CONCLUSIONS: FUTURE LOOK

Contributors

Nicolas Van Aken, University of Liège, Belgium
Alessandra Arcuri, Erasmus University of Rotterdam, The Netherlands
Chris Backes, Maastricht University, The Netherlands
Janneke Bazelmans, University of Amsterdam, The Netherlands
Erik B. Bluemel, University Law Centre of Georgetown, USA
Kurt Deketelaere, Catholique University of Leuven, Belgium
Javier De Cendra De Larragán, Maastricht University, The Netherlands
Stefano Clò, University of Bologna, Italy
Michael Faure, Maastricht University, Erasmus University of Rotterdam, The Netherlands
Giedre Kaminskaite-Salters, Norton Rose LLP, London, United Kingdom
Onno Kuik, Free University of Amsterdam, The Netherlands
Karen E. Makuch, Imperial College London, United Kingdom
Zen Makuch, Imperial College London, United Kingdom
Frans Oosterhuis, Free University of Amsterdam, The Netherlands
Marjan Peeters, Maastricht University, The Netherlands
Marijke Schurmans, Catholique University of Leuven, Belgium
Stefan Weishaar, Maastricht University, The Netherlands
Edwin Woerdman, University of Groningen, The Netherlands

Abbreviations

AAU	Assigned Amount Unit
AB	Assembly Bill
AG	Advocate General
AUS ETS	Australian Emissions Trading Scheme
BAT	Best Available Technique
BERR	Department for Business, Enterprise and Regulatory Reform
BNA International Environment Daily	Bureau of National Affairs International Environment Daily
BRC	Better Regulation Commission
BREFS	Best Available Technology Reference Documents
BVerwG	Bundesverwaltungsgericht
CA ETS	Californian Emissions Trading Scheme
CCA	Climate Change Agreement
CCAP	Center for Clean Air Policy
CCL	Climate Change Levy
CCS	Carbon Capture and Storage
CCX	Chicago Climate Exchange
CDM	Clean Development Mechanism
CEPS	The Centre for European Policy Studies
CER	Certified Emission Reduction
CERT	Carbon Emissions Reduction Target
CETM	Confederación Española de Transporte de Mercancías
CFI	Court of First Instance
CGE	Computable General Equilibrium
CGM	Compagnie Générale Maritime
CH_4	Methane
Chicago Convention	1944 Convention on International Civil Aviation
CHP	Combined Heat and Power
CIRED	International Research Center on Environment and Development
CITL	Community Independent Transaction Log
CJEG	Cahiers Juridiques de l'électricité et du gaz
CMA	Compagnie Maritime d'Affrètement

CNSD	Consiglio Nazionale degli Spedizionieri Doganali
CO_2	Carbon Dioxide
CO_2e	Carbon Dioxide Equivalent
COM	Commission
CPUC	California Public Utility Commission
CSE	Centre for Sustainable Energy
CT	Carbon Trust
Czech Rep./Cz Rep	Czech Republic
DART	Dynamic Applied Regional Trade
DEFRA	Department for Environment, Food and Rural Affairs
DER	Dwelling Emission Rate
dETS	domestic Emissions Trading Scheme
dnc	declared net capacity
DOE	Department of Environment Northern Ireland
DP	Direct Participant
DTI	Department for Trade and Industry
EC	European Community
ECJ	European Court of Justice
ECR	European Court Reports
EDLE	European Doctorate in Law and Economics
EEA	European Environment Agency
EEC	European Economic Community
EELR	European Energy and Environmental Law Review
EFTA	European Free Trade Area
EHA	Enhanced Capital Allowances
EII	Energy Intensive Industries
E.L.R.	European Law Review
EP	European Parliament
EPA	Environmental Protection Agency
EPRI	Electric Power Research Institute
EPS	Emission Portfolio Standards
ERU	Emission Reduction Unit
ESS	Energy Supply Sectors
EST	Energy Savings Trust
ETF	Environmental Transformation Fund
ETG	UK Emissions Trading Group
ETR	Emissions Trading Registry
ETS	Emissions Trading Scheme
ETUC	European Trade Union Confederation
EU	European Union
EUAs	European Union emission allowances
EU ETS	European Union's Emissions Trading Scheme

FEEM	Fondazione Eni Enrico Mattei
GAD	Global and Atmospheric Division
GATT agreement	General Agreement on Tariffs and Trade
GHG emissions	Greenhouse Gas emissions
GLA	Greater London Authority
HAP	Horticulture Assistance Package
HFC	Hydrofluorocarbon
H.R.	House of Representatives
IBGE	Institut bruxellois pour la gestion de l'environnement
ICAO	International Civil Aviation Organisation
ICAP	International Carbon Action Partnership
IEA	International Energy Agency
IFIEC	International Federation of Industrial Energy Consumers
INECE	International Network for Environmental Compliance and Enforcement
IPCC	United Nations Intergovernmental Panel on Climate Change
IPPC	Integrated Pollution Prevention and Control
IPTS	Institute for Prospective Technological Studies
ISO	Independent System Operator
ISTAS	Instituto Sindical de Trabajo, Ambiente y Salud
ITL	International Transaction Log
JEEPL	Journal for European Environmental & Planning Law
JI	Joint Implementation
JV ETS	Japanese Voluntary Emissions Trading Scheme
KP	Kyoto Protocol
lCER	long-term CER (Certified Emission Reduction)
LEZ	Low Emission Zone
LSE	Load-Serving Entity
LULUCF	Land use, Land-Use Change and Forestry
Lux./Lux	Luxembourg
MAC	Marginal Abatement Cost
METRO	Maastricht European Institute for Transnational Legal Research
MS	Member States
Mt.	Million tons
MW	Megawatt
MWh	Megawatt hours
N_2O	Nitrous Oxide
NA	Negotiated Agreement
NAP	National Allocation Plan
NBER	National Bureau of Economic Research

NCCR	Swiss National Centre of Competence in Research
NERA	National Economics Research Associates
NFFO	Non-Fossil Fuel Obligation
NGO	Non Governmental Organization
NI-NFFO	Northern Ireland NFFO (Non-Fossil Fuel Obligation)
NL	Netherlands
NO_x	Nitrogen Oxide
NRP	Dutch National Research Programme on Global Air Pollution and Climate Change
NSW GGAS	The New South Wales Greenhouse Gas Abatement Scheme
NZ ETS	New Zealand ETS
OCC	Office of Climate Change
OECD	Organization for Economic Co-operation and Development
OfGEM	Gas and Electricity Markets Authority
OJ	Official Journal
OTC	Ozone Transport Commission
OTH	Other Demand Sectors
PCT	Personal Carbon Trading
PFC	Perfluorocarbon
PJM	Pennsylvania-New Jersey-Maryland
PNA	Plan National d'Allocation
POLES	Prospective Outlook on Long-term Energy Systems
PPC	Pollution Prevention and Control
PRIMES	Price Induced Model of the Energy System
PSR	Performance Standard Rate
R&D	Research and Development
RECLAIM	Regional Clean Air Incentives Market
RFF	Resources for the Future
RGGI	Northeast Regional Greenhouse Gas Initiative
RILE	Rotterdam Institute of Law and Economics
RJEP	La revue juridique de l'entreprise publique
RMU	Removal Unit
ROS	Renewables Obligation (Scotland)
RPS	Renewable Portfolio Standard
RSA	Royal Society for the encouragement of Arts, Manufacturers and Commerce
RTF	Renewable Transport Fuel
RTFO	The Renewable Transport Fuel Obligations Order
RTO	Regional Transmission Organization
RuG	Rijksuriversiteit Groningen

RWE	Rheinisch-Westfälische Elektrizitätswerke AG
SCM agreement	Agreement on Subsidies and Countervailing Measures
SDA	Social Development Agency
SF6	Sulphur Hexafluoride
SIC	Standard Industrial Classification
SMEs	Small and Medium sized Enterprises
UBA	Umweltbundesamt
UK	United Kingdom
USA	United States of America
t	Ton
tCER	temporary CER (Certified Emission Reduction)
TER	Target Emission Rate
TS	Trading Sectors
UKCIP	UK Climate Impacts Programme
UNFCCC	United Nations Framework Convention on Climate Change
WCI	Western Climate Initiative
WRCAI	Western Regional Climate Action Initiative
WTO	World Trade Organization
yr	year
ZfE	Zeitschrift für Energiewirtschaft
ZuG	Zuteilungsgesetz

PART I

Introduction to the book

1. Introduction

Michael Faure and Marjan Peeters

1. PROBLEM DEFINITION: REASONS FOR THIS BOOK

Emissions trading can no longer be seen as just an interesting theoretical exercise: this market-based approach has developed an increasingly important role, first within the environmental law framework of the USA and later also within that of the EU. The instrument of emissions trading has been applied in order to combat significant environmental problems like acid rain, ozone-depleting substances and climate change. Regarding the two latter problems, the instrument is applied both on the international level as well as on national levels.

Notably for the greenhouse gas emissions problem, emissions trading seems to be very much suited to reaching the necessary reductions in a cost-effective way. In Europe there is now some experience with emissions trading as a result of the implementation of the greenhouse gas Emissions Trading Scheme (EU ETS).[1] The EU ETS is the biggest regional emissions trading system established thus far. The first trading period started on 1 January 2005 and finished on 31 December 2007; the second trading period, during which this book will be published, runs till 2013 and thus comprises five years. In the meantime, only three years after the start of the first trading period, the European Commission released on 23 January 2008 a proposal for a major revision of the EU ETS, which should change the system from 2013 onwards.[2] This proposal includes challenging new topics, like auctioning of allowances, an additional and gradually declining free allocation of allowances on the EU level, and a specific provision for industries facing international competition. The experience with the EU ETS had already

[1] Directive 2003/87/EC of the European Parliament and of the Council of 13 October 2003 establishing a scheme for greenhouse gas emission allowance trading within the Community and amending Council Directive 96/61/EC, OJ L 275/32 25.10.2003.

[2] Proposal for a directive of the European Parliament and of the Council amending Directive 2003/87/EC so as to improve and extend the greenhouse gas emission allowance trading system of the Community, COM(2008)16, Brussels 23.1.2008.

started before 2005, as important decisions regarding the distribution of the tradable allowances to the covered industries needed to be taken before the start of the first trading period. Moreover, the design of the legislative framework necessary for emissions trading was an interesting exercise too, leading to all kinds of new questions. Strikingly enough, those questions, which were in fact quite new for the European governments because there was thus far hardly any experience with this market-based instrument, needed to be answered in an extremely short time period because of the firm deadline set by the politicians aiming to have the EU ETS established before the start of the first commitment period of the Kyoto Protocol.

Moreover, the EU intends to expand its current greenhouse gas emissions trading regime, thereby indeed stressing that this instrument is the core climate change instrument for the EU.[3] Certain member states, like the UK and The Netherlands, intend to adopt domestic measures for applying the instrument to other sources and other pollution problems. In the same vein, the idea of citizens' budgets for carbon emissions is also emerging.[4] Meanwhile, in the USA several initiatives for greenhouse gas emissions trading have been taken at a regional level. In addition, industries initiate voluntary emissions trading activities, not least to prevent future liability claims. In addition, the setting up of a legal framework for trustworthy voluntary emission offsets needs to be considered as well.

The first European experiences with trading of greenhouse gas allowances have thus led to a lot of questions from various perspectives.[5] In this respect it is worthwhile analysing the experience with the ETS in a critical way, aiming to answer the question of what can be learned from this experience at theoretical and policy level, and what lessons thus can be learned for the future application of the instrument. The purpose of this book is to focus on the domestic applications of the emissions trading instrument, especially for greenhouse gases, thereby learning from fresh experiences, critically examining the current practice, and looking to the future for new challenges for the instrument. It may be clear that both lawyers and economists have already questioned the effectiveness of the ETS from various perspectives. For example, lawyers have been critical with regard to the rush for adopting the instrument, and have questioned the flexibility allowed as far as the national

3 See about EU climate change policy Bothe and Rehbinder (2005) (part II of the book); Deketelaere and Peeters (2006).

4 Starkey and Anderson (2005).

5 See an earlier examination of the US and European greenhouse gas emissions trading developments Hansjürgens (2005). See for a specific examination of allocation issues: Ellerman et al. (2007). A description of the development and content of the initial EU ETS has been elaborated on in Delbeke (2006).

allocation plans are concerned, pointing to possible distorting effects for competition and thus for the internal markets. Economists have critically questioned whether the current cap-and-trade system implied in the ETS can be considered as a cost-effective, let alone efficient, tool to reach the targets of reducing climate change. Moreover, the book should not only take into account these critical perspectives on the ETS from a legal and economic perspective. There are, in addition, experiences with emissions trading in other legal systems (like the US) which can be usefully taken into account in rethinking the effectiveness of this ETS. Indeed, the goal of this book is not only to analyse the effectiveness of the ETS, but equally to see what the current experience with the ETS can teach the existing literature with respect to emission trading. In addition, at the policy level, the book also aims to collect some lessons for the future design of the instrument.

Hence, the book will discuss the regulatory schemes for greenhouse gas emissions within the EU and the US. The design and implementation of the legal framework for emissions trading still raises important questions. First of all, we examine why different choices have been made in setting up the current schemes, and what those differences mean for the legal and economic effects in practice. Secondly, we question whether design options thus far only discussed in literature should be applied in practice, like the concept of auctioning, and the benchmark and trade option. It could even be asked whether the emissions trading system is indeed a better solution than the other highly recommended instrument of taxation. In a broader context, one should not forget that emissions trading is part of a comprehensive environmental law system. In that respect, the question emerges of how the instrument relates to other important instruments of environmental law, like integrated licensing.

The book has a theoretical and a policy perspective. The experience with the ETS can usefully be applied to existing theories on emissions trading. Thus, this experience can constitute a fruitful test case to examine to what extent the predictions in the literature concerning the effectiveness of emissions trading have materialized as a result of the ETS. Moreover, the actual experience with the ETS may also allow the refinement of existing theoretical insights and the procurement of more detailed knowledge about the optimal shape and structure of this particular environmental instrument. Indeed, some of the weaknesses of the ETS may thus contribute to a better design of emissions trading in the future. The latter point immediately shows that this book also has a clear policy objective since, in equal measure, it aims at formulating suggestions for improving the current emissions trading scheme concerning greenhouse gases.

2. METHODOLOGY

2.1 Multidisciplinary

As we already indicated, the whole concept of emissions trading is essentially an invention by economists.[6] However, the effectiveness of the emissions trading scheme may to a large extent depend upon the specific way in which the system has been put into a legislative framework. In that respect particular legal aspects, for example concerning the procedure and method of the allocation mechanism, the way in which trading is controlled or the enforcement, are of particular importance. Furthermore, case law, as well, can influence the operating of the scheme in particular cases. Hence, a book that aims at analysing the effectiveness of the European emissions trading scheme for greenhouse gases inevitably has to choose a multidisciplinary approach. Combining a legal and economic approach is also useful since it allows many contributors to use the so-called 'law and economics' methodology to analyse specific aspects of the emissions trading scheme. Indeed, this particular methodology has analysed to what extent legal rules can be considered as promoting efficiency and has equally indicated under what kind of particular conditions one can expect emissions trading to be welfare improving.

The economic approach chosen by various contributors to this book combines classic environmental economic analysis with the previously mentioned law and economics approach. For example, to some extent economic insights are used to analyse the economic consequences of the choice for grandfathering as allocation mechanism rather than auctioning. Other contributors use economic tools to compare, for example, predictions made before the entry into force of the emissions trading scheme with the actual development of the scheme (inter alia looking at prices) after the scheme had been functioning for some time.

This multidisciplinary approach, combining a legal and economic perspective, thus allows a few modest conclusions on the relative effectiveness of the emissions trading scheme. However, as the contributions in the book make clear, one has to be very cautious about drawing policy conclusions on the basis of an analysis of, for instance, the development of the price of a ton of CO_2. This development alone does not necessarily provide hard proof that the emissions trading scheme was either effective or ineffective in reaching

---

6 There is an ample economic literature about emissions trading. See for instance the important work of Tietenberg (1985) and for a further overview of economic literature his website http://www.colby.edu/personal/t/thtieten/tradable_permits.htm. See for a concise overview of law and economics literature Faure (2008).

particular policy goals (more particularly the reduction of CO_2 emissions as agreed to in the Kyoto Protocol). The reason is that it remains often difficult to show that particular effects are necessarily the direct consequence of a policy instrument chosen, in this particular case emissions trading. Another reason to be careful in this respect is that even if it could be shown on the basis of economic data that emissions trading would have had the effect of reducing emissions this does not necessarily imply that it is henceforth also an optimal instrument. The latter would imply that a comparison with other instruments, like taxation, is also made. Some contributors in this book hint at other possible instruments to achieve emission reductions (like inter alia taxation), but these remarks unavoidably remain largely speculative since (at least within the European Union) there is no empirical evidence concerning the effectiveness of a tax system which could be used to analyse the comparative effectiveness of taxation as a policy tool to achieve emission reductions.

2.2 Legal Interdisciplinary

Also within the legal discipline itself many approaches have been chosen within this book to analyse the effectiveness of the emissions trading scheme. For example, some authors used the traditional environmental legal literature with respect to instrument design to analyse the effectiveness of the current design of the emissions trading scheme. An important point of view to analyse the emissions trading scheme is the role that legal principles could play. In that respect, for example, the question arises whether the allocation method of grandfathering chosen in the ETS is in conformity with the polluter-pays principle. More broadly the question also arises whether generally legal principles could serve as a tool in guiding the policy maker when making difficult distributional choices in climate change policy.

The effects of an emissions trading scheme obviously go far beyond environmental law. Hence, the question not only arises to what extent the emissions trading scheme is, given its particular legal design, able to reach the policy goals given. Particular choices also have important implications from a competition law perspective. Hence, the question, for example, arises as to whether the choice for a particular allocation mechanism (more particularly grandfathering) can be reconciled with EU rules concerning state aid.

Moreover, the analysis of the legal aspects of the emissions trading scheme can of course not be limited to an analysis of the legal framework by merely analysing the contents of the EU directive and related EU policy documents and guidelines. More particularly given the importance of legal principles, the question arises as to what extent the judiciary can play its important role in, on the one hand, guaranteeing the effectiveness of the emissions trading scheme and, on the other hand, guaranteeing that the emissions trading scheme still respects basic legal principles following from the rule of law. The question is of course not merely theoretical, since both with the EU directive itself as well as in the decisions at the level of the national member states (by means of national allocation plans and national allocation decisions) decisions may have been taken that to a large extent can affect the rights of actors involved. If they feel that where room for interpretation resulting from ambiguity is possible as well, they will inevitably call on the court system in an attempt to correct decisions which they experience as unfair. Indeed, both at the level of national member states as well as at EU level, interesting case law has meanwhile emerged that provides answers to some of these and other questions. An analysis of the emissions trading directive therefore necessarily also needs to address the question of to what extent the court system has been able to interpret the emissions trading scheme as developed in the directive in such a way that its environmental effectiveness is optimized, whereas on the other hand the interest of actors involved is not jeopardized in an unreasonable way. The question of course also arises whether courts, when asked to answer this necessarily vague question, call for examples of legal principles as an interpretation guideline.

Finally, the legal perspective should not only address the regulatory framework and case law, but also pay attention to the dynamic perspective, thus addressing the question of to what extent the policy maker (and in this particular case more particularly the national member states deciding on allocation plans, or, following the proposal to revise the directive, the Commission or EU legislator itself) is entitled to adapt policy decision concerning the allocation of the tradable allowances to changing circumstances. It is this dynamic perspective which is included in the complicated question mostly referred to as the admissibility of so-called *ex post* adjustments. The latter question is of particular interest since the opinions concerning its admissibility seem to be quite diverging, at least when one compares the opinion of the European Commission (largely negative towards *ex post* adjustments) with opinions in some member states (and recently also supported by case law).

2.3 Comparative Approach

This book clearly chooses not only a multidisciplinary, but also a legal interdisciplinary approach. It places emphasis on legal comparison as well. The need to do so when addressing emissions trading seems obvious: this book largely focuses on the European emissions trading scheme as developed within the framework of the EU. However, the particular implementation of the initial EU ETS depends to a large extent on the way in which member states deal with the emissions trading directive and more particularly via national allocation plans. There interesting differences may appear, also resulting from differences in case law, for example with respect to the mentioned issue of *ex post* adjustments.

However, a comparison should not only take place between EU law and the (varying) approaches in some member states. It also seems interesting to take one particularly interesting member state and devote an entire chapter to it. This is particularly the case for the UK. As the chapter on the UK will show, this legal system is of particular importance, not only for being one of the first to establish a (national) emissions trading scheme (hence giving rise to interesting questions concerning the integration between the EU and the national emissions trading scheme), but also because of a wide experience as well with tools other than emissions trading as instruments to fight climate change. More particularly, the seeming success story concerning so-called climate change agreements made it worthwhile paying specific attention to the UK. The chapter also illustrates how difficult the design of climate change policy becomes: the comprehensiveness between EU law and national law, and between different applicable regulatory instruments, is a complicated issue for the legislative institutions.

A comparison with the United States was interesting as well since the US has some regional greenhouse gas trading regimes where (given the absence of a federal trading scheme) specific problems arise of so-called emissions leakage. The original solutions worked out in several of the regional US regimes are, within a comparison with Europe, highly interesting as well. Moreover, the issue of carbon leakage is also one of the core points of attention within the major revision of the EU ETS, as the proposal includes a specific regime for the energy-intensive sectors or sub-sectors being exposed to significant risks of carbon leakage However, the determination of these sectors, and the design of the specific approach, are yet to be done.

3. FRAMEWORK

The project originated within the Maastricht European Institute for Transnational Legal Research (METRO) to which the two editors of this book and many of the authors are connected.[7] Many of the European researchers who contributed to the book also participate within the transboundary environmental law programme of the Ius Commune Research School.[8] The Ius Commune Research School is a collaboration between the Universities of Amsterdam, Leuven, Maastricht and Utrecht and focuses on the role of law in integration processes.

---

7 See www.rechten.unimaas.nl/metro.
8 See www.iuscommune.eu.

Many researchers connected to both METRO and the transboundary environmental law group of the Ius Commune Research School are interested in environmental law and more particularly climate change issues. The current book is in that respect building upon earlier projects with Edward Elgar. For example, after a conference on 'Institutions and Instruments to Control Global Climate Change' held in Maastricht in June 2001, resulting in a publication (M. Faure, J. Gupta and A. Nientjes (eds), *Climate Change and Kyoto Protocol. The Role of Institutions and Instruments to Control Global Change*, Edward Elgar Publishing, Cheltenham, 2003) subsequent projects focused on the role of environmental law in developing countries, more specifically paying attention to the role of market-based instruments (M. Faure and N. Niessen (eds), *Environmental Law in Development. Lessons from the Indonesian Experience*, Edward Elgar Publishing, Cheltenham, 2006) and on EU climate change policy (M. Peeters and K. Deketelaere (eds), *EU Climate Change Policy. The Challenge of New Regulatory Initiative*, Edward Elgar Publishing, Cheltenham, 2006). The current book focuses specifically on the European emissions trading scheme, thus to a large extent builds upon this earlier research.

4. STRUCTURE OF THE BOOK

As the table of contents shows, the book is divided into four parts and fourteen chapters. This first part contains this editorial foreword drafted by the editors, followed by a general introduction concerning the legislative choices within the European greenhouse emissions trading scheme by Marjan Peeters in chapter 2.

Part 2 discusses the greenhouse gas emissions trading system in the EU from a critical economic and legal perspective. Javier De Cendra de Larragán addresses the allocation of greenhouse gas allowances in the EU from the perspective of legal principles and addresses the issue of harmonization (chapter 3). Nicolas Van Aken discusses (in chapter 4) the possibilities of going to court in the case of emissions trading, followed by an analysis of already-existing case law. Edwin Woerdman, Stefano Clò and Alessandra Arcuri discuss the present design of the EU ETS and more particularly its compatibility with the polluter-pays principle from a legal and economic perspective (chapter 5). Next, Stefan Weishaar discusses the relationship between the EU greenhouse gas emissions trading scheme and competition law (chapter 6). The complicated issue of the admissibility of *ex post* interventions in the present EU ETS is addressed by Chris Backes, Kurt Deketelaere, Marjan Peeters and Marijke Schurmans in chapter 7. They compare the position of the European Commission concerning *ex post* interventions with the way some case law in member states has dealt with it as well as with the important ruling of the Court of First Instance of 7 November 2007. The last paper in this part, by Onno Kuik and Frans Oosterhuis, provides some preliminary elements of the economic impacts of the EU ETS (chapter 8).

Part 3 pays attention to several new developments at the EU level and also discusses a few alternatives and specific case studies. Erik B. Bluemel discusses regional emissions trading initiatives, thereby specifically addressing means for preventing GHG leakage in the US (chapter 9). Karen E. MacDonald and Zen Makuch introduce us to the components of a domestic climate change regulatory and policy framework, by elaborating on the package of climate change policy initiatives in the UK in their discussion (chapter 10). An interesting question from a legal perspective is also the possible linking of different domestic or regional emissions trading schemes, for instance the linking between the EU ETS and regional emissions trading schemes within the US. This complicated issue is addressed by Janneke Bazelmans in chapter 11. Finally a few recent evolutions are discussed, one of them being the expansion of the EU emissions trading scheme to emissions resulting from aviation. Particular problems that arise when applying the EU ETS to aviation emissions are discussed by Giedre Kaminskaite-Salters in chapter 12. Given the fact that the European Commission in its latest proposals provided for auctioning as an allocation mechanism for greenhouse gases, one specific chapter is devoted to the design issues related to the auctioning of greenhouse gases. Stefan Weishaar thus addresses both legal and economic questions relating to the use of auctioning in chapter 13.

Part 4 provides for a few conclusions and an outlook to the future and contains chapter 14 with concluding remarks from the editors.

5. CONTRIBUTORS

As we mentioned above, many of the contributors have worked together either on previous projects or with the editors. Javier De Cendra De Larragán, Michael Faure, Marjan Peeters and Stefan Weishaar are all connected with the Maastricht European Institute for Transnational Legal Research (METRO). They all participate in the Ius Commune Research School as well. The same is the case for other contributors who are connected with partners within the Ius Commune Research School like Kurt Deketelaere and Marijke Schurmans (Catholique University of Leuven), Nicolas Van Aken (Liège) and Janneke Bazelmans (University of Amsterdam); Giedre Kaminskaite-Salters is a solicitor at Norton Rose LLP (London) undertaking Ph.D. research at METRO under supervision of the editors of this book. Karen E. McDonald and Zen Makuch are connected with Imperial College London; Erik B. Bluemel with

University of Denver Sturm College of Law. We also want to mention that several contributors are connected with member institutions of the IUCN Academy of Environmental Law. Both Maastricht University (METRO), The Catholic University of Leuven and Imperial College London are members of this worldwide organization aimed at the further development of environmental law.[9] The editors have worked together on other projects with Alessandra Arcuri (Erasmus University Rotterdam), Edwin Woerdman (University of Groningen) and Stefano Clò (University of Bologna), as well as with Onno Kuik and Frans Oosterhuis (Free University of Amsterdam).

A complete list of contributors and their affiliation is provided following the table of contents.

6. WORD OF THANKS

As editors of this book we are grateful to all contributors for their willingness to participate in this highly interesting and challenging project and for meeting the stringent deadlines we imposed upon them.

The METRO Institute has for many years received support from a consortium of industries for carrying out research into the legal and economic aspects of emissions trading.[10] Moreover, The Netherlands Ministry of the Environment (VROM) sponsored a research team which evaluated the reform of environmental law in The Netherlands (structurele evaluatie milieuwetgeving – STEM) in which other partners inter alia the Free University of Amsterdam (to which Onno Kuik and Frans Oosterhuis are connected), also participated.[11] Some of the papers presented in this book, like chapter 6 on the compatibility of the EU greenhouse gas emissions trading scheme with competition law, chapter 7 on *ex post* interventions and chapter 13 on auctioning, are at least partially a follow-up to research performed earlier for this consortium of industries. We are grateful for the financial support provided and more particularly for the fact that our partners always allowed us to (which may seem obvious but is unfortunately not always) execute our research in full academic independence. A special word of thanks in this respect we owe to Mr. Vianney Schyns (of USG) for his never-ending efforts to support our research initiatives and provide us with challenging feedback on our research results.

---

[9] See www.iucnael.org.

[10] See for further information the METRO website, www.rechten.unimaas.nl/metro under contract research.

[11] See the website, in Dutch; www.evaluatiemilieuwetgeving.nl.

We owe thanks as well to Chantal Kuijpers and Yleen Simonis of the secretariat of the Maastricht European Institute for Transnational Legal Research (METRO) for editorial assistance in the preparation of this book for publication. We owe special thanks to our research assistants Franziska Weber and Escada Kerckhoffs who reviewed the footnotes and the referencing. Finally we are most grateful to our publisher Edward Elgar for their kind professional and efficient support in the publication of this book.

The texts were finalized in April 2008, thus developments after that date could not be taken into account.

Michael Faure and Marjan Peeters
Maastricht, June 2008

REFERENCES

Bothe, M. and E. Rehbinder (eds.) (2005), *Climate Change Policy*, Eleven international publishing.

Deketelaere, K and M. Peeters (eds.) (2006), *EU Climate Change Policy: The Challenge of New Regulatory Initiatives*, Cheltenham, Edward Elgar.

Delbeke, J. (2006) (ed.), *EU Energy Law, Volume IV: EU Environmental Law: The EU Greenhouse Gas Emissions Trading Scheme*, Leuven, Claeys & Casteels.

Ellerman, D., A. Barabar, K. Buchner and C. Carraro (2007) (eds.), *Allocation in the European Emissions Trading Scheme, Rights, Rents and Fairness*, Cambridge, Cambridge University Press.

Faure, M. (1998), *Environmental Regulation, Encyclopedia of Law and Economics*, http://users.ugent.be/~gdegeest/2300book.pdf, version 1998, update forthcoming in 2008 (website visited 8 June 2008).

Hansjürgens, B. (2005), *Emissions Trading for Climate Policy*, Cambridge, Cambridge University Press.

Starkey, R. and K. Anderson (2005), *Domestic Tradable Quotas, a Policy Instrument for Reducing Greenhouse Gas Emissions from Energy Use*, Tyndall Centre for Climate Change Research, Norwich, UK

Tietenberg, Th.H. (1985), *Emissions Trading: an Exercise in Reforming Pollution Policy,* Washington D.C., Resources for the Future.

www.rechten.unimaas.nl/metro.

www.iuscommune.eu.

www.iucnael.org.

http://www.colby.edu/personal/t/thtieten/tradable_permits.htm.

www.evaluatiemilieuwetgeving.nl.(in Dutch)

PART II

Greenhouse gas emissions trading in the EU

2. Legislative choices and legal values: considerations on the further design of the European greenhouse gas Emissions Trading Scheme from a viewpoint of democratic accountability

Marjan Peeters

1. INTRODUCTION

1.1 Aim of This Chapter

Emissions trading is now widely acknowledged as the major instrument for regulating greenhouse gas emissions. The *effective and efficient* control of greenhouse gas emissions through the issuance of a restricted amount of tradable permits is increasingly seen as an attractive approach. However, *the specific design* of this instrument, for which different models are available, raises many questions from an economic and legal perspective. This book focuses on how the emissions trading instrument is being applied and will be applied for regulating greenhouse gases in the European legal order. It has become clear that Europe too is seeking the correct modeling for the instrument: the European Commission already proposed a drastic revision only a few years after the initial greenhouse gas emissions trading scheme started to operate in 2005.[1] Following this proposal of the Commission of 23 January 2008, important legislative decisions need to be undertaken by the Council and the European Parliament.[2] We are however still at the stage of building understanding of the different design options and the related economic effects and

[1] Directive of the European Parliament and of the Council of 13 October 2003 establishing a scheme for greenhouse gas emissions allowance trading within the Community and amending Council Directive 96/61/EC, OJ L 275/32 25.10.2003.

[2] Proposal for a directive of the European Parliament and of the Council amending Directive 2003/87/EC so as to improve and extend the greenhouse gas emissions allowance trading system of the Community, COM(2008)16, Brussels 23.1.2008.

legal aspects of the instrument.[3] Moreover, emissions trading is not to be seen as a superior instrument, but as an attractive option that needs to be examined and to be compared with other approaches, like command and control and taxation. It would indeed be wrong to assume that emissions trading would be the unique approach to be applied as the one and only world-wide regulatory approach.[4] Other instruments, for instance border tax adjustments, instruments like labeling and taxation, are interesting to examine as well.

Within the EU context, it is nevertheless clear that, with the adoption of directive 2003/87, emissions trading has become a key instrument of EU climate change policies, and literature in principle supports the idea of applying emissions trading for greenhouse gas emissions in this regional context. Nonetheless, the optimal design of the emissions trading instrument for specifically the EU has not been crystallized yet, and this chapter aims at enhancing the understanding of designing an emissions trading scheme for greenhouse gases within the EU context. After having emphasized the need for a legal analysis of the emissions trading instrument, it will present some important design options for that instrument. Subsequently, it will discuss the proposed major revision of the present scheme from one specific aspect that was raised in the 'reform' literature by Bruce A. Ackerman and Richard B. Stewart, which is what they call the 'democratic case' of emissions trading, which we will take further and call democratic accountability.[5] We will elaborate on their argument that emissions trading in fact contributes to the democratic accountability of environmental law, and will review how this argument can be understood in view of the present proposal to change the initial European greenhouse gas emissions trading scheme. By doing so, we introduce a value for assessing the design of the emissions trading instrument that has been under-explored thus far in the literature concerning European greenhouse gas emissions trading. Remarkably, this exercise shows us that the initial greenhouse gas emissions trading directive facilitating national governments to allocate tradable rights is not that bad at all from the perspective of democratic accountability. On a more general level, this discussion shows that we are still building a framework of criteria to assess the emissions trading instrument, in which different economic and legal perspectives need to be balanced.

---

3 Also in the USA there is an ongoing debate about the design of emissions trading models and moreover the additional use of technological standards for air pollution notably by SO_2, NO_x and PM, see Brian Potts (2007).

4 As argued by Geert van Calster (2008).

5 Ackerman and Stewart (1988).

1.2 Increasing Attention to Emissions Trading and the Need for Analysis

While the European greenhouse gas emissions trading system is operating in its current form, some developments already indicate a possible broader use of emissions trading in future European environmental policy. Firstly, as will be discussed in this book, the Commission has proposed the expansion of the present greenhouse gas emissions trading scheme to other greenhouse gases and to other sectors, like aviation. Secondly, following the Green Paper presented by the Commission called 'Market-based instruments for environment and related policy purposes', a general discussion was held among the European institutions concerning the use of market-based instruments, including emissions trading, in different environmental policy areas within the EU. The Green Paper mentions the possible use of emissions trading not only for air pollution (in a broader sense than climate change) but also for habitat policies.[6] The recently revised directive on ambient air quality and cleaner air for Europe already specifically mentions the possibility of reducing air pollution through the use of economic instruments, such as taxes, charges and emissions trading.[7] Such schemes could be developed on the national level, although a transnational approach could be attractive because of larger economic benefits. At the same time, we already see some national applications, like the NO_x emissions trading regime for industries that has been operating in the Netherlands since 2005.[8] Such national applications are however limited because the covered industries need to comply with the permit requirements as requested by the IPPC-directive, which means that the use of the best available technology needs to be followed.[9] Thirdly, the instrument of tradable permits has also emerged within the renewable energy policies of some

---

6 European Commission, COM(2007)140.

7 Directive of the European Parliament and of the Council on ambient air quality and cleaner air for Europe. Annex, B(3)(g) (publication in Official Journal pending during writing this article, see for the legislative procedure http://www.europarl.europa.eu/oeil/file.jsp?id=5287672).

8 See for a description: European Environmental Agency, Market-based instruments for environmental policy in Europe, Technical report 8/2005, http://reports.eea.europa.eu/technical_report_2005_8/en, p. 23.

9 Directive 2008/1/EC of the European Parliament and of the Council of 15 January 2008 concerning integrated pollution prevention and control (codified version), OJ 29.1.2008, L24/8, article 9(4) says that permit conditions in the form of emission limit values and equivalent parameters and technical measures shall be based on the best available techniques, without prescribing the use of any technique or specific technology, but taking into account the technical characteristics of the installation concerned, its geographical location and the local environmental conditions.

member states, specifically in the form of tradable certificates that represent investments in renewable energy. Such certificates can be used by electricity producers in order to comply with the commitment to deliver a certain percentage of the total electricity production from renewable sources. In this field, the instrumental approach has not been harmonized within the EU, because it is still not sufficiently clear whether this instrument is to be preferred above a feed-in tariff system. The experience with on the one hand quantity-based permit-trading and on the other hand price-based instruments (subsidies, feed-in tariffs) does not, according to the Commission, determine which instrument should be preferred, as both kinds of instruments are expected to have the same economic efficiency.[10] Hence, the proposal for a directive on the promotion of the use of energy from renewable sources, also released on 23 January 2008, contains the flexibility of leaving each member state the choice of whether to have a national-based support scheme, or to trade on the basis of such certificates, which are called Guarantees of Origins (certificates proving the renewable origin of energy).[11]

The real challenge with emissions trading is to go further than just mentioning the possible use of emissions trading, as is being done in the Green Paper on market-based instruments. How the instrument could be designed for a specific problem needs to be explored, and how it then can be judged against other possible regulatory approaches like subsidies, taxes, or classical command and control instruments. When such a design for emissions trading is be taken up, the following core points will need to be reviewed:

- the level of environmental protection to be ensured;
- the identification of the tradable permit and related aspects (content of the permit, duration, legal status, ownership, tradability);
- the choice of the model and the design of implementing procedures for distributing the tradable permits, including the establishment of necessary administrative competences for the allocation of the rights and competences to intervene within the market;
- the fine-tuning of the emissions trading approach with other regulatory approaches, especially when local effects of the environmental problem are at stake;
- the establishment of a reliable monitoring scheme together with an effective enforcement mechanism.

---

10 European Commission, Commission Staff Working Document, The support of electricity from renewable energy sources, accompanying document to the Proposal for a directive on the promotion of the use of energy from renewable sources, Brussels, 23.01.2008, p. 14.

11 European Commission, COM(2008)19 Final.

Each of these core points can be designed in different ways, and the abundant (predominantly economic) literature is addressing these modalities both in an instrumental way – reviewing the effectiveness and efficiency of the design options – and in a contextual way, thereby analysing how these core decisions are made by the legislator, and which influences are relevant for the ultimate outcome of the political process. Most of the studies are ex ante assessments, but there is also emerging literature on ex post assessments of emissions trading schemes. [12] In 2004 an OECD report delivered ex post assessments of different schemes, thereby showing that the acid rain allowance trading scheme has been successful, meaning that it had both a cost-effective and an environmental effective outcome, while other applied emissions trading schemes did not deliver the expected outcome.[13] However, this OECD study pays hardly any attention to legal aspects. It indeed appears that assessments from a legal approach are quite scarce compared to economic and political science literature.

1.3 The Legal Perspective

From economic literature, we learn that the efficiency and the effectiveness of the regulatory approach through emissions trading are attractive factors.[14] It is exactly stemming from these instrumental characteristics that emissions trading has become so popular in the field of climate change policy: it simply saves money when using this instrument. When we put the emissions trading instrument, which is in fact an economic instrument, into a legal perspective, a hurdle has to be jumped. It is quite obvious that economists are much more familiar with emissions trading than are lawyers. In general terms, lawyers seemingly feel traditionally more confident with standards and with prescribing behavior through permit conditions instead of letting the market do the work, leaving private operators quite some discretion to decide. This might explain why emissions trading has had less attention by lawyers compared to economists, and why the instrument is perhaps less favored by lawyers. There is however an important job to be done through legal analysis. The economic-oriented studies naturally under-explore core legal values like democratic

---

[12] We refer here to assessments with respect to emissions trading schemes that already existed before the establishment of the EU ETS. 'EU ETS' is the abbreviation of the 'European Union Emissions Trading Scheme'.

[13] OECD, Tradable Permits: Policy Evaluation, Design and Reform, Paris, 2004.

[14] See the important work of Tietenberg (1985), and for a further overview of economic literature on his website http://www.colby.edu/personal/t/thtieten/tradable_permits.htm. See for an overview of literature (law and economics) Michael Faure, *Environmental Regulation, Encyclopedia of Law and Economics*, http://users.ugent.be/~gdegeest/2300book.pdf, version 1998, update forthcoming in 2008 (website visited 8 June 2008).

accountability and the legal control of administrative decisions, transparency and public participation, the role of principles like legal certainty and equal treatment, human rights, access to courts, and legal aspects of the design of an adequate compliance and enforcement mechanism.

From a legal perspective, the analysis of the emissions trading instrument can be made on two dimensions.

First, which legal claims are made through court procedures, and how they have been solved can be analysed. The quite numerous court procedures that have occurred in the first phase of the EU ETS, both at the European courts and the national courts show that quite a few legal questions were posed, mainly by industries and member states. By analysing those cases, we develop a better understanding of the legal concerns of interested parties, and how those were addressed by courts. From such an analysis, recommendations for improving the implementation of the legislative framework, or even improvements of the legislative framework itself might be deduced.

Secondly, apart from the court procedures that show 'hard core legal problems', there are important values that can less easily, or even not at all, be tested by court procedures. This does not mean however that they should be overlooked. For instance, the democratic accountability, already mentioned, of the emissions trading instrument has up till now been under-explored in the debate about the EU ETS, and we will try to stimulate such a discussion in Section 4. Such legal analysis concentrates on the foundations that underpin the legislative framework and the legal systems in which the emissions trading system will be applied. It aims to contribute to the understanding of and to comment on the decision-making as being undertaken in practice.

1.4 The Importance of a Mature Legal Framework

This chapter does not discuss legal aspects of emissions trading on the international level, like international emissions trading among states, and the project-based mechanisms known as Joint Implementation and the Clean Development Mechanism. These emissions trading concepts have been provided by the Kyoto Protocol, and they should be analysed within the context of the specific framework of international law. It is however increasingly argued that the instrument of emissions trading should preferably be applied within a well-developed legal system, meaning that the basic obligation that no pollution will be caused unless this is covered by a tradable permit (or credit) is ensured through an adequate enforcement regime.[15] The need for

---

15 This has also been recognized for instance by the International Network for Environmental Compliance and Enforcement (INECE), See also: http://inece.org/emissions.

monitoring and enforcement of emissions trading within the EU context has already been discussed in literature, and will remain an important factor for consideration.[16] The international legal system is still weak with regard to compliance and enforcement. From this perspective of compliance, it has even been argued that on the international level a harmonized tax system should be preferred to that of a carbon trading system.[17] Also, for developing countries, one can make some reservations when considering economic instruments for their domestic environmental law policies. It seems a wiser approach to experiment with the emissions trading instrument first in relatively well developed legal systems, in order to get a better understanding of their effects and possible improvements before applying the instrument in legal orders that are less mature.[18]

1.5 Structure of this Chapter

Section two sheds a light on the important task of the legislator to choose the right model of emissions trading. The section will specifically focus on auctioning and the model known as credit and trade (or PSR trading). Section 3 presents the major revisions to the current emissions trading scheme as proposed in January 2008 by the European Commission. Section 4 firstly debates the argument that emissions trading promotes the democratic accountability of environmental law, and will then review the current state of affairs and the major revisions from this perspective. We will take a modest approach, in the sense that we aim to enhance a discussion about this value and how it interrelates with the current and proposed design of greenhouse gas emissions trading specifically in Europe. In Section 5 a conclusion will follow.

2. THE CHOICE OF THE LEGISLATOR REGARDING THE EMISSIONS TRADING MODEL

2.1 Cap and Trade through Free Allocation or Auctioning

When reviewing the actual emissions trading scheme as being applied in the EU, it is important to emphasize that the model as shaped by the political process is quite different from the ideal model being presented in the literature. The most striking difference is that the original EU ETS lacks a meaningful

16 Peeters (2006a); Peeters (2006b).
17 Hovi and Holtsmark (2006). See before already Victor (2001).
18 Faure, Peeters and Wibisana (2006).

role for auctioning. This is in line with current applications of emissions trading in the USA: the main option used is a free allocation of allowances along administrative criteria, like the level of the historical pollution. There is however agreement among economic scholars that auctioning should in principle be preferred. The concept of emissions trading as presented by J.H. Dales in 1968 already started from the idea of auctioning tradable rights. This basic idea is hugely supported in the literature, as far as it concerns emissions trading in a domestic legal scheme, and for a regional scheme as within the EU.[19] A. Denny Ellerman et al. (2007) for instance state: 'There is hardly an economist who does not deplore the limited use of auctioning and the concomitant extensive use of free allocation in the EU ETS (as well as in other cap-and-trade systems).' Also Jonathan R. Nash, who examined emissions trading in view of the 'polluter pays principle', concluded that specifically for emissions trading on a national level auctioning should be recommended.[20] An auction provides for the most efficient initial distribution of the tradable permits, has fewer governmental costs compared to grandfathering, and, moreover, according to Nash, fits best with the polluter-pays principle.

The question of how specifically for carbon policies the auctioning of tradable permits could be done has also been explored. Peter Cramton and Suzi Kerr discussed a possible auction of carbon permits, which they restricted to CO_2 emissions because of monitoring problems with other greenhouse gases. They proposed that the auctioning would happen regularly, suggesting a quarterly basis. They found that the model for auctioning carbon permits would not be very complex, which predominantly follows from the nature of the CO_2 emissions: for this specific pollution problem neither the source nor the timing of the emissions is important.[21] Indeed, when considering the possible application of emissions trading, the specific characteristics of the problem to be regulated is of course very relevant. In general, emissions trading is seen as very suitable for environmental problems without local effects, like greenhouse gases. As far as greenhouse gases other than CO_2 that cause local effects ('hot spots'), those need to be taken into account within the regulatory package. This makes as such the emissions trading instrument less attractive, but not necessarily unattractive. It depends on the question of how the emissions trading approach can be combined with the locally based regulatory concerns. The acid rain emissions trading program in the USA is, for instance, a trading

[19] Dales (1968), republished by Edward Elgar in 2002.

[20] Nash (2000, p. 508). However, for the international level he foresees that the differences in wealth between the participating countries will cause problems when auctioning emissions rights. A free allocation would then be the second-best alternative, on which more easily commitment will be reached.

[21] Cramton and Kerr (1998).

scheme that runs together with technology-based standards that aim at avoiding serious local effects.[22]

Cramton and Kerr furthermore explored through which specific auction procedure the bids can be done.[23] Cramton and Kerr state moreover that even in an upstream approach, where energy producers and other big operators would be obliged to surrender permits, market power of these permit traders would not become a problem. They examine in this respect the situation in the USA, where more than 1700 possible permit-buyers would be covered. Following this observation, it is fair to assume that competition problems due to market power do not seem to become a concern in the case of auctioning carbon permits within the EU.

Despite the clear recommendations from the literature, auctioning has thus far not been favored by legislators, even though this model delivers revenue to the governmental budget through the sale of the permits. Maybe we can assume that industries thus far have succeeded in their lobby against the financial burden of buying allowances for their environmental pollution. However, the positive societal effect designating the revenues must not be overlooked: one of the attractive aspects of auctioning is called the double dividend, which means that the revenue can be used for lowering taxes, notably labor taxes. According to Cramton and Kerr, the revenue from auctions would be refunded through tax cuts to all citizens of the nation. In their view, this effectively means that *polluters are buying the right to pollute from the public.*

However, the introduction of auctioning also means that the legislator prefers to rely on the functioning of the market. One important feature of auctioning is that even more than compared to the free allocation, important decision-making will essentially be left to the market. This concerns the main question: who is going to emit and how much? While under a free allocation program as is running in the EU the government heavily determines the initial distribution of the permits, an auction program excludes the government from the distribution (unless specific additional arrangements are to be made).

It is nevertheless important to note that Dales did not envision within his auction model that the government should not intervene at all any more.[24] He did foresee that the government should guide the market process, in particular if the price were to increase or decrease more than preferred. He thus foresaw a quite active role for the government to intervene into the market: if prices were to increase too much, additional permits should be put on the market by the government, and in case of a price fall, the government should be ready to

---

22 Pring (2006).

23 Cramton and Kerr recommend the ascending clock auction. The non-paper and the Report under the project 'Review of the EU Emissions Trading Scheme' refer also to the sealed-bid uniform price auction. See these papers for the technical details.

24 See above, note 19.

buy permits in order to increase the price again and thus make technological innovation more attractive again.

However, one needs to recognize that the emissions trading scheme designed by Dales was only of a quite limited scale, because it concerned emissions trading for water pollution in a certain water basin area. The greater scale of a possible auctioning scheme in the EU falls far from this first idea, as many permit-buyers and sellers would be in the market, which makes the need for governmental intervention in the market process less on an assumption. However, elaborating on Dales' idea, we need further analysis about the extent and form of possible governmental intervention in case of an auctioning and unexpected market functioning. It would be more of a risk if the possible role of governments after the start of the market were to be ignored by the legislator when adopting an emissions trading scheme.

In sum, the legislative choice for auctioning, which, as will be discussed in Section 3 is now part of the proposal of the Commission to amend the initial EU ETS, means that crucial decisions need to be made on the legislative level, including the definition of the tradable permit, the auction method, and the coverage of the model (who needs to buy, who does not?). Individual decision-making by the administration seems hardly needed anymore, except for instance in case of unexpected market functioning or force majeure cases. In addition, it can be anticipated that in the context of enforcement some individual fine-tuning could probably occur to restore unjustifiable outcomes of the auctioning model, if any. We expect that an active brokerage function would not be needed in the EU context, because of the large scale, but this should first be analysed in a more elaborate way. But, in general, an auctioning scheme is characterized by (1) legislative decisions concerning the coverage and the auction procedures, (2) hardly any administrative work except for monitoring and enforcement, and (3) much decision-making through the market process, by industries; it is this market process that will determine which sources substantially reduce their emissions and which sources will decide to buy allowances.

2.2 The Option of Credit and Trade

Besides the well-known cap-and-trade mechanism, either through auctioning or through the free allocation of permits as applied in the initial EU ETS, there is another emissions trading model that could be considered by the legislator. This mechanism is called 'credit and trade'. Alternative names for this system are 'benchmark and trade' or 'Performance Standard Rate trading' (PSR). Within such a system there is contrary to cap and trade no absolute cap identifying the total amount of allowed emissions which will be divided into tradable permits to be allocated either through auctioning or a gratis distribution.

Instead, a relative approach is taken by establishing a general performance standard, indicating the allowed amount of emissions *per unit of production,* or *per unit of fuel.* If an industry were to produce fewer emissions than indicated by this relative standard, it could sell these credits to other industries or, when the legislator so allows, reserve these credits for future use. When an industry exceeds the relative standard, it is obliged to cover the extra emissions by means of an emission credit, bought from another industry, or taken from its own reserve. The total amount of pollution will be steered by adjusting the benchmark: if the total amount of pollution is higher than expected, the benchmark can be adjusted, meaning that the performance standard will be set at a lower level. Compared to cap and trade, the total amount of pollution will thus be regulated ex post, through an additional policy decision by the legislator or the delegated administrative institution. While with a cap-and-trade approach the total amount of pollution will be set ex ante, the credit-and-trade approach initially allows for increases of pollution above the preferable maximum amount.

One basic comment on the credit-and-trade mechanism is that it lacks this ex ante cap on the total amount of emissions: emissions may grow if the total amount of production were to grow. Given that its effectiveness is uncertain, some economists express their preference for a cap-and-trade approach, even though this would be grandfathering with a gratis allocation.[25] On the other hand, however, the PSR can be found politically attractive, especially within carbon policies if a threat of carbon leakage were to become a real concern. This might happen if in a specific country or regional organization the government wants to introduce carbon policies, while other important countries in the rest of the world hesitate to do so. Indeed, a system of cap and trade would be less effective if European industries were to decide to relocate their activities to other parts of the world where a cap is lacking or where less costly carbon policies might exist.[26] Moreover, a distinction can be made between, on the one hand, the total cap to be reached within a nation (or a regional organization), and, on the other hand, the commitment to be reached by a specific sector, like a global competing industrial sector. Here, the legislator could decide to approach the sector with a credit-and-trade approach, while ensuring the total amount of emissions by using one or more compensating options, which are: demanding higher efforts from other sectors, offsetting the surplus emissions by buying credits on the international emissions trading market, or, alternatively, adjusting the cap for the international competing sector. Industries arguing in favor of credit and trade stress the latter option, meaning

---

25 See furthermore the chapter in this book by Edwin Woerdman, Alessandra Arcuri and Stefano Clò.

26 Weishaar (2007).

the control of the total amount of pollution through the ex post adjustments of the benchmark. A study conducted for the International Federation of Industrial Energy Consumers, IFIEC, has argued that the credit-and-trade approach with a single fuel-specific benchmark for electricity production would lead to a cheaper electricity price compared to grandfathering and auctioning.[27] This results from the fact that allowances will be allocated free of charge (no auctioning), without incurring any opportunity costs as is the case with cap and trade with gratis allocation. The limited effect of credit and trade on electricity costs minimizes the risk of emissions increase outside the EU due to replacement of industrial production (insofar as this leakage would be a real threat, which of course needs to be assessed too). Moreover, the method allows a less complicated entrance of (clean) newcomers compared to auctioning and cap and trade with gratis allocation. The IFIEC study proposes the adjustment of the benchmark in future years, if in earlier years the carbon emissions have been higher than expected. This method in fact entails a borrowing of emissions of future years, which will be compensated through a more stringent benchmark in those later years, if this provision indeed can be effectively applied by the legislator or the delegated administration.

The IFIEC study explains that a cautious approach towards achieving the preferred cap would be to take a high electricity production scenario, which could even mean that an overachievement would occur (an even lower total amount of carbon emissions than ex ante determined). An alternative approach would be to set up additional policies, like the renewable energy policy, in order to stimulate the transition towards a low-carbon energy society. One effect of credit and trade for the electricity sector would be that fewer incentives for low-carbon options will be provided for other sectors outside the emissions trading scheme other than by auctioning and grandfathering. This is a consequence of the resulting lower electricity price compared to auctioning and grandfathering. Here, the main question to be answered by the legislator is whether such incentives should come from the EU ETS (thereby taking also into account that energy consuming industries covered by the EU ETS would face higher electricity prices) or from other regulatory measures compelling or stimulating those sources outside the EU ETS to reduce their carbon emissions.

It is obvious that the credit-and-trade option is to the advantage of the industries that propose this method. This underlines observations in the literature on private interest theory of regulation, meaning that industries, realizing that environmental regulation is unavoidable, will cooperate in the development of

[27] Ecofys (Bart Wesselink, Sebastian Klaus, Alyssa Gilbert and Korneli Blok). The IFIEC method for the allocation of CO_2 allowances in the EU Emissions Trading Scheme. A review applied to the electricity sector, March 2008.

the regulation and try to change the contents to their advantage.[28] However, this does not mean that the method as such should be overlooked by literature and, moreover, the responsible legislative institutions.[29]

When discussing credit and trade, it should be acknowledged that the definition of the credit baselines (the benchmarks) could be a demanding task. This is because of the differences among sources, even within the same sector.[30] In our assumption, this effort of setting the benchmark needs only to be done once, for each sector, at the start of the credit-and-trade system, followed probably with later fine-tuning when necessary. This would of course introduce uncertainty (possible adjustments of the benchmark) and some administrative costs (for instance when for certain industries specific arrangements need to be taken in case the common benchmark were to be found unreasonable, being disproportional to the specific installation).

3. THE MAJOR REVISION OF THE EU ETS: TOWARDS HARMONIZATION

3.1 Introduction

The European Commission launched on 23 January 2008 a far-reaching revision of the current EU ETS.[31] It means that the emissions trading scheme would be extended to other major industrial emitters (main new sectors are specific non-combustion sources in the chemical industry and the aluminum and ammonia industry). Meanwhile, a proposal to extend the scheme to the aviation sector has already been made.[32] The EU ETS would also cover greenhouse gases other

28 See Faure (2008), referring to Maloney and McCormick (1982, pp. 99–123).

29 The continuing opinion of the Commission to forbid ex post arrangements is in fact not convincing, see the chapter in this book by Chris Backes, Kurt Deketelaere, Marjan Peeters and Marijke Schurmans, 'The underestimated possibility of ex post adjustments: some lessons from the initial greenhouse gas emissions trading scheme'.

30 Ellerman et al. (2007).

31 See the Commission of the European Communities, Proposal for a Directive of the European Parliament and of the Council amending Directive 2003/87/EC so as to improve and extend the greenhouse gas emissions allowance trading system of the Community; see also Press release European Commission, 'Boosting growth and jobs by meeting our climate change commitments' IP/08/80, 23 January 2008, and 'Questions and Answers to the Commission's proposal to revise the EU Emissions Trading System', Memo 08/35, Brussels 23 January 2008.

32 Proposal for a Directive of the European Parliament and of the Council amending Directive 2003/87/EC so as to include aviation activities in the scheme for greenhouse gas emission allowance trading within the Community, COM(2006)818.

than carbon dioxide, namely nitrous oxide and PFCs (perfluorocarbons). In fact, the Commission has concluded that emissions trading will remain the key approach within European climate change policy, but that another design is needed. The initial model, especially its allocation procedure with national allocation plans, has indeed suffered criticism, because of its complexity, vagueness, and competition-distorting concerns.[33] One of the major criticisms of the current emissions trading scheme is however that it is not yet effective enough. Indeed, the data concerning the first two years show that more allowances have been distributed than industries needed according to their emission records. Although the fact that not all allowances have been used to cover emissions could in theory mean that the instrument has stimulated technological and other innovation, causing emission reductions, the overall conclusion is however that the member states have been (too) generous in allocating allowances in the first phase to the EU ETS sector. When analysing this aspect of over-allocation, one needs to take into account that the Kyoto commitment period (2008–2012) was not yet applicable during the first operating years of the EU ETS. This means that the member states were not yet bound to an emission reduction obligation. Indeed, for the period 2008–2012 the member states as well as the European Community are bound to the overall greenhouse gas emission targets stemming from the Kyoto Protocol and the so-called Burden Sharing Agreement.[34] This legally different situation is relevant and thus should be taken into account when assessing the effectiveness of the EU ETS in its first years (2005–2007). Moreover, the Commission has stressed the fact that the first allocation and trading period (2005–2007) was to be seen as a learning phase.[35] For the period 2008–2012 the legal situation is different, as then the member states as well as the Community need to comply with the emission reduction commitments following from the Kyoto Protocol. For reviewing compliance, the Kyoto Protocol has established reporting duties. Furthermore, Council Decision 280/2004/EC provides a

33 Communication of the Commission, Building a global carbon market – report pursuant to Article 30 of Directive 2003/87/EC, COM(2006)676, 13.11.2006.

34 Council Decision 2002/358/EC of 25 April 2002 concerning the approval, on behalf of the European Community, of the Kyoto Protocol to the United Nations Framework Convention on Climate Change and the joint fulfilment of commitments thereunder. See further Marc Pallemaerts, Rhiannon Williams, 'Climate Change: the International and European Policy Framework', in Deketelaere and Peeters (2006a, pp. 22–50). See for a quantification of the respective emission levels of the member states in terms of tonnes of carbon dioxide equivalent Commission Decision of 14 December 2006, OJ L 358, 16.12.2006, including an extra assignment of so-called assigned amount units to Denmark.

35 One might question whether it is to be supported by experimenting with legislation, but this will not be debated here.

mechanism for monitoring greenhouse gas emissions within the member states.[36] It for instance establishes a mechanism for monitoring all anthropogenic emissions by sources and removals by sinks of greenhouse gases, and evaluating progress towards meeting commitments in respect of these emissions by sources and removals by sinks. The Commission can of course take infringement procedures towards member states that do not comply with such monitoring provisions, and, moreover, if a member state were to breach the emissions reduction commitment. Furthermore, the Commission may also start infringement actions if a member state were to be in breach of the monitoring, reporting, verification and enforcement prescriptions of the greenhouse gas emissions trading directive. As such, important provisions have been established in order to monitor and to enforce the greenhouse gas reduction commitments, but whether they will be effective enough remains to be seen.

However, the initial EU ETS suffers from a distributional problem, since the determination of the total amount of emissions to be given to the EU ETS-covered installations is the subject of sensitive debate. The initial EU ETS does not indicate exactly how many emissions should be ultimately reduced by the covered sectors in each country, as this is left to a decision by member states in their national allocation plans, to be reviewed by the European Commission, which could be followed by a procedure at the Court of First Instance, to be followed by an appeal at the European Court of Justice. Most of the member states were not able to finalize the procedures for the national allocation plans for the period 2008–2012 in time, which thus means that these member states were not able to comply with the deadline for surrendering allowances to the industries before 1 March 2008. It is clear that the national allocation is as such a delicate part of the current scheme, and some legal conflicts between member states and the Commission have occurred regarding the decision-making of the Commission according the approval of the national allocation plans.[37]

3.2 The Revision

Important proposed amendments are:

- The establishment of an EU-wide cap for the covered installations. In addition, an eight-year trading period is envisioned, which results in a

[36] Decision no. 280/2004/EC of the European Parliament and of the Council of 11 February 2004 concerning a mechanism for monitoring Community greenhouse gas emissions and for implementing the Kyoto Protocol of 11 February 2004 OJ L 49, 19.2.2004.

[37] See the overview in the chapter written by Nicolas Van Aken.

third trading period running from 2013 to 2020, and a fourth trading period from 2021 to 2028.

- The new scheme should lead to an emissions reduction in 2020 of 21% compared to 2005 levels. The Commission proposes that the EU ETS sector can deliver a reduction larger than 20% reduction in 2020 when compared to 1990, as it is cheaper to reduce emissions in the ETS sector.[38] Other sectors should deliver an emissions reduction of around 10% in 2020 compared to 2005.
- There will be a linear reduction of the total amount of allowances for the EU ETS sector, with a factor of 1.74% per year.[39] This linear reduction will apply beyond the end of the trading period 2013–2020; the Commission shall review the linear factor no later than 2025.[40] In case of an international agreement on climate change leading, by 2020, to mandatory reductions of greenhouse gas emissions exceeding the minimum reduction levels agreed upon by the European Council, the linear factor shall likewise increase, in order to ensure that the Community quantity of allowances in 2020 will be decreased.[41]
- Auctioning will be the main allocation method, to start in 2013 with the power sector, and gradually including the other sectors, resulting in an overall auctioning by the year 2020.[42] The Commission estimates that around 60% of the total number of allowances will be auctioned in 2013, and this proportion will increase gradually. The auctioning will be done by the member states, and the Commission proposes to regulate the maximum amount of allowances that may be sold by them. From a viewpoint of solidarity, a part of this amount will be redistributed among member states.[43] The Commission proposes a redistribution of a part of the amount of allowances to be auctioned from member states with an average level of income per head that is more than 20% above the EU average.

---

[38] European Commission, Questions and Answers on the Commission's proposal to revise the EU Emissions Trading System, memo/08/35, Brussels, 23 January 2008, p. 3.

[39] Compared to the average annual total quantity of allowances issued by member states in accordance with the decisions of the Commission on their national allocation plans for the period 2008 to 2012, see art. 1 of the directive proposal, replacing art. 9 of the initial directive.

[40] According to the new version of art. 9 as being proposed by the Commission.

[41] According to the new version of art. 28 as being proposed by the Commission.

[42] According to the new version of art. 10 as being proposed by the Commission, and art. 10a(7) (new version as being proposed by the Commission).

[43] According to the new version of art. 10 as being proposed by the Commission.

- From 2013 and in each subsequent year up to 2020 there will be an allocation of allowances free of charge to installations which are exposed to a significant risk of carbon leakage.[44] It is the task of the Commission to determine the relevant sectors. There are several conditions that should be respected in adopting these measures, such as the rule that that no free allocation will be made for the electricity sector (which would mean that the credit-and-trade system for this sector would be excluded), that the maximum amount of allowances to be given to the installations shall not exceed the verified emissions of those installations in the first trading period, and that 5% of the total amount will be available for new entrants (but electricity production by new entrants will not be given gratis allowances).
- Moreover, the Commission will be required to deliver a report to the European Parliament and the Council containing an analysis of the situation of the energy-intensive sectors or sub-sectors that have been determined as being exposed to significant carbon leakage.[45] This needs to be done not later than June 2011 and in view of the light of international negotiations and the extent to which these lead to global greenhouse gas emission reductions. The report should be accompanied by relevant proposals when necessary, which could mean the adjustment of the proportion of allowances given free, or to include within the scheme importers of products produced by the sectors or sub-sectors which are determined to be exposed to significant risks of carbon leakage.[46] This could mean that the exemption from auctioning for EU ETS sectors will be withdrawn. The envisioned inclusion could mean that importers of certain products would be required to surrender allowances. It is obvious that such a system would need careful consideration in view of the United Nations Framework Convention on Climate Change (UNFCCC), and the WTO agreement. It is explicitly stated that any binding sectoral agreements which lead to global emissions reductions of the magnitude required to effectively address climate change, and which are monitorable, verifiable and subject to mandatory enforcement arrangements shall also be taken into account when considering what measures are appropriate.
- Furthermore, the Commission proposes that a certain percentage of the proceeds from the auctioning of the allowances, including the allowances that are redistributed from a solidarity perspective among

---

44 According to art. 10a (8) as being proposed by the Commission.
45 According to art. 10b as being proposed by the Commission.
46 According to art. 10b as being proposed by the Commission.

member states, *should* be earmarked for investments in carbon-friendly investments, for instance concerning renewable energy and energy efficiency.[47] The proposal doesn't say 'shall', but expresses only a wish about the spending of the revenues. A binding determination of the way in which the revenues will be spent by the member states conflicts with the EC Treaty, as there is no competence for doing so.

- Harmonization of the allocation procedures to be followed by the member states, like the date at which ultimately allowances must be issued and what in this respect should be done in case of closure of an installation. It is for instance proposed that installations that 'cease' their operation shall receive no further allowances.[48]
- The opting in and opting out of industries. Opt-in was possible in the initial EU ETS and will be expanded following the Commission's proposal. Opt-out was only possible for the first period (2005–2007) but should be possible again from 2013 onwards.[49]
- Harmonization of the conditions for using JI and CDM.[50]

In order to execute the EU ETS, the Commission should develop some important legislative and administrative competences, which are:

- the determination of the sectors (or presumably also sub-sectors) that will not be covered by the method of auctioning, at the latest by 30 June 2010 and every three years thereafter; the proposal gives a quite extensive list of conditions to be respected by the Commission in its decision-making;[51]
- the adoption of a Regulation by 31 December 2010 to ensure that auctioning by member states will be done in an open, transparent and non-discriminatory manner;[52]
- the adoption of 'Community-wide and fully-harmonised implementing measures' for the free allocation of allowances;[53]
- the Commission shall adopt a regulation for the monitoring and reporting of emissions and, where relevant, activity data, from the

---

47 According to art. 10(3) as being proposed by the Commission.

48 According to the new version of art. 11 as being proposed by the Commission.

49 See for opt-out art. 27 as being proposed by the Commission.

50 According to art. 11a as being proposed by the Commission. The connection with international emissions trading will not be discussed in this contribution.

51 The new art. 10a(9) as being proposed by the Commission.

52 Art. 10(5) as being proposed by the Commission.

53 Art. 10a (1) as being proposed by the Commission.

activities listed in Annex I which shall be based on the principles for monitoring and reporting set out in Annex IV and shall specify the global warming potential of each greenhouse gas in the requirements for monitoring and reporting emissions for that gas.[54] The Commission should also adopt a regulation for the verification of emission reports and the accreditation of verifiers specifying conditions for the accreditation, mutual recognition and withdrawal of accreditation for verifiers, and for supervision and peer evaluation as appropriate. In the present directive, it has only been regulated that the Commission should adopt guidelines for monitoring and reporting emissions.[55]

- Specific decisions concerning the opting in and opting out of industries.[56]

These regulations and other decisions shall be adopted following the procedure being referred to in art. 23(3) of the directive, which is known as the comitology procedure. In addition, the Commission has a major influence on the future content of the EU ETS as it has the sole competence to propose amendments of the adopted legislative framework.

The legal framework for the EU ETS is an important but not the only part of EU climate policy. In addition, the Commission has proposed a legal framework for carbon capture and storage, which will be connected to the EU ETS.[57] Moreover, it released a proposal on effort sharing to meet the EU's greenhouse gas reduction commitment in sectors not covered by the EU ETS (such as buildings, services, smaller industrial installations, transport, agriculture and waste). Moreover, the proposal for renewable energy already mentioned provides a legal framework for a major increase of renewable energy production by member states.

---

54 According to the new version of art. 14 as being proposed by the Commission.

55 Art. 14 of directive 2003/87.

56 See concerning the experiences with opt-in possibilities under the acid rain program Ellerman (2004), thereby specifically referring to the difficulties and consequently rather high administrative costs in setting the right baseline for opting in.

57 European Commission, Proposal for a directive of the European Parliament and of the Council on the geological storage of carbon dioxide and amending Council Directives 85/337/EEC, 96/61/EC, Directives 2000/60/EC, 2001/80/EC, 2004/35/EC, 2006/12/EC and Regulation (EC) No 1013, COM(2008)18, Brussels 23 January 2008.

3.3 Shifts of Decision-making to the EU Level

This concise overview of the major revision of the EU ETS scheme shows that there will be a major shift from decision-making on the national level to the EU level. The national allocation plans and subsequent national allocation decisions will no longer be part of the future emissions trading scheme as proposed by the Commission, as there will be an EU-wide cap for the installations covered by the EU ETS. Moreover, the coverage of the scheme will be expanded. These arrangements would heavily influence the national climate change policies of the member states, as they will be restricted in their policy on reaching commitment with their national greenhouse gas emissions reduction targets. In the former national allocation plans national governments had a considerable degree of discretion regarding the industries covered by the EU ETS. The proposed approach takes away such discretion. The Commission explains that this decentralized system implied an incentive for member states to favor their 'own' industries.[58] A 'race to the top' would in fact occur (meaning ample room for emissions) if member states aim to allocate as many allowances as possible. However, it would be wrong to assess this effect without taking into account the fact that the member states would need to comply with the national emissions reduction target if industries were to be allocated a relatively high share. In other words, an advantageous policy for the EU ETS sector would mean that the reduction efforts would be higher in other sectors. Another additional option for member states to follow is to compensate the generous approach for their industries by making use to some limited extent of the Kyoto mechanism, and thus buying credits abroad in order to compensate part of the emissions caused by the industries. The shift of the distribution of the room for emissions to industries to the EU level takes away the effect of member states having a generous approach to their industries, but raises the question of what the consequences are for national policies and thus the other relevant greenhouse gas emitting sectors. It has already been argued that the harmonized cap for the EU ETS sector would imply a disincentive for member states to impose measures on this sector that go further, as those measures would lead to a 'leakage' of allowances, and thus emissions, to other member states.[59] If, for instance, additional emissions reduction measure were to be applied by a member state concerning an industrial sector covered by the EU

[58] European Commission, Questions and Answers on the Commission's proposal to revise the EU Emissions Trading System, memo/08/35, Brussels, 23 January 2008, p. 3.

[59] Netherlands Environmental Assessment Agency, Consequences of the European Policy Package on Climate and Energy, Bilthoven, 2008, available at www.mnp.nl.

ETS, the available room for emission given by the EU ETS cap would still be used but then in other countries where such additional measures are lacking. As a consequence of this leakage effect, member states would focus their climate policies on other sectors, not covered by the EU ETS. In other words, member states could explore, in view of article 176 EC Treaty the adoption of measures that go further for the EU ETS industries, but in practice this is unlikely to happen in case of the proposed EU ETS cap, because of the leakage effect. Extra reductions in one member state might simply be nullified by fewer reductions in other member states.[60] This shows that the establishment of an EU-wide cap does not only limit the national legislators in addressing the EU ETS sector, but also influences the intensity of national policies concerning the other sectors.

One other major revision is the shift towards auctioning. This auctioning of emissions rights was clearly not preferred by the European legislative institutions at the time of the adoption of Directive 2003/87.[61] It is even prescribed by the initial EU ETS directive that the member states should allocate most of the allowances free during the first and second trading period. The initial directive does not furthermore contain any provision indicating the possible design of an auction, like whether auctioning shall be open to potential bidders from outside the national legal system.[62] The political preference for a free allocation is stimulated by the nature of the environmental problem at hand: climate change is a world-wide problem that suffers from the fact that some states with important economies have not yet committed themselves to legally binding emissions reduction commitments. This means for Europe that submitting industries to auctioning costs while in other parts of the world meaningful emissions reduction obligations are lacking was not a preferable option. In this respect, one can see that the design of the initial European greenhouse gas emissions trading model has been influenced by the lack of progress towards binding commitments on the international level. Nevertheless, the possibility of auctioning has now been tabled by the Commission. This means that the EU will move from the present, as such, cumbersome grandfathering process towards the, in essence, simpler auctioning method. Instead of the (thus rather complicated) task for authorities to decide on allocation criteria that lead to a justified allocation of tradable permits free of charge, the auctioning method seems to require fewer governmental costs because it leaves the distribution of the allowances to the discretion of industries. These industries shall

60 The Netherlands Environmental Assessment Agency calls this the 'waterbed effect', p. 38.

61 Art. 10 of directive 2003/87.

62 Non-paper on the use of auctioning for allocating emissions trading allowances in the second trading period 2008–2012 and further on.

then determine how many allowances they need for a certain price, taking into account their marginal cost of abatement of the greenhouse gas. However, it remains to be seen whether auctioning will make it through the legislative process, and, if so, exactly how auctioning will be designed and implemented, and which legal problems will have to be solved in this respect. Moreover, the Commission does not only propose the auctioning method, but also proposes what should be done with an important share of the revenue from the auctioning.

The proposal of the Commission thus implies that the European legislator would develop a further influence on the national climate change policies. It herewith responds to the broadly heard wish to harmonize the system. This harmonized approach does not only relate to the new environmental target to be reached in 2020, because this could also have been provided by decision-making regarding the member states' emissions reductions targets for 2020 or even further. The current proposal also further harmonizes how that environmental target should be complied with, and takes away national discretion for important industrial sectors and the power sector. The option of an EU-wide cap for the covered industries, the expansion of the scheme, and the choice for obligatory auctioning with earmarked revenues and the choice for a separate treatment for sectors exposed to significant risks of carbon capture are the headlines. The initial directive already introduced a far-going harmonization of monitoring, reporting and notably also enforcement, thereby prescribing some administrative sanctions like a penalty and naming and shaming. This approach has been upheld in the proposal of the Commission.

4. EMISSIONS TRADING AND DEMOCRATIC ACCOUNTABILITY

4.1 Introduction

The literature has already extensively discussed the emissions trading instrument, and this needs to be continued regarding new or adapted emissions trading models. The previous section already mentioned a study that explores the effect of the cap for the EU ETS sector as proposed by the Commission, which would imply that member states would be reluctant to impose additional greenhouse gas reduction obligations on their industries. In addition to studies that examine the possible economic and policy effects of several models, legal analysis can contribute to a further discussion about the chosen regulatory options and their alternatives. As stated in Section 1, hard core legal studies will focus on concrete legal claims and possible legal solutions, like compensation for a specific industry that could face a disproportional administrative decision. In addition, a broader discussion about how and to what extent the regulatory option fits into the legal system and its main values is also worthwhile. This section will explore the current state of affairs, including the Commission proposal, from one specific value that has been identified in the early literature supporting the concept of emissions trading within the American context, which is that of democratic accountability. This specific fundamental value has been put at the forefront in the economics-oriented literature about emissions trading. Already by 1987, Bruce Ackerman and Richard B. Stewart took up the debate over the seeming conflict between the market and democracy, but argued that the use of emissions trading would facilitate a governmental discussion about core values to be reached through environmental regulation.[63] Because one of the main questions of the present European emissions trading scheme is the level on which the most important decisions should be made, it is interesting to explore the current developments in view of the specific value of democratic accountability and to link this further with the former discussion about how emissions trading is related to this concept.

4.2 Emissions Trading Submits the Core Environmental Question to the Political Debate

Ackerman and Stewart are known to be supporters of a market-based environmental law. In their publication 'Reforming Environmental Law' published in 1984 they made a plea for introducing incentive-based regulatory instruments, notably emissions trading.[64] Of course, the possible use of market-based instruments is extensively debated, and there is also literature critically discussing the expected positive effects of emissions trading and stressing the relevance of command- and control-like options.[65] We here only focus on the specific argument used by Ackerman and Stewart in their subsequent article 'Reforming Environmental Law: the Democratic Case for Market Incentives', where they focus on the positive role that emissions trading, in their opinion, would play in view of democratic accountability.[66] It is important to note that these authors start their discussion in the light of the command-and-control approach as being developed in the 1980s within the United States, which was primarily based on the application of the best-available control technology. Such technology standards were largely determined through uniform central regulations, adopted by the federal Environmental Protection Agency. This

63 Ackerman and Stewart (1988).
64 Ackerman and Stewart (1984).
65 Latin (1985), and, later, Driesen (2003).
66 Ackerman and Stewart (1988).

involves a complex scientific, engineering and economic assessment to be done by the government, with a demanding information effort. They also provide a fertile ground for litigation, 'producing reams of technical data, complex adversary rule-making proceedings and protracted judicial review'.[67] Ackerman and Stewart state that the instrument of emissions trading would move away from technical issues, and would enhance the democratic debate on the core environmental goals to be reached, which they summarize in the following question: 'during the next *n* years, should we instruct the EPA [Environmental Protection Agency] gradually to decrease (or increase) the number of pollution rights by *x* percent?'[68]

They propose an environmental control system that is pollution based instead of technology based, and that within the core political process the maximum amount of emissions will be determined. It would release the government from making complicated technical assessments about setting the right permit conditions, and the authors refer to the extensive hearings the EPA had to undertake in order to set the right technology standard for a certain industrial sector, followed by needing to defend the criterion before the courts. Emissions trading as being proposed by J.H. Dales would be an important improvement, where the technological decisions will be left to the industries, and where the cap on the total amount of emissions is representing the maximum amount of pollution that would be allowable during a certain time in a certain period. In their view, auctioning should be chosen as the distributional method. From the moment the existing permit of an industry expires, the industries would be able to buy a new one through an auction.[69]

Of course, a pollution-based control system can also be achieved by other regulatory approaches, notably the environmental quality standards that determine the maximum acceptable pollution level in a certain area. The crucial question is, however, how such quality standards might be achieved, and therefore different regulatory options can be examined. One can suggest a tax, although it seems difficult to set the tax at such a level that the set maximum amount of level as indicated by the environmental quality standard will not be exceeded. One can suggest the classical permit approach, with emission limit values based on technology criteria, but then the administration would have a burdensome task to set the limits in such a way that, first, the environmental quality standard will be achieved and, secondly, that the specific costs and circumstances for industries will be taken into account. Ackerman and Stewart argue that this complicated debate is time-consuming and thus moves away from the core question about the environmental quality to be reached. The

67 Ackerman and Stewart, p. 174.
68 Ackerman and Stewart, p. 189 (clarification within brackets done by author).
69 Ackerman and Stewart, pp. 180–81.

alternative market-based approach through emissions trading would allow the political process to concentrate on the core environmental question.

Within carbon policies, the total amount of greenhouse gas concentrations is already at the core of the debate, and we already know that on the international level, through the conclusion of the Kyoto Protocol, important but sensitive decision-making occurred regarding the maximum allowable amount of pollution for developed countries within the years 2008–2012, accompanied by the international emissions trading instruments. The set emissions reduction targets as included in the Kyoto Protocol are indeed followed by the choice within the EU for an emissions trading scheme; first there were initiatives in the United Kingdom, Denmark and The Netherlands for setting up domestic emissions trading schemes, but then there was the adoption in 2003 of the initial greenhouse gas emissions trading directive. When we analyse the initial EU ETS from a perspective of democratic accountability as discussed by Ackerman and Stewart, we first need to understand that this emissions trading scheme only covers a particular part of all greenhouse gas-emitting sources in the EU. The core question how much could be polluted in each member state was thus already determined through the EU burden-sharing agreement for the former 15 member states, and the Kyoto Protocol for the new member states.[70] Furthermore, we see that the initial directive left the question of how much pollution might be caused by a certain sector to the member states, as they needed to decide within their national allocation plans on the total quantity of allowances to be allocated to the EU ETS sector, with (only) a review by the Commission. The member states could no longer change the binding emissions reduction commitment applicable for the state, but were able, to a certain extent, to decide on the distribution of the maximum amount over the sectors, taking into account the possibility to buy emissions rights abroad, which would then increase the total amount of emissions within the country. This means that the political debate about the core question of how much pollution may be created, in particular by *whom*, did not occur in any single democratic debate, but went through a multi-level system.

However, the proposal of the Commission amending the current emissions trading directive implies that the total allowable quantity of pollution for the selected European industries will be determined an the EU level. This means that the decision about the total amount of allowances available for a sector will be decided for a substantial part on the EU level. The proposal needs to go through the co-decision procedure, enabling the Council and the European Parliament to discuss the matter and to propose amendments. Obviously, the democratic process within the EU has to be judged on its own merits, and it is

70 Except for Malta and Cyprus.

different from those on the national levels. The democratic character of the EU system, or in other words its democratic deficit, is precisely one of the current sensitive issues. The European Parliament obviously has a less mature and less powerful role than the national parliaments in the EU generally have.[71] Regarding climate change policies, it can be expected that the national parliaments will try to influence the decision-making regarding climate change on the EU level, predominantly in the Council through, for instance, the link of the ministerial accountability to parliament.[72] Through the informal (lobby) process and the formal processes at the national level, the total amount of reduction to be achieved by the respective country together with the method of distributing the allowances and other crucial distributional elements might be part of the political national debate, with the aim of influencing the EU decision-making process.

In sum, in the specific context of (European) climate change, the political debate not only focuses on the question 'how much pollution should be allowed during a certain time in a certain area' but also, given the broad nature of the problem and the many different greenhouse gas-emitting sources (not only industries, but also agriculture, transport, land use, households), on the question 'who needs to reduce and how much'? Within climate change policies the second question is indeed a challenging one, as there are so many and different greenhouse gas-emitting sources. Political disagreements and legal conflicts about this distribution are easily possible. Up till now, there has not been such a distributional effort within European environmental law as within the climate change dossier.

Precisely with regard to this distributional effort, the current proposal of the Commission implies an important shift of decision-making, as the efforts to be made by the EU ETS sector will also be prescribed by setting the EU cap. Moreover, the method of distribution will be amended and further harmonized, leaving hardly any room for a political debate at the national level. However, the debate at the EU level on the distribution of the total amount of allowable emissions cannot be comprehensive, as the member states retain a considerable degree of discretion (also depending on other EU measures) concerning other sources.

The envisioned idea that emissions trading would emphasize the democratic accountability within environmental law would thus indeed seem

---

71 See for a detailed discussion about how to improve the accountability within the EU Verhey, Broeksteeg and Van den Driesche (2008).

72 See concerning (other) possibilities through which national parliaments (being multiple-actor institutions) might effectively influence the European decision-making process P. Kiiver (2007), Europe in Parliament, towards targeted politization, (Dutch) Scientific Council for Government Policy, Web publications 23, www.wrr.nl.

applicable in the context of environmental law in the USA, for instance concerning air pollution, as argued by Ackerman and Stewart. However, the matter is quite a lot more complex in the case of climate change, especially in the current state of affairs in the EU. Following the proposed future climate policy measures in the EU, both the European decision-making and the national decision-making procedures will be important for determining how much pollution may be caused by specific sectors.[73] Nevertheless the national parliaments would not be relevant any more specifically for the EU ETS sector if the current proposal were to be adopted in this form, despite their presumed attempts to influence the European legislative institutions. As noted in the previous section, the EU-wide cap even discourages national governments from considering policies for the EU ETS sector additional to the EU-wide emissions trading cap.

One can question whether it is really necessary for the EU to determine the amount of pollution to be allocated to specific sectors. We assume that from an economic point of view, such an approach would indeed be seen as an optimal approach. However, one could question from the democratic point of view whether it would be wise to exclude national governments from the debate about the distribution of the climate change efforts. A policy approach through which the EU would only adhere to the (in the international negotiations also to be discussed) core question of how many emissions would be acceptable during a certain period, while leaving discretion to member states to decide about the regulatory instruments and whether distributional choices should not be excluded from the considerations at all. In that case, which is more or less done through the initial EU ETS, there is quite a clear distinction regarding democratic accountability: on the EU level the commitments of states contributing to the overall reduction target of for instance minus 20% or 30% in 2020 would be discussed, while the national governments and parliaments would decide about the distribution of the efforts and the tools. In this form, democratic accountability seems more transparently arranged: the EU level will then focus on commitments among states, while the states themselves focus on commitments among sectors. The EU could even facilitate the setting up of facultative regulatory instruments, like a facultative emissions trading tool, leaving it to the member states to decide whether or not to participate. As long as the EU will take up its responsibility in holding the member states compliant with their monitoring and reporting duties, the fear for a race to the bottom of national climate policies is not convincing, as ultimately states need to comply with their emissions reductions targets and thus anyway need to

---

73 This is however not only a characteristic of climate change, as with other air pollution problems like SO_2 many different sources contribute while the EU only partly harmonizes the relevant legislation.

adopt regulatory measures. However, the argument that such a decentralized system would harm the 'level playing field' has strong support within the EU, which consequently leads to the current proposal to take away from member states an important part of their climate change policy.[74]

4.3 Lack of Attention to Democratic Accountability in Reports of the Current System

As far as is known, there is no comprehensive study available that discusses the present emissions trading scheme from the perspective of democratic accountability, especially on the national level where the national allocation plans need to be concluded. There is as such a lack of information about how parliaments were involved in the national allocation plans and decisions. It would be interesting to know to what extent and how the national democratic debate concentrated on the setting of the cap for the EU ETS sector, and on the specific distribution of the allowances. An analysis of how democratic decision-making and moreover how public participation have influenced the emissions trading system and the distribution of commitments among sources inside and outside the EU ETS would contribute to an understanding of the specific functioning of greenhouse gas-emissions trading. Furthermore, the influence of interest groups and lobbying could be clarified too.

The official review of the current greenhouse gas-emissions trading system also under-explores these aspects. The formal report for the year 2007 for instance hardly mentions any of them. Even the involvement of the public regarding the procedure for establishing the national allocation plans has hardly been mentioned, let alone the possible effects of such an involvement.[75] Indeed, neither article 21 nor article 30 of the greenhouse gas-emissions trading directive indicates that the review of the EU ETS should focus on the national decision-making procedures, or on access to information, public participation and access to courts. The 2007 report gives some information regarding access to information, but it does not elaborate on the effect of the legal provision that information is available to the public: the report does not show to what extent this information has led to public comments or to other involvement of the public.[76]

---

[74] See for an interesting view on the advantages of having only European set environmental quality standards together with different national approaches for compliance with them Faure (2001).

[75] European Environment Agency, Application of the Emissions Trading Directive by EU Member States, reporting year 2007, delivered in 2008 (www.eea.europa.eu).

[76] European Environment Agency, p. 75.

The 2007 report however shows another important outcome of the initial scheme, which is the large cost of the national allocation procedures. It is exactly this outcome that illustrates the demand for an alternative emissions trading model. It is true that the shift towards the EU level makes the allocation process much better to understand, as it is hardly undoable to read and review 27 different national allocation plans and allocation procedures. The democratic accountability of the regime will thus be improved as far as the transparency of the process is involved: it is much better to follow one single European decision-making procedure compared to 27 different national procedures, with 27 decisions by the Commission regarding the NAP. However, we must not overlook the fact that some important lobbying will take place regarding the co-decision procedure, and thus that even within the centralized and thus much easier to oversee democratic procedure on the EU level many hidden arguments need to be found in the national discussions and forthcoming influences to members of the European parliament and the council.

4.4 Additional Tools for Democratic Decision-making: Public Participation[77]

One important difference between the present integrated permit-system in the EU (following the so-called IPPC-directive) and the greenhouse gas-emissions trading system is that in the first approach the public concerned has a broad possibility to participate in the administrative decision-making procedure where it concerns an individual activity in a certain place. Such public participation can be seen as a kind of democratic involvement within administrative decision-making, as the administration cannot take a decision before hearing the view of the public concerned. The participation of the public (in a broad sense) can also be facilitated in case of other administrative decisions, like the adoption of plans and even legislation. However, we need to be careful with the importance of public participation as a means of providing an additional tool for enhancing the democratic value of administrative decision-making: we do not clearly know yet how worthwhile and influential this public participation may be. Further examination of the role of public participation in the integrated permit procedures needs to pay attention to the question of whether the public is able to submit meaningful and relevant views in cases of highly complex and technical matters. How is it, for instance, really possible for citizens and environmental NGOs to criticize effectively a best-available technology?

[77] We connect here to the terms as being used in the Aarhus Convention on access to environmental information, public participation in public decision-making, and access to justice in environmental matters, http://www.unece.org/env/pp/documents/cep43e.pdf.

Moreover, we need also to examine whether the public is indeed interested in participating in permit procedures. Maybe some environmental problems, like problems that do not cause a direct threat to the backyard, will be ignored by the public. Moreover, in the specific context of the IPPC-directive we may be concerned about the way that the so-called best-available technology documents (BREFS) are developed – the democratic aspect of the procedure is a critical point. In sum, the classical permit procedure as being established by the IPPC-directive allows the public and environmental NGOs to participate in the decision-making procedure, even though we cannot judge the value of it because we need to have a more detailed examination of the question of what this opportunity really means. Nonetheless, the important difference between emissions trading and classical permission is that in the latter public participation is in principle possible regarding the specific amount of pollution to be caused at a certain place. The carbon trading instrument, concentrating on the global effect, doesn't need this local approach. The core element, the reduction of greenhouse gases, can indeed be determined by the legislator. Notably, the question of how the amount of pollution is to be distributed among society will be heavily decided by the democratic institutions by setting the cap and the method and coverage of the trading scheme. Public participation on the micro-level (the amount of emissions to be done by the specific installation) is as such not needed, except in the situation where the greenhouse gas also has a local effect. Public participation however then concentrates on the local effect, not on the greenhouse gas effect. Compared to a traditional permit instrument, there is with emissions trading a shift from public participation regarding individual permit procedures to the political debate concerning the content of legislation, setting the amount for allowed greenhouse gases, the coverage and the allocation method.[78] The market will then govern where exactly pollution will be done.

Following this observation, the citizens can no longer influence the individual decision-making by industries through legal means, as within a classical permit procedure. As the operators of firms are free to decide on either establishing greenhouse gas reducing opportunities or buying allowances, this means that the public and NGOs have no legal opportunity to influence the decision-making of this installation. The lack of involvement of citizens has been criticized by Robert Baldwin. He stated that with emissions trading the public, notably consumers, have no or only little information regarding the emissions abating efforts of suppliers and manufactures, or the locations at which any abatement efforts are being made.[79] The market on the basis of

[78] See the critical discussion by MacDonald and Makuch (2006).

[79] R. Baldwin (2008), p. 24. See also MacDonald and Makuch (2005) discussing the (lack of) desire of the public to participate in the emissions trading decision-making process.

which industries make their decisions hides, in fact, information which may be crucial to environmentally conscious consumers for deciding what product to buy. When buying products or services, consumers cannot take account of the carbon output of the product or service, and thus, they cannot influence the carbon-behavior of industries. This lack of information, and thus lack of opportunities for citizens to behave in a carbon-friendly manner, can maybe be solved through additional measures, like labeling or even an additional carbon tax, although these also need to be considered on their merits. Moreover, labeling would help citizens who want to conduct carbon-friendly behavior and thus want to know about the carbon load of products and services.

Any attempt to repair the lack of citizen involvement within emissions trading means that an additional policy approach is needed. For instance, the taxation of the carbon load would mean that products and services would be double targeted: first by the emissions trading instrument, and secondly by the carbon tax. Such a combination of instruments reduces the main argument that emissions trading is a least-cost approach. Nevertheless, if an additional approach through labeling and taxation would enhance citizen involvement and thus would imply an additional force for steering towards low carbon options, it would still be worthwhile to consider it.

In general, it is indeed recognized within environmental law literature that the combination of different instruments might be a worthwhile approach.[80] However, here too we are in the learning stage: with what instruments could emissions trading possibly be combined, and how? In view of enhancing citizen involvement for steering society to low-carbon consumption labeling could be considered.[81] However, the question of how labels could or should be designed needs to be raised, and, moreover, whether labels would be effective and whether taxation would perform better instead. One of the main questions is whether it would be wise to set up a *carbon* label, or whether a *broader environmental* labeling scheme should be preferred? If, again, a carbon-focused approach were to be established, as an additional tool to greenhouse gas-emissions trading, the sole concentration on carbon could stimulate shifts to other environmental problems. Such effects need of course to be studied, not only when considering labeling (and taxation), but already at the start of policies focusing on reducing carbon dioxide and other greenhouse gases.

Another question would be whether the labeling scheme should be based on voluntary approaches, or whether a labeling scheme should be prescribed by law. One can indeed imagine that industries might establish a voluntary labeling scheme, but then the question must still be answered about how the

---

80 See the detailed discussion in Gunningham and Gabrosky (1998).

81 OECD Environmental Outlook to 2030, summary in English, OECD 2008, p. 13.

trustworthiness of such a system is guaranteed. Some governmental involvement in order to guarantee the reliability of the labels, is to be recommended here. It is obvious that an obligatory labeling system or the governmental inspection of voluntarily established labels would entail additional costs for industries and governments. However, looking at it from a democratic perspective in a broad sense, meaning citizen involvement, it could be attractive to consider additional approaches that enable citizens to have a choice as to what products and services to buy. As both (carbon) taxation and (carbon) labeling of products and services concerns primarily need to be considered at the EU level because of internal market concerns, it is first up to the Commission whether to choose to examine and eventually to propose such additional tools.[82]

4.5 Democratic Accountability in the Long Run

Moreover, the climate change problem will not be solved over a short period as a huge period of transition is required. This can be illustrated by the perceived need to settle long-term emissions reduction standards, running up until 2050. This time-dimension raises an interesting question from a democratic accountability perspective, as it touches upon the responsibility of the present generation, represented by their chosen government, to the latter generations, who in effect have no present-day representation. On the other hand, a far-reaching binding environmental commitment would bind later governments and later generations, who could well have a different view on the matter and also about the stringency of the environmental target. It would be interesting to explore further this 'sustainable development' aspect from a view of 'accountability' of the present government to later governments. Maybe it is preferable to set long-term policy targets, and shorter legally binding targets, enabling each future democratically chosen legislator to take up its responsibility and to adopt its decision-making guided by, possibly, newly arisen insights and circumstances.

5. CONCLUSION

The current state of affairs is that the Commission, following the plea from member states, has proposed the harmonization of the EU ETS. It wants to

---

[82] Of course, one needs to review the meaning of the already adopted taxation and labeling legislation, but it seems that these measures have only a very limited meaning.

introduce auctioning, an EU-wide cap, and a further harmonization of the method of free allocation and the use of the flexible mechanisms. This would imply an important shift to decision-making on the EU level, with some far-reaching competences for the Commission, removing an important part of climate policies from the national governments. The new election of the European Parliament in spring 2009 seems to imply an incentive for a hasty process, aiming at having measures adopted before these elections, and thus probably not giving enough careful consideration in the legislative process to main questions concerning the design of the scheme. As with establishing the legal framework for emissions trading, the most important choices need to be made by the legislator. Where a critical choice needs to be made among different models, we can only hope that the legislative process will be as careful as possible, thereby not only taking into account the economic optimal solution, but also discussing different options from the perspective of legal values.

With emissions trading, we are however still in a learning phase regarding the possible design of the instrument and its related economic and other effects. Although the acid rain emissions trading system has been proven successful, this is no guarantee that other emissions trading models will be successful as well. In addition, not only instrumental effects like effectiveness and efficiency count: legal values also need to be considered when reviewing the merits and pitfalls of an instrument. We indicate the need for further assessments of the functioning of the different forms of emissions trading schemes, focusing not only on macroeconomic effects and administrative costs, but, moreover, also with regard to legal values. Not only the analysis of case law, but moreover a discussion of emissions trading against the backdrop of core legal values is needed in order to understand the system and to assess further improvements of the system.

One of such values is democratic accountability. In this article, we have elaborated on the specific view that emissions trading would allow the legislator to concentrate on the core question of environmental law. However, we have also stated that within particularly the climate change dossier, which concerns so many and such different sources, the core question 'what may be polluted during a certain period in a certain region?' is to be complemented with another one, which is 'which sector needs to reduce how much?' For comprehensive climate change policy-making, and as long as the emissions trading instrument cannot or will not cover all greenhouse gas-emitting sources, the legislator thus needs to discuss this distributional question too. For the specific legal system of the EU, with the concept of shared competences and the principle of subsidiarity, it needs furthermore to be considered on what level those questions should be answered. The initial EU ETS includes quite a clear distinction regarding both questions, and it would as such have been possible to proceed with that in the forthcoming years: on the

EU level the commitments of states contributing to the overall reduction target of for instance minus 20% or 30% in 2020 would then be discussed, while the national governments and parliaments would then decide about the distribution of the efforts between the EU ETS sector and other sectors. In this form, the democratic accountability then seems transparently arranged: the EU level will then focus on commitments among states, while the states themselves focus on commitments among sectors. The proposal of the Commission to revise the EU ETS implies however an important shift of decision-making from member states to the EU-level, which means that the distributional question can no longer be comprehensively addressed by a single legislator: both the European and the national legislators will be responsible, each for their part, for the distributional choices to be made. The push for a level playing field thus seems to win, from getting a broader experience with different national approaches for allocation issues, and, moreover, from leaving important distributional choices to national democratic processes.

In this vein, we also need to consider what the consequences of a further harmonization through the emissions trading instrument would be for public participation and, in a broader sense, the specific position of citizens. It is a pity that this aspect has yet been under-explored in the assessment of the initial emissions trading scheme. However, we still need to explore what the possible role of public participation and other forms of citizen involvement might be even within climate change policies. Thereby we need to recognize that precisely this market-based type of regulation could have as a consequence that citizens might feel even more lost, because the emissions trading instrument prevents them from getting involved in the decision-making of private companies. When the market and not the government lead decisions, there is less room for public participation compared to the traditional permit system. Therefore, the idea of a label revealing the carbon load of a product and service seems a complementary instrument to emissions trading, especially if the emissions trading instrument were to be further harmonized at the EU level. Labeling could probably make the citizen feel connected again with climate change but in an alternative and possibly even better way: the citizen can then act as a critical consumer of carbon products and services. This consumer behavior can have an influence on market decisions taken by the industries and service providers, and that influence might be even more powerful than public participation. Of course, labeling also is surrounded by challenging questions: would it be wise to think of a carbon label, or would it be better to consider an environmental label as this would prevent the trade-offs between the carbon problem to other environmental concerns? How can it be fitted into the international trade law framework? Should a labeling system be mainly voluntary, combined with some governmental checks on the credibility of the labels, or should the labels as such be obligatory? Should a tax maybe be preferred over labeling? Is the consumer indeed a critical consumer aiming at reducing carbon emissions by buying low-carbon products? Such questions should of course be taken up when exploring the usefulness of labeling (or taxation), and this instrument is as such thus surrounded by some challenging questions. However, the market-based character of emissions trading together with the shift of decision-making from national democratic processes to decision-making on the EU level creates in fact a need to investigate which complementary instrument will facilitate citizens having an alternative say within climate change policies.

REFERENCES

Ackerman, B.A. and R.B. Stewart (1984), 'Comment: Reforming Environmental Law', *Stanford Law Review*, 37 (5), 1333–65.

Ackerman, B.A. and R.B. Stewart (1988), 'Reforming Environmental Law: The Democratic Case for Market Incentives', *Columbia Journal of Environmental Law*, 13, 171–99.

Baldwin, R. (2008), 'Regulation Lite: The Rise of Emissions Trading', *Law Society Economy Working Papers*, www.lse.ac.uk/collections/law/wps/wps.htm and http://ssrn.com/abstract=1091784.

Van Calster, G. (2008), 'Against Harmonisation – Regulatory Competition in Climate Change Law', *Carbon & Climate Law Review*, 2 (1), 89–95.

Cramton, P. and S. Kerr (1998), 'Tradable Carbon Permit Auctions: How and Why to Auction Not Grandfather', *Discussion Paper 98-34*, May, Resources for the Future.

Dales, J.H. (1986), *Pollution, Property and Prices: an Essay in Policy-making and Economics*, Toronto, University of Toronto Press, republished by Edward Elgar in 2002.

Driesen, D.M. (2003), *The Economic Dynamics of Environmental Law*, Cambridge, MA, MIT Press.

Ecofys (Wesselink, Bart; Klaus, Sebastian; Gilbert, Alyssa and Blok, Korneli) (March 2008), *The IFIEC Method for the Allocation of CO_2 Allowances in the EU Emissions Trading Scheme. A Review Applied to the Electricity Sector.*

Ellerman, A. Denny (2004), 'The US SO2 Cap-and-Trade Programme', in OECD, *Tradeable Permits, Policy Evaluation, Design and Reform*, pp. 86–88.

Ellerman, A.D., B.K. Buchner and C. Carraro (2007), 'Unifying Themes', in A. Denny Ellerman, Barbara K. Buchner and Carlo Carraro (eds), *Allocation in the European Emissions Trading Scheme: Rights, Rents and Fairness*, Cambridge University Press, 2007, p. 352.

Faure, M. (1998), 'Environmental Regulation, Encyclopedia of Law and Economics', http://users.ugent.be/~gdegeest/2300book.pdf.

Faure, M. (2001), 'Regulatory Competition versus Harmonisation in EU Environmental Law', in D. Esty and D. Geradin (eds), *Regulatory Competition and Economic Integration*, Oxford, Oxford University Press, pp. 263–86.

Faure, M., M. Peeters and A. Wibisanan (2006), 'Economic Instruments: Suited for Developing Countries?', in Michael Faure and Nicole Niessen (eds), *Environmental Law in Development. Lessons from the Indonesian Experience*, Cheltenham, UK and Northampton, MA, Edward Elgar.

Gunningham, N. and P. Gabrosky (1998), *Smart Regulation, Designing Environmental Policy*, Oxford, Clarendon Press.
Hovi, J. and B. Holtsmark (2006), 'Cap-and-Trade or Carbon Taxes? The Feasibility of Enforcement and the Effects of Non-compliance', *International Environmental Agreements*, 6, 137–55.
Kiiver, P. (2007), Europe in Parliament, towards Targeted Politization, (Dutch) Scientific Council for Government Policy, Web publications 23, www.wrr.nl.
Latin, Howard A. (1985), 'Ideal versus Real Regulatory Efficiency, Implementation of Uniform Standards and "Fine-Tuning" Regulatory Reform', *Stanford Law Review*, 37, 1267–1332.
MacDonald, K. and Z. Makuch (2006), 'Emissions Trading and the Aarhus Convention: a Proportionate Symbiosis?', in Marjan Peeters and Kurt Deketelaere (eds), *EU Climate Change Policy. The Challenge of New Regulatory Initiatives*, Cheltenham, UK and Northampton, MA, Edward Elgar, pp. 125–52.
Maloney, M.T. and R.E. McCormick (1982), 'A Positive Theory of Environmental Quality Regulation', *Journal of Law and Economics*, 25, 99–123.
Nash, J.R. (2000), 'Too Much Market? Conflict between Tradable Pollution Allowances and the 'Polluter Pays Principle'', Harvard Environmental Law Review, 24, 465–535.
Pallemaerts, M. and R. Williams (2006), 'Climate Change: the International and European Policy Framework', in Marjan Peeters and Kurt Deketelaere (eds), *EU Climate Change Policy: The Challenge of New Regulatory Initiatives*, Cheltenham, UK and Northampton, MA, Edward Elgar, pp. 22–50.
Peeters, M. (2006a), 'Enforcement of the EU Greenhouse Gas Emissions Trading Scheme', in Kurt Deketelaere and Marjan Peeters (eds), *EU Climate Change Policy: The Challenge of New Regulatory Initiatives*, Cheltenham, UK and Northampton, MA, Edward Elgar, pp. 169–87.
Peeters, M. (2006b), 'Inspection and Market-based Regulation through Emissions Trading: the Striking Reliance on Self-monitoring, Self-reporting and Verification', *Utrecht Law Review*, 2 (1), 177–95.
Potts, Brian H. (2007), 'Trading Grandfathered Air, a New, Simpler Approach', *Harvard Environmental Law Review*, 31, 115–62.
Pring, G. (Rock) (2006), 'A Decade of Emissions Trading in the USA: Experiences and Observations for the EU', in Marjan Peeters and Kurt Deketelaere (eds), *EU Climate Change Policy. The Challenge of New Regulatory Initiatives*, Cheltenham, UK and Northampton, MA, Edward Elgar, pp. 188–204.
Tietenberg, Thomas H. (1985), *Emissions Trading: an Exercise in Reforming Pollution Policy*, Washington D.C., Resources for the Future.
Verhey, L., H. Broeksteeg and I. Van den Driesche (2008), *Political Accountability in Europe: Which Way Forward?*, Groningen: Europa Law Publishing.
Victor, David G. (2001), *The Collapse of the Kyoto Protocol and the Struggle to Slow Global Warming*, Princeton, NJ, Princeton University Press.
Weishaar, S. (2007), *Law and Economics Analysis of the European Greenhouse Gas Emissions Trading System: Allocation and Competition*, PhD thesis, Maastricht University, The Netherlands.

3. Too much harmonization? An analysis of the Commission's proposal to amend the EU ETS from the perspective of legal principles

Javier De Cendra De Larragán[1]

1. INTRODUCTION

The EU has adopted a strong political stance towards climate change. In recent years it has taken a number of important steps in an attempt to develop a comprehensive climate change policy,[2] at which core lay the EU burden-sharing agreement established by Council Decision 2002/358/EC[3] and the emissions trading scheme (EU ETS) established by Directive 2003/87/EC[4] (hereafter Directive 2003/87/EC or the directive). The latter directive covers the key elements of the EU ETS, including the basic rules for the allocation of allowances. However it adopts a highly decentralized approach, affording Member States a significant leeway to make important regulatory choices. This decentralized approach has seemingly brought numerous problems, reducing the capacity of the EU ETS to achieve the main objectives of the directive, namely environmental effectiveness, cost-effectiveness and economic efficiency, and the sound functioning of the internal market by avoiding distortions of competition.[5] Article 30 of the directive mandates the

---

1 Ph.D Researcher, METRO, University of Maastricht.

2 However, EU climate change policy cannot be considered to be comprehensive yet. See Ludwig Kramer, 'Some reflections on the EU mix of instruments on climate change', in Peeters and Deketelaere (2006).

3 Council Decision 2002/358/EC concerning the approval, on behalf of the European Community, of the Kyoto Protocol to the United Nations Framework Convention on Climate Change and the joint fulfilment of commitments thereunder, OJ L 130, 15.5.2002, p.1.

4 Directive 2003/87/EC of the European Parliament and of the Council establishing a scheme for greenhouse gas emission allowance trading within the Community and awaiting Council Directive 96/61/EC, OJL 275, 25.10.2003, p. 32.

5 Article 1, recital 7, recital 18.

Commission to undertake a review of the directive which may lead to amendments in order to improve the functioning of the scheme. Pursuant to this mandate, the Commission has recently issued a proposal to amend the directive. This proposal introduces a highly harmonized scheme in which Member States are left with virtually no choices when implementing its crucial elements.[6]

Against this background, this article seeks to examine closely the reasons given by the Commission for adopting such a highly harmonized approach, particularly with respect to the allocation methodology, from the perspective of three key legal principles of Community law, subsidiarity, polluter pays principle, and equality. These principles are selected because they are frequently relied upon to request further harmonization of EU (environmental) law.[7] The specific question this chapter wants to examine is whether the approach proposed by the Commission can be supported and/or required by those principles, and if so to what extent.

The term 'allocation methodology' or 'allocation process' is used to refer to three closely related elements: (1) the determination of the cap; (2) the allocation to sectors; (3) the allocation to individual installations.[8]

2. STRUCTURE

The chapter starts by outlining the core elements of an emissions trading scheme and the concept of harmonization. Second, it introduces the legal principles which will be used for the analysis. Third, it examines the allocation methodology in Directive 2003/87/EC from the perspective of the concept of harmonization. It reviews the problems to which that allocation methodology has given rise, from the perspective of legal principles, and considers the solutions which have been proposed to solve them in the recent Commission's proposal to amend the Directive. Last, it analyses whether the particular level of harmonization proposed by the Commission seems justified on the basis of those principles.

---

6 COM(2008)30 Final.

7 See in general, for the polluter pays principle, Jans and Vedder (2007, p. 43). In the context of the EU ETS, see for instance Hepburn et al. (2006); for the principle of subsidiarity, see for instance Kramer (2007, pp. 17–20); for the principle of equality, see Tridimas (2006, pp. 95 *et seq.*). In the context of the EU ETS, see Del Río (2006, p. 464).

8 See for instance Zapfel (2007). See also Ecofys (2006, p. 6).

3. EMISSIONS TRADING SCHEMES AND THE ABSTRACT APPEAL FOR (FULL) HARMONIZATION

3.1 Emissions Trading Schemes

An emissions trading scheme (ETS) consists of a market in which companies trade among each other tradable emission rights representing each a certain amount of emissions, with the objective of achieving reductions of emissions at the lowest possible cost. Such a market allows in theory the promotion of three important policy objectives: (1) environmental effectiveness; (2) cost-effectiveness and economic efficiency; (3) equity. When the amount of tradable emission rights is set ex ante by the regulator, the environmental effectiveness is in theory defined by the cap. Further, the functioning of the market promotes cost-effectiveness in achieving the target, by promoting allocative efficiency. Lastly, it is possible to promote equity through the allocation methodology.[9]

While the concept of an ETS is in theory very simple, its actual design and implementation has an important influence on the degree to which the three objectives are actually achieved. The edifice of an ETS is built upon six main pillars: (1) the mechanism to set the cap; (2) the rules to set the scope (or coverage) (3) the methodology to allocate emission rights to companies participating in the scheme; (4) the mechanism for monitoring, verifying and reporting the emissions; (5) the market itself where allowances can be traded; (6) the system of penalties.

A too-lenient cap, an incomplete coverage in terms of competitors, allocation rules which distort competition within the market, lack of competition in the market, defective measuring, verifying and reporting rules (or their defective application), and inadequate and/or insufficient penalties may all constitute factors that reduce the beneficial effects of the scheme. The risks that any of these problems arise increase when an ETS is introduced at a multi-jurisdictional level, with each jurisdiction enjoying competence to make important design choices on those pillars. Hence, in these cases, a highly harmonized approach could seem in principle preferable.

3.2 Harmonization in Community Law

The concept of harmonization in Community law is closely linked to the principles of conferral,[10] subsidiarity and proportionality. In the context of

---

[9] Daly (1992).

[10] As amended by the Treaty of Lisbon amending the Treaty on European Union and the Treaty establishing the European Community, signed at Lisbon, 13 December 2007 OJ C 306, 17.12.2007.

environmental legislation, the competence to regulate is shared between the Community and the Member States. Hence Community action must be in conformity with the principles of subsidiarity and proportionality.[11]

The Community can, on the basis of the principles of subsidiarity and proportionality, introduce harmonized regulation covering only certain aspects of a particular field, leaving others to be regulated by the Member States, or may decide to totally harmonize all the relevant aspects of that particular field, leaving Member States with limited discretion to adopt different approaches. In the latter cases, the freedom of Member States to regulate the field will depend both on the content of the Community legislative Act and on the legal basis on which that legislative Act was based.[12] The choice therein is between partial or complete harmonization. Harmonization can be, aside from partial or complete, minimum, maximum, or anywhere between both extremes. When harmonization is complete, Member States cannot in principle go beyond the stringency level set in Community regulation. Complete harmonization is found mainly in those fields of environmental law where there is a strong relation with the free movement of goods, normally with legal basis on Article 95 EC.[13] However, in the ambit of environmental law with a legal basis on Article 175, Article 176 EC guarantees that Member States can always set more stringent environmental objectives than those laid down in Community legislation, but not different policies.[14]

Community environmental law is generally introduced through directives while regulations are exceptional.[15] Directives are mandatory as regards ends, but leave Member States freedom as regards the means. However, while some directives may introduce partial and minimum harmonization, others may introduce complete harmonization, thus resembling regulations. Whereas complete harmonization is generally necessary in the context of product standards, to avoid Member States introducing measures which amount to barriers to trade, in the context of emission standards complete harmonization is generally not required. Here, Member States and companies may have an interest in pushing for more harmonization at EU level in order to avoid distortions of competition in the internal market stemming from the differing stringency of domestic standards.

---

11 EC Treaty, Article 3b.
12 Kramer (2007, p. 126).
13 Jans and Vedder (2007, p. 94).
14 Kramer (2007, p. 127). See also Jans and Vedder (2007, p. 103).
15 Kramer (2007, p. 56); Jans and Vedder (2007, p. 87).

3.3 The Degree of Harmonization in Directive 2003/87/EC

Directive 2003/87/EC has introduced the basic architecture of the EU ETS. Member States have a large degree of freedom when implementing the directive in their domestic legal systems. The Commission must develop guidance to assist Member States in that process in order to ensure that its objectives are achieved, and can influence the choices made by Member States mainly through the evaluation of national allocation plans (NAPs).[16] Hence, the directive has introduced a complex mode of harmonization. The main issue analysed in this chapter is whether the principles of subsidiarity, polluter pays, and equality may require a higher level of harmonization than is currently the case and whether the level proposed by the Commission is the adequate one to ensure respect for those principles.

4. THE ANALYTICAL FRAMEWORK BASED ON LEGAL PRINCIPLES

Legal principles of Community law can be roughly divided into three groups: (1) constitutional principles; (2) general principles of Community law; (3) institutional principles, or principles regulating the relation between institutions and Member States.[17] Legal principles fulfill generally three functions: (1) guide the legislator in the law-making process; (2) guide the interpretation of legislation by courts; (3) serve to test the legality of legislation and administrative decisions.[18] In this contribution the focus lies particularly on constitutional principles and general principles of Community law and functions (1) and (3).

4.1 The Principle of Subsidiarity

Article 3b of the Lisbon Treaty[19] on the principles of subsidiarity and proportionality reads as follows:

> In areas which do not fall within its exclusive competence, the Union shall act only if and insofar as the objectives of the proposed action cannot be sufficiently

---

16 See Articles 9, 14, 29, and Annex III of the directive.

17 Tridimas (2006, p. 4).

18 Ibid.

19 Treaty of Lisbon amending the Treaty on European Union and the Treaty establishing the European Community, signed at Lisbon, 13 December 2007. O JC 306, 17.12.2007.

achieved by the Member States, either at central level or at regional and local level, but can rather, by reason of the scale or effects of the proposed action, be better achieved at Union level [. . .] Under the principle of proportionality, the content and form of Union action shall not exceed what is necessary to achieve the objectives of the Treaties.

The principle of subsidiarity sets three requirements for the Community legislator to gain competence: (1) that a certain issue cannot be sufficiently dealt with by Member States; (2) that it can be dealt with more effectively by the Community; (3) a proportionality test. The principle of subsidiarity in Community law does not seem to have been designed as a mechanism to preserve and promote (a certain level of) autonomy of lower levels within societies.[20] The EC Treaty puts the accent on promoting cooperation between the Community and Member States in order to achieve shared objectives rather than on establishing a sharp differentiation of competences. So, the question of which precise level of harmonization would ensure that the main objectives of an ETS are achieved while leaving as maximum freedom as possible to Member States to pursue their own regulatory choices does not have a precise legal answer.

The Protocol on the subsidiarity principle[21] introduces a number of formal requirements to ensure that the principle is taken seriously. The Commission must consult widely before proposing legislative acts, must make its case on the basis of qualitative, and wherever possible, quantitative indicators, and must ensure that the burdens for administrations stemming from the proposal are minimized and remain commensurate with the objective sought. Under the ordinary legislative procedure, if 55% of the members of the Council or a majority of the votes cast in the European Parliament are of the opinion that a proposal made by the Commission is not compatible with the principle of subsidiarity, the legislative proposal will not be considered further. The Court of Justice has jurisdiction to examine whether legislative acts infringe the principle of subsidiarity, and the key issue is the intensity of its review, which so far has been low.[22] Kramer has pointed out that the decision on the level at which regulation should be adopted is ultimately a political one.[23]

Academics within the field of Economics Analysis of Law have sought to find a number of criteria to answer more precisely that question.[24] The starting point is that the EC Treaty gives competence to the Community to

---

20 Barber (2005).

21 OJ C 306/150, 17.12.2007.

22 Craig (2006, p. 425).

23 Kramer (2007, p. 20).

24 See for instance Van den Bergh (1996); Rehbinder and Stewart (1985); Carbonara et al. (2008).

regulate only if and insofar as centralization leads to superior (welfare) effects.[25] There would seem to be three main arguments in favor of a decentralized approach:

1. When legislators compete with each other in the market of laws *a la Tiebout*;
2. When there are informational asymmetries, and hence local governments seem to be in a better condition than central regulators to monitor industries;
3. Competition between regulators may serve as a learning process to achieve better solutions in terms of welfare.

Arguments in favor of (a higher degree of) centralization would be:

1. The existence of transboundary externalities, such as environmental pollution;
2. The existence of economies of scale and of transaction costs;
3. The danger of a race to the bottom, which arises when legislators do not compete as competitive firms in the market, but rather under prisoners' dilemma conditions;
4. The risk of regulatory capture.[26]

Applying this set of criteria may lead to the conclusion that some elements of a regulatory framework should be harmonized while others should or could remain under the competence of Member States. Sweeping harmonization or complete decentralization will rarely turn out to be the best approach. However these criteria will not be systematically assessed in this paper except when evidence can point to their direct relevance for a certain choice.

1.2 The Polluter Pays Principle

The polluter pays principle is defined neither in the EC Treaty nor in the case law of the European courts. Article 175 EC refers to it as one of the guiding principles of Community law, and it is at the basis of the guidelines on environmental state aid.[27] The ECJ related the polluter pays principle to the principle of proportionality in *Standley*.[28] In that case, a number of farmers

---

25 Van den Bergh (1996, p. 364).

26 Van den Bergh (2000).

27 Community Guidelines on State Aid for Environmental Protection, OJ C 82, 1.4.2008.

28 Case C-293/97 *Standley* [1999] ECR I-2603.

challenged several decisions made by the Secretary of State of the UK regarding the implementation in the UK of Council Directive 91/676/EEC of 12 December 1991[29] concerning the protection of waters against pollution caused by nitrates from agricultural sources. The directive required the identification of areas affected by nitrate pollution, and the restriction of economic activity within them. The farmers claimed that the directive imposed a disproportionate burden upon them, because other pollution sources were left unaddressed. According to the farmers, the directive infringed the polluter pays principle, the principle of rectification at source, the principle of proportionality and the fundamental right to property. The British court made a preliminary reference to the ECJ. In relation to the polluter pays principle, the ECJ ruled that the directive does not mean that farmers must take on burdens for the elimination of pollution to which they have not contributed.[30] Further, according to the Court,

> the Member States are to take account of the other sources of pollution when implementing the Directive and, having regard to the circumstances, are not to impose on farmers costs of eliminating pollution that are unnecessary. Viewed in that light, the polluter pays principle reflects the principle of proportionality on which the Court has already expressed its view.[31]

Jans considers that this means that European measures must avoid putting burdens on persons and undertakings for the elimination of pollution to which they have not contributed.[32]

Leaving aside definitional problems with the principle,[33] it has two aims: on the one hand, to ensure that polluters internalize the social costs generated by pollution, thereby reducing production to the socially optimal levels; on the other, to ensure that such internalization is made by all companies within a market, in order to avoid distortions of competition stemming from different environmental regulations.[34]

The socially optimal level is generally equated with the level set by the government, though this does not have to be the case, and there seems to be a trend towards governments seeking to find and prescribe the optimal level of

---

29 OJ 1991 L 375, p. 1.

30 *Standley*, fn. 28, para. 51.

31 *Standley*, above, para. 52.

32 Jans and Vedder (2007, p. 44).

33 De Sadeleer (2002). Two difficult questions must be answered to make the principle operational: (1) who is the polluter; (2) how much must the polluter pay?

34 See the Communication from the Commission to the Council regarding cost allocation and action by public authorities on environmental matters, OJ L 1975, 194/1. See Gaines (1991). See also Jans and Vedder (2007, p. 42).

pollution. This would seem to be the objective of the EU when adopting a (non-legally binding) long-term target for the entire EU of 60–80% emissions reduction by 2050, a reduction considered necessary to ensure a global average increase in temperature below 2 degrees Celsius.[35] On the other hand, there is deep disagreement among economists about what constitutes the optimal level of pollution, and the choice of a certain level may seem ultimately rather arbitrary.[36] From a regulatory perspective, the principle requires ensuring that environmental regulations implement a relatively similar level of internalization across countries in order to avoid distortions of competition. However, it does not require complete harmonization of the content of regulatory approaches.

As mentioned, the polluter pays principle is at the core of the Community guidelines on state aid for environmental protection,[37] which refer explicitly to tradable permit schemes.[38] The guidelines consider that ETS can lead to state aid when a Member State grants allowances to companies below their market value, which is by definition the case when allowances are allocated for free. According to the guidelines, 'when the global amount of permits granted by Member States is lower than the global expected needs of undertakings, the overall effect on the level of environmental protection will be positive'.[39] Hence, state aid will be allowed. Further, at the individual level of each undertaking, if the allowances granted do not cover the totality of expected needs of the undertaking, the state aid will be also allowed. In order to limit distortions of competition, no over-allocation of allowances can be justified and provision must be made to avoid barriers to entry.[40] These criteria apply to the second trading period 2008–2012. But the approach of the guidelines seems legally problematic. How to calculate the expected needs of installations? How to calculate the expected needs of each undertaking? What does over-allocation precisely mean? Which provisions must be made to avoid barriers to entry? The guidelines lead to a substantial amount of legal uncertainty,[41] and this indicates the difficulty of shaping precise rules on the basis of the polluter pays principle.

---

35 COM(2006)676.

36 Risbey (2006). See for the calculation of economics costs of abatement Hope (2006).

37 Community guidelines on state aid for environmental protection, OJ C 82/1, 1.4.2008.

38 Ibid, para. 55.

39 Ibid.

40 Ibid.

41 De Sepibus (2007b), at p. 25.

4.3 The Principle of Equality

4.3.1 Criteria for its application

The principle of equality in Community law functions both as a powerful tool for pursuing ever deeper Community-wide harmonization within the internal market and, in the hands of the European courts, to test the legality of Community and domestic legislation implementing the former. The principle of equality as applied by the European courts requires that competitors are treated similarly, and that differential treatment be objectively justified. In general, two elements are required to determine whether there is a breach of the principle: (1) a set of criteria for comparing undertakings; (2) a set of criteria to decide whether differential treatment is justified or not. According to settled case-law, the ECJ and the CFI will first have to determine whether two products or undertakings are in a comparable situation. In order to do so, the Court will normally have recourse to the criterion of competition.[42] This criterion serves to establish whether two producers or undertakings compete in the internal market. In the case of undertakings, the Court may have regard to their production[43] or to their legal structure[44] in order to determine whether their competitive positions are comparable. For two products or two undertakings to be in a comparable situation it is sufficient, in principle, that they are potentially in competition.[45] The criterion of substitutability or competitiveness is used by the Court as the principal criterion throughout the field of economic law, for instance in relation to Article 28[46] or Article 82 EC.[47]

Difference in treatment between comparable situations is not prohibited where it is objectively justified. The notion of objective justification is not easy to define in the abstract, but depends on the objectives of legislation and the circumstances of the case. Toth has, on the basis of case law, listed the cases where differential treatment of products or undertakings is justified in view of the aims that the Community institutions lawfully pursue as part of Community policy:[48]

[42] Joined Cases 103 and 145/77 *Royal Scholten-Honig v Intervention Board for Agricultural Produce* [1978] ECR 2037, paras. 28–29. See Tridimas (2007, p. 81).

[43] Case 14/59 *Pont-a-Mousson v High Authority* [1959] ECR 215 at 232.

[44] See Joined Cases 17 and 20/61 *Klockner v High Authority* [1962] ECR 325 at para. 345.

[45] Case C-319/81 *Commission v Italy* [1983] ECR 601, para. 16.

[46] Case C-391/92 *Commission v Greece* [1995] ECR I-1621.

[47] Case 27/76 *United Brands v Commission* [1978] ECR 207.

[48] A.G. Toth. *The Oxford Encyclopaedia of European Community Law* (Oxford University Press, 1990), Vol I, pp. 191 et seq.

1. Where its purpose is to obviate special difficulties in a sector of industry;
2. Where it is not arbitrary in the sense that it does not exceed the broad discretion of Community institutions;
3. Where it is based on objective differences arising from the economic circumstances underlying the common organization of the market in the relevant products.

The guiding principle seems to be that the difference in treatment must not be arbitrary, i.e., must be based on rational and objective considerations.[49] However, there are two additional elements that must be considered in deciding whether Community secondary law is in breach of the principle.

4.3.1.1 *The right of the Community to legislate step-by-step* The ECJ has held that a harmonization measure which is intended to standardize previously disparate national rules may produce different effects depending on the previous state of the national laws but that does not amount to discrimination, provided that the measure applies equally to all Member States.[50] Incremental harmonization may lead to differences in treatment since certain legal relations may be subject to Community rules whereas other comparable relations remain subject to national laws. Such difference in treatment may be provisional or more permanent, where the Community does not intend, or for one reason or another, is unable to harmonize an area of law. The harmonization process is by nature a complex exercise surrounded by legal and political difficulties which determine the time in which harmonization measures are adopted as well as their content.[51] If in *Francovich II* the Court had reached the opposite result, it would have elevated the principle of non-discrimination to an autonomous source of harmonization.[52] Thus, in coordinating national laws, partial, step-by-step, harmonization is the preferred, indeed the only feasible, Community policy.[53] In *Safety High-Tech*,[54] the ECJ stated that Article 174.1 EC does not oblige the Community to protect the environment in its entirety, from which it can be implied that it is obliged to protect it at least partially. Winter argues that the Court could have gone further by stating that, in situations where a complex set of problems must be solved urgently,

---

49 Case 139/77 *Denkavit v Finanzant Warendorf* [1978] ECR 1317.

50 (Case C-331/88 *Fedesa and Others* [1991] ECR I-4023).

51 (See also Case C-36/99 *Ideal Tourisme* [2000] ECR I-6049).

52 Case C-479/93 *Francovich v Italian Republic* [1995] ECR I-3843, at 1837.

53 In Case C-63/89 *Assurances du Credit v Council and Commission* [1991] ECR I-1799. Case C-479/93 *Francovich v Italian Republic* [1995] ECR I-3843, (*Francovich II*).

54 Case C-284/95, Judgment of 14/07/1998, *Safety Hi-Tech / S. & T.* [1998] ECR I-4301.

but are difficult to handle because of limited instrumental and administrative capacity, all issues need not be solved all at once, and public authorities could go step-by-step by singling out individual actors (or substances), as long as this would be part of a broader plan providing for systematic action in the future.[55] In *Standley*, the Court could have asked whether that directive was part of a broader framework to tackle all nitrate sources, but instead it satisfied itself with an isolated consideration of the contribution of one source, farming, to pollution by nitrates. In the context of legislation adopted *ex novo* with legal basis on Article 175 EC, harmonization often needs to proceed step-by-step, and may give rise to distortions of competition. For instance Directive 2003/87/EC has adopted a highly decentralized approach, while including in Article 30 the possibility to adopt a more harmonized approach in the future. Some authors have denounced that such an approach leads to distortions of competition within the internal market and have pleaded for a higher level of harmonization.[56]

Moreover, the Court has stated that the obligation of the administration to ensure as far as possible equal competitive conditions to all market participants is conditioned by the need to implement structural objectives by means which are 'achievable and verifiable in practice' and which do not give rise to undue administrative and supervisory difficulties.[57]

4.3.1.2 *The consideration of the discretion of Community institutions* In *Wuidart and Others*,[58] the ECJ found that, in taking policy decisions which require the assessment of a complex economic situation, the Community legislature enjoys wide discretion and the Court will only engage in marginal review. The Court cannot substitute its own assessment for that of the Community legislature, but must confine itself to examining whether the assessment made contains a manifest error or constitutes a misuse of powers or exceeds clearly the bounds of the legislature's discretion. In such cases, the discretion of the institution which wrote the act extends also 'to a certain extent, to the findings as to the basic facts, especially in the sense that it is free to base its assessment, if necessary, on findings of a general nature'.[59]

4.3.2 Limits to its applicability

Even if full harmonization were to be achieved, Community legislation would still impact differently different undertakings. As AG Jacobs has noted, 'the

55 Winter (2004, p. 27).
56 Ray Schmitt (2006); De Sepibus (2007a).
57 See *SAM Schiffahrt*, fn. 59.
58 Joined Cases C-267-85/88 *Wuidart and Others* [1991] ECR I-435.
59 Joined Cases C-248-9/95 *SAM Schiffahrt and Stapf v Germany* [1997] ECR I-4475, paras. 24–25.

principle of equality cannot preclude the legislature from adopting a criterion of general application [...] it may affect different persons in different ways, but beyond certain limits any attempt to tailor the legislation to different circumstances is likely only to lead to new claims of unequal treatment'.[60] The same can be said in respect of the principle of proportionality.[61] The principle of equal treatment as a criterion for burden sharing constitutes one particular version of the principle of proportionality in respect of burden sharing of the costs imposed by regulation, since it mandates equal burdens for comparable individuals or firms. Hence, the reasoning made by AG Jacobs would seem to apply also more generally to the principle of proportionality. Furthermore, if we consider that the polluter pays principle constitutes a particular implementation of the proportionality principle, as suggested by the ECJ in *Standley*, then the same conclusion can be extended to it. The polluter pays principle can be used as a criterion for the distribution of burdens of regulation, with the criterion being the amount of pollution emitted by each polluter. Hence, a measure implementing the principle will inevitably impact different persons differently. Seeking to tailor legislation to each individual circumstance will likely lead to new claims of unequal treatment. This is important in relation to allocation methodologies. An allocation methodology which is based upon general criteria will by definition give advantage to some companies and damage others. By contrast, an allocation methodology which is based on the specific circumstances of each installation will tend to give each installation exactly what it needs in order to cover its emissions. But as this is impossible, since by definition the total amount of allowances allocated must be lower than actual emissions – if the emissions reduction target is to be fulfilled, in fact some installations will receive fewer allowances vis-à-vis other installations. Which installations are benefited and which are damaged will depend precisely on the criteria employed for the distribution. Those criteria should not breach the polluter pays principle, which means that, in general, the criterion for allocation must not have the effect that the worst polluters receive more allowances relative to the more efficient polluters. In this way, the principle of proportionality and the polluter pays principle reinforce each other. The principle of equality is a particular version of the former. Indeed, from the perspective of the polluter pays principle, two undertakings that compete in the internal market are only comparable when, in addition to being competitors, they emit similar amounts of pollutants per input or output. Only in that case would they be entitled to a similar amount of allowances. Hence, the polluter pays principle constitutes the relevant criterion to introduce allocation methodologies that treat different undertakings differently.

---

60 Joined Cases C-13-16/92 *Driessen and others* [1993] ECR I-4751 at p. 4780.
61 Tridimas (2006, p. 78).

4.3.3 Equality and harmonization

The principle of equality cannot become an autonomous source used by the Court to further harmonization, but constitutes a powerful tool in the hands of the Commission to strive for an ever higher level of harmonization. The reason traditionally used is the need to ensure a 'level playing field'. This concept is closely related to that of 'fair play'. The question is what should be understood by the concept of a 'level playing field'? Clearly it cannot mean total equalization between competitors, for if this were the case, competitors themselves would have to be turned into equals, and the basis for competition, comparative advantage, would entirely disappear. A less radical answer is to suggest that all elements which do not define competitors *qua* competitors must be 'leveled'. But this is unrealistic, since many factors are simply beyond the control of the regulator. A more modest approach would be to consider that competitors should compete under identical regulatory frameworks. But on reflection this appears wholly unattainable in reality. Moreover, there is no intrinsic reason why this would be preferable to having several regulatory frameworks alongside each other. After all, different situations and actors demand different regulatory solutions for their particular problems, and the principle of subsidiarity aims to guarantee autonomy to adopt different solutions to similar problems. Hence, a last definition would require simply that competitors may compete under fair conditions which permit them pursuing their competitive strategies in the market. This would require at a minimum that they can sell their products in other jurisdictions in similar conditions to those of national producers. However, this rationale only weakly applies to emission standards. Fairness is not the same as the legal principle of equality, and should not be conflated with it. The application of the principle of equality requires adopting a conception of fairness. The European courts do not endorse any particular theory, and the principle does not seek to advance any particular idea of the social good.[62] Indeed, the principle of equality does not dictate a single result, and more than one policy may be compatible with it. Hence, in the hands of the Court it does not seek to resolve issues but to act as a check on decision makers, requiring them to justify their policy choices in a consistent and rational manner.[63]

Once the Community legislator has decided to harmonize a field to a certain extent, the principle of equality can be used by the European courts to check the legality of Community secondary law. As seen above, harmonization as such cannot ensure equal treatment. Some criteria for comparison are required. Equality as a principle of distribution constitutes one version of

---

62 Tridimas (2006, p. 62).
63 Ibid.

proportionality, and since there are not two identical firms, it needs to be complemented with substantive principles, such as proportionality and the polluter pays principle. To conclude this section, we must recall that the intensity of review applied by the courts is not very high, and therefore it is not likely that it will overturn a directive on the basis of a breach of the principle of equality.[64]

5. CURRENT METHODOLOGY FOR ADOPTING DISTRIBUTIONAL CHOICES IN DIRECTIVE 2003/87/EC AND PROBLEMS

Directive 2003/87/EC adopts a highly decentralized approach to the six pillars of an ETS. This highly decentralized regime leaves Member States ample discretion to regulate those pillars as they see fit, and has led to concerns about the inadequacy of the EU ETS to achieve its objectives. The analysis will consider the allocation methodologies and the scope.

5.1 The Setting of the Cap

Article 9 Directive 2003/87/EC mandates Member States to fix the cap that applies to their industries. Criteria 1 to 4 and 12 of Annex III of the Directive guide the setting of the cap. The main requirements are compliance with the Kyoto Protocol, proportionality in the distribution between covered and non-covered sectors, and consistency with the national climate change programme. Annex III has been interpreted further by the Commission in a non-paper and in two guidance documents.[65] However, those criteria are sufficiently general as to allow Member States a substantial margin of discretion when determining the share of the domestic cap to be achieved by the sectors covered in the EU ETS. In the first trading period (2005–2007) a number of factors played a role in ensuring very lenient caps and hence over-allocation: (1) general lack of experience with the scheme; (2) lack of data availability of historical emissions of covered installations; (3) general use of emission projections despite their unreliability; (4) fear of damaging the competitive positions of national industry, and pressure from industry.[66] The Commission approached the issue of over-allocation in the NAPs from the perspective of state aid rules, considering

64 See also Boute (2007).

65 European Commission (2003a); European Commission (2003b); European Commission (2005b).

66 Ellerman et al. (2007a).

that allocation over expected needs could not be authorized. However, the Commission depended for its assessment on the projections prepared by Member States. The Commission in fact acknowledged that there was a degree of over-allocation in the NAPs of the first trading period (2005–2007). In the second trading period (2008–2012), the Commission adopted a much more stringent approach to caps, once some NAPs had already been delivered,[67] thanks to the availability of data on 2005 emissions and to the availability of EU-wide projections on economic growth and on improvements in energy efficiency. As a result, the aggregate cap for the second trading period is below business as usual emissions. A number of Member States from Eastern Europe have challenged the Commission decisions in the ECJ, and it remains to be seen whether the court upholds their views.[68] Member States have considered the differences in ambition levels between different Member States as the most important element in creating distortions in the internal market, since those levels were directly translated into individual allocations. The problem was not so much the lack of legitimate reasons to adopt different ambition levels, but rather that the bases on which to do so were not well defined, which may have led to a race to the bottom in addition to an element of regulatory capture.[69]

5.2 The Coverage

Annex I of Directive 2003/87/EC establishes the installations that participate in the scheme. Three issues must be examined: (1) the coverage and definition of combustion installation; (2) the rules for the expansion of the coverage; (3) the treatment of small installations.

First, the assessment of the first round of NAPs made evident that Member States had interpreted Annex I differently, leading to different coverage in different Member States. The Commission sought to solve this issue by providing an interpretation of the term 'combustion installation' in its guidance and by requiring a certain level of harmonization.[70] However, some differences remained.

---

67 European Commission (2006a).

68 Case T- 32/07, *Slovakia v Commission*, OJ C 69 of 24.03.2007, p. 29, Action brought on 7 February 2007; Case T-183/07, *Poland v Commission*, OJ C 155 of 07.07.2007, p. 41, Action brought on 28 May 2007; Case T-194/07, *Czech Republic v Commission*, OJ C 199 of 25.08.2007, p. 38, Action brought on 4 June 2007; Case T-221/07, *Hungary v Commission*, OJ C 199 of 25.08.2007, p. 41, Action brought on 26 June 2007.

69 Ellerman et al. (2007a, pp. 348–49).

70 European Commission (2005b).

Second, Annex I provided for partial harmonization. Member States can, according to Article 24, seek to introduce additional activities and gases within the scope of the directive. The Commission must approve this through the comitology procedure, applying the following criteria: (1) potential effects on the internal market; (2) potential distortions of competition; (3) the environmental integrity of the scheme; (4) the reliability of the planned monitoring and reporting system. Thus, Member States are not left totally free to make that decision. In the first trading period, the Commission has accepted all requests from Member States to expand the scheme. Moreover, Article 24.4 mandates the Commission to study, in the event Member States would be allowed to expand the coverage, whether the scope of Annex I could be amended to introduce such an expansion at EU level.

Third, the implementation of Annex I showed that many installations which are covered by it are rather small and have negligible emissions, therefore many consider that it is not worthwhile to have them in the scheme.[71] The directive allowed Member States to exempt them under certain conditions only during the first trading period. The Commission adopted a more flexible approach in its second guidance document, without accepting their exclusion, since this is not possible according to Article 27 of the current directive.[72]

5.3 The Allocation Rules

The discussion above has shown the link between the cap and the individual allocations. We now turn to the allocation rules. There are in theory three different methods to allocate allowances: allocation for free on the basis of historical emissions (grandfathering), benchmarking and auctioning. Article 10 does not prescribe any allocation methodology. In fact, the only thing it does is to limit the share of auctioning up to 2012, to prevent the possibility that some Member States would allocate allowances for free while others would auction them. This provision was a concession to industry, which was afraid of the potentially serious distortions of competition that would follow from it.[73] Annex III does not prescribe any specific methodology either, but rather introduces a number of conditions to guide the process. Article 30 Directive 2003/87/EC mandates the Commission to explore the need for further harmonization of the allocation methodology, and mentions explicitly auctioning and EU-wide benchmarks.

71 Ellerman et al. (2007a, p. 341).
72 European Commission (2005b).
73 Robinson et al. (2007, p. 66).

The reason for this very low level of harmonization seems to be that Member States wanted freedom to explore the best manner to deal with the allocation rules together with their industry, which represents a distributional exercise of a considerable magnitude, and to compensate for the differences in the stringency of Member State caps resulting from Council Decision 2002/358/EC.[74] The Commission produced two documents to guide the allocation process. The majority of Member States adopted grandfathering as the main allocation method for incumbents, while using a benchmarking approach for new entrants. This approach has been followed generally in the two allocation rounds.[75] The main reason not to use benchmarks for incumbents was that benchmarks are data intensive, groups must be created, and their application would lead to deviations from recent emissions which were too large to become acceptable, due to significant source heterogeneity.[76] The manner in which grandfathering was implemented differed widely between Member States. In general most used assumptions of future growth rates combined with a compliance factor to respect the Kyoto target. However, approaches differed strongly as regards the treatment of early action, new entrants, closures and transfers, ex post allocation, CHP, and myriad other rules to differentiate between industrial sectors in order to take account of their potential to abate emissions and their needs.[77]

Economists have highlighted the problems that grandfathering gives rise to.[78] First, grandfathering tends to keep the status quo, hence adopting a static approach which reduces dynamic efficiency. Second, grandfathering tends to favor incumbents against new entrants, because the latter do not have historical emissions and therefore under an absolute cap may be treated less well. Third, grandfathering coupled with updating, which seems unavoidable, generates perverse incentives to increase emissions in order to receive more allowances in the future and reduces the dynamic efficiency of the scheme. Since updating may give incentives to companies to increase their emissions instead of decreasing them, in order to receive more allowances in future periods, it goes against the aim of the polluter pays principle. Fourth, it reduces the incentives of installations to trade, since they receive all the allowances they need. Fifth, coupled with an absolute cap, it damages the position of companies which gain market share, since they must purchase allowances from their competitors. Sixth, grandfathering has been in a number of national

---

74 Zapfel (2005, pp. 29–30).

75 Betz et al. (2006, p. 373). See also Ellerman et al. (2007a, pp. 351–55).

76 Ibid.

77 See Ellerman et al. (2007a, pp. 360–62). See also Ray Schmitt (2006), explaining Germany's approach.

78 Grubb and Neuhoff (2006).

allocation plans coupled with special rules for new entrants, closures and ex post allocation, which can be extremely complex and non-transparent, therefore reducing legal certainty and creating tensions with rules on free movement.[79] The majority of Member States have introduced these types of rules in one way or another. The reason for introducing those rules is to compensate for the fact that grandfathering tends to perpetuate the status quo.[80] This rationale, which acknowledges the existence of stranded costs, will logically weaken as time goes by. Hence, while acceptable as a first step in a learning process, in order to gain acceptance of the ETS by incumbents, it is not clear whether it constitutes the best-possible approach to allocation. On the one hand grandfathering makes it harder to set stringent targets since a link is established between the cap and individual allocations, and companies receive a strong incentive to lobby for generous allocations;[81] on the other it allows participants to accept it by minimizing distributional effects within Member States. Importantly, it may be used to avoid disproportionate impacts of the ETS on certain sectors or individual participants. Determining which allocation method is the best requires making a comparison with other allocation methodologies on the basis of a set of criteria (such as environmental effectiveness, cost-effectiveness and equity). It should be taken into account that in reality different allocation methodologies are applied alongside each other, and can be applied at different levels (Community or Member State level); hence the exercise of comparison becomes even more difficult. It is in this light that the remainder of the contribution must be read.

6. ANALYSIS OF THE RECENT COMMISSION'S PROPOSAL FROM THE PERSPECTIVE OF HARMONIZATION

The Commission has issued a proposal to amend Directive 2003/87/EC.[82] The changes proposed are far reaching and pervasive, but here the analysis is limited to those changes related to the allocation methodology and the scope.

6.1 The Setting of the Cap

A background study prepared for the Commission in order to prepare the review of the directive considered three levels at which the cap and the allocation could

---

79 Peeters, Weishaar and de Cendra de Larragán (2007).
80 Ibid.
81 Robinson et al. (2007, p. 65).
82 European Commission (2008b).

be harmonized: (1) both the cap and allocation could be set at EU level; (2) the cap could be set at EU level and the allocation at Member State level; (3) the cap could be set at Member State level and the allocation rules could be harmonized at EU level.[83] Each alternative would require adopting further decisions. The report concluded that an EU-wide cap, or an EU-wide sectoral cap, together with a more harmonized approach to allocation was to be preferred for reasons of keeping incentives to reduce emissions, enhancing transparency, limiting distortions in the internal market, and practical feasibility.

As regards the cap, the Commission has proposed the introduction of an EU-wide cap that would be set at EU level, on the basis of allocated volumes for the period 2008–2012. This cap would be reduced each year by 1.74% until 2020, in order to ensure compliance with the 2020 Community target of reducing Community-wide emissions by 20%. This approach introduces complete harmonization of the cap at EU level. Member States lose all their discretion to set national caps, including the possibility to impose stricter caps upon their companies. The Commission adduces four reasons for introducing an EU-wide cap:[84]

1. National caps do not provide sufficient guarantees that EU emission reduction targets will be achieved;
2. Setting caps at national level is not likely to allow the achievement of cost-effectiveness;
3. An EU-wide cap introduces a long-term perspective thereby increasing predictability, especially when coupled with trading periods of 8 years rather than of 5 years;
4. It sends a clear message to investors about the long-term Community policy on climate change.

These reasons will be considered below under the heading on subsidiarity and the cap.

6.2 The Coverage

The proposal introduces a number of amendments in relation to the coverage. First, it inserts an explicit definition of the term combustion installation and couples it with an explicit list of activities. Second, it expands the coverage of the EU ETS by including new sectors and gases, with the aim of including as

---

[83] Ecofys (2006).
[84] European Commission (2008b, p. 7).

many sectors as possible in order to achieve a clear and undistorted carbon price applicable throughout the EU. Further, the proposal states that when the Commission authorizes a Member State to unilaterally include additional sectors or gases, it may at the same time authorize other Member States to do the same. Third, the proposal seeks to provide a solution to the case of small installations, by combining a threshold on installed capacity with one on emissions from individual installations as the criterion for exclusion. Installations with an installed capacity between 20 and 25 MW and less than 10,000 Tco2/yr can be excluded from the scheme, provided that they are subjected to equivalent measures, Member States apply to the Commission, and the latter does not object. The rule however does not resolve completely the potential distortions of competition that may be generated when one competitor is covered by the EU ETS and other is not. For instance, an installation emitting 11,000 Tco2/yr and having 24 MW would be included in the scheme, while a competitor emitting 9,500 tons and having 24 MW would be excluded. Even clearer is the case of an installation with 19 MW of installed capacity and 11,000 Tco2/yr, and a competitor with 20 MW and 11,000 Tco2/yr. In sectors where this is a realistic situation the current proposal will not solve the problem.

6.3 The Allocation Rules

The most far reaching amendment to the current directive, together with the introduction of an EU-wide cap, is the change in the allocation rules. The proposal adopts auctioning, i.e. paying the full price of the allowances, as the basic allocation principle. This choice is very much influenced by the choice of an EU-wide cap. The Commission gives a number of reasons to choose auctioning:[85] (1) is the allocation methodology which best ensures efficiency; (2) ensures best transparency and simplicity; (3) avoids undesirable distributional effects; (4) best complies with the polluter pays principle;[86] (5) rewards early action to reduce emissions. According to the report prepared for the Commission, choosing grandfathering would require either assuming the same share in emissions under the cap or applying a growth rate, possibly combined with an EU-wide factor. The former solution ignores differences between Member States as well as differences in trends, technological reduction potentials and cost between different sectors. The latter requires the use of growth rates with associated uncertainties.[87]

85 European Commission (2008b, p. 7).
86 Ibid.
87 Ecofys (2006, p. 15).

Auctioning will be introduced gradually. From 2013 onwards, it will be fully applied to electricity producers and to carbon capture and storage (CCS). In 2020 it will apply across the entire ETS. The Commission will adopt a Regulation on timing, administration and other aspects of auctioning to ensure that it is conducted in an open, transparent and non-discriminatory manner. Auctioning constitutes the simplest approach and avoids distortions, but gives rise to two important effects: on the one hand it may lead to leakage of production and emissions; on the other, it generates substantial revenues that governments may use for a variety of purposes.

In order to tackle the risk of carbon leakage (the displacement of greenhouse gas-emitting activities out of the EU thereby increasing global emissions), the proposal differentiates between the power sector, which does not face international competition, and the other industrial sectors. The former will have to purchase all the allowances it needs through auctions or in the market. For the latter a transitional period will apply. In 2013 they will receive 80% of the allowances free, but this percentage will decrease linearly towards full auctioning in 2020. However, energy-intensive industries which are determined by the Commission to be exposed to a significant risk of carbon leakage could receive up to 100% of the allowances free of charge. Alternatively, an effective carbon equalization system could be introduced in order to level the international playing field, which would need to comply with the principle of common but differentiated responsibilities and with the WTO rules (GATT and SCM agreement). The Commission must determine the energy sectors or sub-sectors which are likely to be subject to carbon leakage by 2011; until that time a significant number of installations will not know which allocation regime will apply to them, which notably seems to reduce legal certainty, taking into account that one of the objectives of the EU ETS is to give incentives to industry to invest in low-carbon technologies.

As regards the use of the revenues obtained through auctioning, the proposal foresees that 90% of the allowances will be distributed to Member States according to their relative share of 2005 emissions in the EU ETS (or the average of 2005 and 2006 emissions).[88] The remaining 10% will be redistributed from Member States with an average level of income per head that is more than 20% above the EU average towards the other Member States, on the basis of fairness and solidarity. The proposal foresees that the auctions will generate significant revenues and proposes the precautionary principle as the guiding principle to spend them. The proposal is that 20% of the proceedings should be destined for various measures, including mitigation, R&D, renewable sources of energy, CCS, the Global Energy Efficiency and Renewable

88 European Commission (2008b, p. 8).

Energy Fund, deforestation, adaptation in developing countries and social aspects such as possible increases in electricity prices in lower and middle incomes. While the Commission invokes the precautionary principle as the guiding rationale for the spending of revenues, the measure arguably has a fiscal nature, especially taking into account the volume of revenues which would be generated, around 50 billion annually until 2020 according to the Commission.[89] Article 175 EC requires that provisions primarily of a fiscal nature be adopted by unanimity, hence it is foreseeable that Member States will not accept this earmarking of revenues.

For sectors other than the power sectors and CCS, the proposal gives to the Commission the power to set through comity Community-wide and fully harmonized transitional rules for free allocation. It is remarkable that the setting of allocation rules, which constituted the single most controversial issue in the negotiation and implementation of Directive 2003/87/EC, is now proposed to be solved through the comitology procedure. According to the proposal, the allocation methodology should take into account a large number of factors such as the most efficient techniques, the possibility of substitution, alternative production processes, use of biomass and CCS, and the need to avoid perverse incentives to increase emissions. These requirements logically point towards the introduction of EU-wide benchmarks to the largest possible extent. However, in some cases benchmarking may not be feasible, and the Commission will have to face similar problems to those currently faced by Member States when developing their NAPs. Member States need to take into account the particular situation of thousands of installations in the internal market. The proposal does not say anything about how to solve this problem, and about the degree of discretion that the Commission will have to exert in implementing the allocation methodology. The question that arises is whether the Commission has the capacity to perform this task, and whether adopting a simple approach in order to avoid going through the nightmare represented by the elaboration of NAPs will ensure respect for the principles of equal treatment and proportionality. None of these issues is dealt with in the proposal.

The proposal also foresees the creation of an EU-wide new entrant reserve, from which allowances should be distributed with the same methodology to that applied to incumbents. Installations that close will not receive further allowances. Hence, the proposal adopts total harmonization both for the setting of the cap and for the allocation methodology.

---

89 ENDS Europe Daily 2482, 12 February.

7. HARMONIZATION AND LEGAL PRINCIPLES

7.1 The Principle of Subsidiarity

7.1.1 The principle of subsidiarity and the cap

There would seem to be in principle a presumption in favor of further harmonization of the cap. First, emissions that lead to climate change constitute a transboundary externality, requiring a coordinated approach between Member States and at international level, for which the Community has comparative advantage. Second, the experience of the first two allocation rounds has shown that a low degree of harmonization cannot ensure sufficiently stringent caps, due to evidence of regulatory capture which has led to an inevitable race to the bottom.[90] The consequence of this race to the bottom affects only sectors within the ETS, therefore the burden is passed to sectors outside the ETS and to tax payers. Hence, reasons of fairness would support this harmonization also. Moreover, providing certainty to investors also seems to require a long-term EU-wide target.

However, the reasons provided by the Commission for harmonizing the cap do not appear to be sufficiently convincing to shift the establishment of the cap to the EU level, in particular given the fact that the Commission has ensured that allocation volumes in the second allocation round are below business as usual. The first allocation round was a period of learning by example, and Members States on the one hand lacked experience with the ETS, and on the other did not wish to subject their companies to stringent targets that could place them at a disadvantage vis-à-vis competitors from other Member States. But with the initial problems, such as data limitations and the setting of methodologies, largely solved, and with more experience gathered on the functioning of the ETS, it is not immediately obvious that Member States are incapable and/or unwilling to comply with EU emission reduction targets, or that the Commission is unable to ensure compliance. Indeed, the Commission is now in a better position to check the adequacy of targets than in the first allocation round; more stringent targets will lead to higher prices of allowances, hence enhancing emission reductions and trade of allowances; further, there would also be lower variations in relative stringency between national allocation plans, since a common methodology will be applied. Hence the main problems have already been dealt with.[91] Thus, whereas it is possible that harmonization indeed achieves more transparency, effectiveness and equality than does the current approach, it would seem that the Commission's proposal does not sufficiently make the case.

---

90 Betz et al. (2006).
91 Kruger et al. (2007, pp. 126–27).

7.1.2 The principle of subsidiarity and the scope

The proposal introduces three changes to the scope:

1. A definition of combustion installation coupled with a list;
2. The expansion of the coverage is dependent upon the possibility of accurately monitoring emissions;
3. A change to the rules determining the coverage of small installations, as seen in section 6.2 above.

The first change does not amount to harmonization, but is rather a requirement of legal certainty and transparency. As regards the second, whether this provision respects the principle of subsidiarity can be discussed, since it preempts Member States from regulating non-covered sectors as they see fit. On the other hand, allowing the introduction into the ETS of gases and installations which cannot be adequately monitored would reduce its environmental effectiveness. The third seeks to reduce administration costs, and to achieve equality and fairness. However, as we have seen above, it may give rise to new inequalities. Hence it is not clear that in this respect a Community approach is really necessary, as the report carried by Ecofys acknowledged.[92]

7.1.3 The principle of subsidiarity and the allocation rules

As mentioned above, allocation rules can be used to promote a certain conception of fairness. However, the way in which they are designed can violate the polluter pays principle and the principle of equal treatment within the internal market. Hence, the case for harmonization must be analysed mainly under those principles.

From the perspective of subsidiarity, the question which must be asked is which objectives of the ETS related to the allocation rules cannot be sufficiently achieved by Member States and can be better achieved by the Community. One can think of two: avoiding distortions of competition within the internal market and avoiding barriers to the right of free establishment. The former will be analysed under the principle of equal treatment. The second would justify enough harmonization to prevent allocation rules from being designed so as to favor national industries vis-à-vis foreign ones. This is currently the case with closure and transfer rules such as those introduced in the German NAP for the first trading period. According to those rules, operators that close installations in German territory can keep their allowances as long as they open a similar installation in German territory and during a certain period of time. This rule may violate Article 28 EC on the free movement of

92 Ecofys (2006, p. 18).

goods and Article 43 on the right of establishment protected by Article 43 EC. In *Germany v Commission*, the CFI mentioned but did not examine this issue since it did not constitute the core of the case.[93] Further harmonization must depend on the aim of achieving a 'level playing field', which is a notoriously difficult concept to deal with, and which could therefore cover almost all the reasons that the Commission adduces to push for auctioning, such as the need to increase simplicity and transparency, while reducing distributional effects. If the argument for complete harmonization is that 27 Member States regulating the allocation rules will as a matter of fact yield a less simple and transparent approach than when the Community does so, then this suggests that the principle of subsidiarity is not fully taken seriously. A strong argument can be made that once environmental effectiveness and economic efficiency are ensured, Member States may retain competence to ensure their own understanding of fairness within their own jurisdictions, subject to compliance with the treaties and with general principles of Community law.

7.2 The Polluter Pays Principle

7.2.1 The polluter pays principle and the cap

As seen above, it has been traditionally considered that the polluter pays principle was complied with by any standard set by the regulator, whether optimal or not from an economic perspective. This view has evolved and there is now a trend towards finding and enforcing the optimal level of pollution. A few comments are in order. First, if there are different regulators, each aiming at setting the optimal level of pollution, each may find a different optimum level. The geographical distribution of environmental impacts does not depend on the location of emissions, since all gases mix uniformly in the atmosphere, but it certainly may have an impact on the incentives of countries to set emission reduction targets. Countries facing less damage will have an incentive to adopt lower targets, which may be the optimal response from their perspective but not from the perspective of other countries. Hence coordination may be necessary in order to apply the polluter pays principle evenly throughout the Community. Second, if there is a risk of a race to the bottom, legislators will tend to set domestic caps that are too low. For these reasons, the polluter pays principle would justify setting the cap at EU level.

7.2.2 The polluter pays principle and the scope

The polluter pays principle cannot achieve its aim of avoiding distortions of competition if it does not cover all competitors within a market. The polluter

---

93 Case T-374/04 *Germany v Commission*, OJ C 315 of 22.12.2007, paragraph 146.

pays principle does not prescribe per se a specific regulatory instrument, since many can in principle be used to implement it (including different allocation methodologies). Hence, different Member States could implement the principle through different instruments, and as long as the cap and the coverage were to be comparable, the principle would be complied with. Obviously, some instruments would be more efficient and cost-effective than others. If the focus is placed on maximizing these variables within the internal market, it follows that the (in theory) most cost-effective and efficient instrument should be applied to all competitors within the market.[94] But this is not legally required by the polluter pays principle.

7.2.3 The polluter pays principle and the allocation rules

The polluter pays principle requires the internalization of social costs into production costs. The principle is complied with through the setting of a cap that applies to each and every installation. However, different allocation rules will distribute that cap differently among polluters. The criteria used by each rule will tend to benefit some polluters over others. For instance, grandfathering based on historical emissions gives more allowances to the worst polluters, which can then sell them at a profit in the market. Further it tends to create a disadvantage for early actors. Grandfathering coupled with rules on updating and rules on closures compound this problem. Benchmarking may reduce those problems, although that depends on their specific design. Auctioning ensures equal treatment of all polluters in the sense that it distributes the burden in accordance with only one criteria: the level of pollution of each polluter. On the other hand, companies facing different marginal abatement cost (MACs) will be affected differently by auctioning. When the differences in MACs are large, auctioning may lead to strong impacts on the positions of some companies. If those companies are competitors in the market, auctioning could lead to some companies losing their ability to compete. Hence, the introduction of auctioning must take this potential impact into account in order to comply with the principles of equality and proportionality.

The question of whether allocation rules may violate the polluter pays principle must also be analysed from the perspective of state aid rules. We have already seen the defective approach of the Community guidelines on state aid for environmental protection. Moreover, although the Commission considers that free allocation constitutes in principle state aid, the academic literature seems divided on this issue.[95] For state aid to exist, all the conditions laid

---

94 See in general Nordhaus (2007). See for the EU ETS, Del Río (2006, p. 465).

95 See De Sepibus (2007b); see Lorenz (2004); see Merola and Crichlow (2004); see Weishaar (2007).

down in Article 87 EC must be fulfilled:[96] (1) the conferral of an advantage; (2) granted by the State or through State resources; (3) which distorts or threatens to distort competition; (4) by favoring certain undertakings or the production of certain goods or services (the selectivity criterion); (5) and which affects trade between Member States. Iñigo Sanz et al. consider that allocation of allowances for free does not in principle constitute state aid, since the requirements of Article 87 EC are not complied with.[97] Angus Johnston considers that free allocation constitutes, *prima facie*, state aid.[98] De Sepibus considers that the question of whether certain allocation methodologies constitute illegal state aid cannot be answered conclusively in the light of prevailing case law.[99] In Case T-233/04,[100] the CFI considered that the NO_x emission trading scheme implemented by The Netherlands did not constitute state aid, since, although it fulfilled the conditions of conferring an advantage through state resources, did not constitute a selective measure. The NO_x scheme covered industrial installations with an installed capacity above 20 MW, thereby applying an objective criterion to determine the coverage. Since this is also the approach followed by the EU ETS, the reasoning also applies therein. The case however does not consider whether specific choices within allocation rules may constitute state aid. In its analysis of the second round of NAPs the Commission has sought also to subject allocation rules to state aid scrutiny, when they have the potential to discriminate at sector or installation level. In a number of NAPs, it has considered that a number of choices cannot be justified on the basis of the objectives of Directive 2003/87/EC.[101] Whether this approach is successful remains to be seen. Hence, it is not yet clear whether respect for state aid rules seems to require the harmonization of allocation rules.

7.3 The Principle of Equality

7.3.1 The principle of equality and the cap

The principle of equality is not directly related to the setting of the cap, in the sense that the caps set by Member States do not have a direct impact upon the

[96] Case C-126/01 *GEMO* [2003] ECR I-13769, paras. 21 and 22.
[97] Sanz Rubiales, I, et al. (2007, pp. 250–55).
[98] Johnston (2006, p. 119).
[99] De Sepibus (2007b p. 31).
[100] Case T-233/04 *Netherlands v Commission*, OJ C 128, 24 May 2008.
[101] Decision on the second German NAP, 29.11.2007, at paras. 20–24; Decision on the second Polish NAP, 26.3.2007, para. 23; Decision on the second Spanish NAP, 26.2.2007, paras. 8–9.

individual allocations to installations.[102] Hence, an argument based on equality can only indirectly support the establishment of an EU-wide cap, in order to increase fairness.

7.3.2 The principle of equality and the scope

The principle of equality seeks to ensure that competitors are treated similarly within the internal market. The difficulty arises when extending the scope of the EU ETS could cover some industries while leaving competitors outside. When those extensions are made by the Community legislator, the right to legislate step-by-step and the margin of discretion afforded by the European courts suggest that a breach of the principle of equality is unlikely. The French Council of State has raised a preliminary question with the ECJ on this issue in relation to the French NAP for the first trading period. The Council of State expressed doubts as to whether including the steel sector while leaving out the aluminium and plastic industries is in line with the principle of equal treatment, since they produce products which compete in the internal market.[103] In a similar case, the Belgian Arbitration Court ruled that the Walloon NAP did not breach the principle of equality.[104]

Moreover, complete harmonization cannot ensure equal treatment in relation to small installations, because rules to determine the coverage may actually lead to differential treatment of competitors. Hence in the latter case there is no clear advantage in introducing complete harmonization of the coverage.

7.3.3 The principle of equality and the allocation rules

Once the cap has been set at a level that ensures the attainment of the environmental objective, and a market is introduced that ensures allocative efficiency, the allocation rules may be used to achieve a certain conception of fairness, subject to compliance with the principle of equal treatment and freedom of establishment. Indeed, differential treatment is only the first step in finding a breach of equality, the second being that the difference cannot be justified. The CFI has acknowledged the right of Member States to introduce ex post allocation rules in order to preserve the integrity of the internal market and to avoid distortions of competition, despite the strong opposition of the Commission. This indicates that the principle of equality does not reduce in advance the freedom of Member States to choose allocation rules when they can be justified by the logic of the scheme.[105]

---

[102] Case T-27/07, *US Steel Kosice v Commission*; Case T-489/04, *US Steel v Commission*; Case T-130/06 *Drax Power and Others v Commission*; Case T-387/04, *EnBW Energie Baden-Wurttemberg v Commission*.

[103] French Council of State, Arrest No. 287110 of 8 February 2007, Article 3.

[104] Belgian Arbitration Court, Arrest No. 92/2006 of 7 June 2006, B.15.

[105] Case T-374 *Germany v Commission*, paras. 140–42.

Indeed, the use of different allocation methodologies in different Member States in the first two trading periods led to a situation in which competitors did receive substantially different amounts of allowances. Criterion 5 of Annex III of the Directive requires that NAPs do not discriminate between installations. This criterion sets a test independent from that of state aid. Hence, even if a NAP were to be found to comply with state aid rules, it could still be assessed under this criterion and be found in breach of it. Now, this criteria cannot require complete harmonization. For instance, the Commission has used criterion 5 to require that Germany has the so-called allocation guarantees for new installations, which gave them an advantage over slightly older installations by ensuring them a compliance factor of zero for 14 years.[106] The Commission considered that the starting date could not be used as a criterion to justify discrimination between existing installations.[107] Further, the Commission has considered that allocations on the basis of the emissions in one year may constitute state aid, since it damages installations with particularly low emissions in that year.[108] Moreover it has also criticized the use of benchmarks based on average technology to power producers using fossil fuels while applying benchmarks based on best-available technology to others using gas, since that rule favors the former.[109] The principle of non-discrimination could provide a rationale to eliminate those type of rules without requiring full harmonization of the allocation rules

Nevertheless, it must be recognized that the 'level playing field' argument is consistently used to seek further harmonization, even by Member States themselves, in order to prevent a race to the bottom. Whether this is really a danger does not matter, since often the mere appearance of a danger is enough to justify tackling it. The question then becomes whether harmonization may deliver what it promises.

First, adopting auctioning as the general allocation method may actually discriminate in favor of companies that have more technical and financial capacity to participate in auctions. Small installations may be particularly penalized by this approach, if they cannot participate themselves in the auctions and need to purchase allowances in the secondary market.

Second, Member States can use the revenues from auctioning in very diverse manners, for instance to reduce corporate taxation, or to finance public spending. Since fiscal matters are beyond the competence of the Community, different approaches will have an impact on companies located in different Member States in different ways, and new 'distortions' in the internal market will be generated.

---

106 German NAP of the German Allocation Act, Section 11, para. 1.

107 European Commission, Decision on the second German NAP, 29.11.2006.

108 Decision on the second Polish NAP, 26.3.2007, para. 23.

109 Decision on the second Spanish NAP, 26.2.2007, paras. 8–9.

8. CONCLUSIONS

This paper has sought to find out whether an analysis based on three legal principles of Community law may support and/or require a higher level of harmonization of the ETS. We must now distinguish between the roles that these principles play in the activity of courts and in the activity of the legislator.

In relation to the activity of the courts, the answer seems to be negative. Harmonization is mainly a political exercise, for which the legal principles examined offer limited guidance. The function of the principle of subsidiarity is to require an adequate justification for further harmonization. The polluter pays principle does not seem specific enough to require a higher level of harmonization. The principle of equality applies mainly as a tool for testing the legality of Community secondary legislation and the adequacy to it of implementing legislation, and hence has the capacity to set certain limits to the choices of Member States, but is not an independent tool for introducing a higher level of harmonization. Nevertheless, the role of these and other principles can only be fully appreciated in particular cases, not in the abstract.

Considering the role of legal principles in guiding the legislator towards higher levels of harmonization, the answer may be somewhat different. Harmonization is by essence a political exercise. The principle of subsidiarity can be conceived of as a tool to promote cooperation and as a tool to ensure the involvement of lower levels of (domestic) government. It hence seeks to ensure that a reasonable balance in the distribution of competences is achieved, but drawing a precise line is a problem for which several reasonable answers can be given. There is a clear conceptual relationship between the polluter pays principle, the principle of equality and the principle of proportionality, and all three are particularly important for the allocation problem. Together with the principle of integration they seem to require a climate change policy that is comprehensive in terms of gases and economic sectors, and which aims at setting a carbon price across the economy while taking into account the responsibilities, capacities and needs of different sectors and undertakings. Adequate implementation of these principles across the EU necessarily requires a certain level of harmonization. Deciding the precise level is a political question for which clear legal answers cannot be expected.

Nevertheless, a number of conclusions can be ascertained from this contribution. The principle of subsidiarity may justify the level of harmonization proposed in relation to the cap, but complying with it would seem to require more detailed argumentation on the side of the Commission. It is not so clear however that it would require the proposed harmonization of the scope and the allocation rules. The polluter pays principle would seem to require an EU-wide cap. It could support a harmonized scope together with the principle of cost-effectiveness. In relation to the allocation rules, the principle would require some harmonization, but not as complete as currently proposed by the Commission. The principle of equality does not require an EU-wide cap. Moreover, it does not justify the approach to small installations adopted by the proposal. Finally, it does not require a full harmonization of allocation rules. The harmonization proposed seems to bow to the endless drive for an ever deeper EU-wide centralization. To conclude, the proposal tackles the key problems that arose in relation to the current directive, and improves significantly its ability to promote environmental effectiveness, cost-effectiveness and equity. However, the level of harmonization proposed does not seem totally justified in the light of those objectives and taking into account key principles of Community law.

REFERENCES

Armenteros, M.F. (2005), 'State Aid Issues Raised by the Implementation of Climate Change Policy Instruments', in M. Bothe and E. Rehbinder (eds.), *Climate Change Policy*, Utrecht, Eleven International Publishing, pp. 219–57.

Barber, N.W. (2005), 'The Limited Modesty of Subsidiarity', *European Law Journal*, 11(3), 308–25.

Betz, R., K. Rogge, and J. Schleich (2006), 'EU Emissions Trading: an Early Analysis of National Allocation Plans for 2008–2012', *Climate Policy*, 6(4), 361–94.

Betz, R. and M. Sato (2006b), 'Emissions Trading: Lessons Learnt from the 1st Phase of the EU ETS and Prospects for the 2nd Phase', *Climate Policy*, 6(4), 351–59.

Bothe, M. and E. Rehbinder (eds.) (2005), *Climate Change Policy*, Utrecht: Eleven International Publishing.

Boute, A. (2007), 'Combating Climate Change and Securing Electricity Supply: The Role of Investment Protection', *European Environmental Law Review*, 16(8), 227–49.

Carbonara, E., B. Luppi and F. Parisi (2008), 'Self-defeating Subsidiarity: an Economic Analysis', University of Minesota Law School, Legal Studies Research Paper Series, Research Paper No. 08-15.

Convery, F.J. and L. Redmond (2007), 'Market and Price Developments in the European Union Emissions Trading Scheme', *Review of Environmental Economics and Policy*, 1(1), 88–111.

Craig, P. (2006), *EU Administrative Law*, Oxford, Oxford University Press.

Daly, H.E. (1992), 'Allocation, Distribution, and Scale: Towards an Economics that is Efficient, Just, and Sustainable', *Ecological Economics*, 6, 185–93.

Delbeke, J. (ed.) (2006), *EU Environmental Law*, Leuven: Claeys & Casteels.

Del Río, P. (2006), 'Harmonization versus Decentralization in the EU ETS: an Economic Analysis', *Climate Policy* 6(4), 457–75.

De Sadeleer, N. (2002), *Environmental Principles – from Political Slogans to Legal Rules*, Oxford, Oxford University Press.

De Sepibus, J. (2007a), 'Scarcity and Allocation in the European Emissions Trading Scheme – a Legal Analysis', *NCCR Working Paper No. 2007/32.*

De Sepibus, J. (2007b), 'The EU ETS Put to the Test of State Aid', *NCCR Working Paper.*

Ecofys (2006) 'Harmonization of Allocation Methodologies, Report prepared for the European Commission under the project "Review of EU Emissions Trading Scheme" ', October 2006.

Egenhofer, C., N. Fujiwara, Ahman Markus and Zetterberg Lars (2006), 'The EU Emissions Trading Scheme: Taking Stock and Looking Ahead', *CEPS Policy Briefs*.

Ellerman, D. and B.K. Buchner (2007), 'The European Emissions Trading Scheme: Origins, Allocation and Early Results', *Review of Environmental Economics and Policy*, 1(1), 66–87.

Ellerman, D., B.K. Buchner, and C. Carraro (2007a), 'Unifying Themes', in Denny Ellerman, Barbara K. Buchner and Carlo Carraro (eds.), *Allocation in the European Emissions Trading Scheme*, Cambridge, Cambridge University Press, pp. 339–69.

Ellerman, D., B.K. Buchner and C. Carraro (eds.) (2007b), *Allocation in the European Emissions Trading Scheme*, Cambridge, Cambridge University Press.

European Commission (2003a), Non-paper of 1 April 2003 on allocation mechanisms.

European Commission (2003b), Communication on guidance to assist Member States in the implementation of the criteria listed in Annex III to the Directive 2003/87/EC establishing a scheme of greenhouse gas emissions allowances trading within the Community and amending Council Directive 96/61/EC, and on the circumstances under which force majeure is demonstrated, COM(2003) 830, 7 January 2004.

European Commission (2005a), Report on Competition Policy 2004.

European Commission (2005b), Communication, 'Further guidance on allocation plans for the 2008–2012 trading period of the EU Emission Trading Scheme', COM(2005)703 Final.

European Commission (2006a), Communication from the Commission to the Council, The European Parliament, the European Economic and Social Committee and the Committee of the Regions, Building a global carbon market – Report pursuant to Article 30 of Directive 2003/87/EC, COM(2006)676.

European Commission (2008a), Proposal for a Decision of the European Parliament and of the Council on the effort of Member States to reduce their greenhouse gas emissions to meet the Community's greenhouse gas emission reduction commitments up to 2020, COM(2008)17 Final.

European Commission (2008b), Proposal for a Directive of the European Parliament and of the Council amending Directive 2003/87/EC so as to improve and extend the greenhouse gas emission allowance trading of the Community, COM(2008)16 Final

Faull, J. and A. Nikpay (eds) (2007), *The EC Law of Competition*, Oxford, Oxford University Press.

Foundation for International Environmental Law and Development (Field) (1999), Designing Options for Implementing an Emission Trading Regime for Greenhouse Gases in the EC, in Scoping Paper N. 1. EC Trade and Competition Law Issues Raised by the Design of an EC emission Trading System.

Gaines, S.E. (1991), 'The Polluter Pays Principle: From Economic Equity to Environmental Ethos', *Texas International Law Journal*, 26, 463–96.

Grubb, M. and K. Neuhoff (eds.) (2006), *Emissions Trading and Competitiveness*, London: Earthscan.

Grubb, M., C. Azar and U.M. Persson (2006), 'Allowance Allocation in the European Emissions Trading System: a Commentary', *Climate Policy*, 5, 127–36.

Grubb, M. and K. Neuhoff (2006), 'Allocation and Competitiveness in the EU Emission Trading Scheme: Policy Overview', *Climate Policy*, 6, 7–30.

Grubb, M. and F. Ferrario (2006b), 'False Confidences: Forecasting Errors and Emission Caps in CO_2 Trading Systems', *Climate Policy*, 6(4), 496–501.

Hansjürgens, B. (ed.) (2005), *Emissions Trading for Climate Policy: US and European Perspectives*, Cambridge, Cambridge University Press.

Harrison, D. and D. Radov (2007), 'United Kingdom', in Ellerman et al. (eds), pp. 41–71.

Hepburn, C., M. Grubb, K. Neuhoff, F. Matthes and M. Tse (2006), 'Auctioning of EU ETS Phase II Allowances: How and Why?', *Climate Policy*, 6(1), 137–60.

Hope, C.W. (2006), 'The Social Cost of Carbon: What Does It Actually Depend On?', *Climate Policy* 6(5), 565–72.

Jans, H.J. and H.H.B. Vedder (2007), *European Environmental Law*, Europa Law Publishing.

Johnston A. (2006), 'Free Allocation of Allowances under the EU Emissions Trading Scheme – Legal Issues', *Climate Policy* 6, 115–36.

Kramer, L. (2007), *EC Environmental Law*, London, Sweet and Maxwell.

Kruger, J., E. Oates Wallace and A.W. Pizer (2007), 'Decentralization in the EU Emissions Trading Scheme and Lessons for Global Policy', *Review of Environmental Economics and Policy*, 1(1), 112–33.

Lefevere, J. (2005), 'Greenhouse Gas Emissions Trading: A Background', in Bothe et al. (eds.), pp. 103–28.

Lorenz, M. (2004), 'Emission Trading – the State Aid Dimension', *European State Aid Law Quarterly*, 3, 399–455.

Martinez, Kim K. and K. Neuhoff (2005), 'Allocation of Carbon Emission Certificates in the Power Sector: How Generators Profit from Grandfathered Rights', *Climate Policy*, 5, 61–78.

Meadows, D. (2006), 'The Emissions Allowance Trading Directive 2003/87/EC Explained', in Delbeke (ed.) (2006), pp. 63–115.

Mehling, Michael A. (2005), 'Emissions Trading and National Allocation in the Member States – an Achilles' Heel of European Climate Policy?', *Yearbook of European Environmental Law*, 5, 113–56.

Merola, M. and G. Crichlow (2004), State Aid in the Framework of the EU Position after Kyoto: an Analysis of Allowances Granted under the CO_2 Emissions Allowances Trading Directive', *World Competition*, 27(1), 25–51.

Neuhoff, K., M. Åhman, R. Betz, J. Claudius, F. Ferrario, K. Holmgren, G. Pal, M. Grubb, F. Matthes, K. Rogge, M. Sato, J. Schleich, J. Sijm, A. Tuerk, C. Kettner and N. Walker (2006a), 'Implications of Announced Phase II National Allocation Plans for the EU ETS', *Climate Policy*, 6, 411–22.

Neuhoff, K., F. Ferrario and M. Grubb (2006a), 'Emission Projections 2008–2012 versus NAPs II', *Climate Policy*, 6(4), 395–410.

Neuhoff, K., K. Keats and M. Sato (2006c), 'Allocation, Incentives and Distortion: the Impact of EU ETS Emisssion Allowance Allocations to the Electricity Sector', *Climate Policy*, 6(1), 73–92.

Nordhaus, W.D. (2007), 'To Tax or Not to Tax: Alternative Approaches to Slowing Global Warming', *Review of Environmental Economics and Policy'*, 1(1), 26–44.

Peeters, M. and K. Deketelaere (2006), *EU Climate Change Policy – The Challenge of New Regulatory Initiatives*, Edward Elgar, Cheltenham UK.

Peeters, M. and S. Weishaar, J. De Cendra De Larragán (2007), 'A Governance Perspective on the Choice between "Cap and Trade" and "Credit and Trade" for an Emissions Trading Regime', *European Environmental Law Review*, 16(7) 191–201.

Ray Schmitt, B. (2006), 'A Level Playing Field? Initial Allocation of Allowances in Member States', in Peeters, M. and Deketelaere, K. (2006), *EU Climate Change Policy*

– *The Challenge of New Regulatory Initiatives*, Edward Elgar, Cheltenham UK, pp. 83–97.

Rehbinder, E. and R. Stewart (1985), 'Legal Integration in Federal Systems: European Community Environmental Law', *33 Am. J. Comp. L.*, 33, 371–446.

Risbey, S.J. (2006), 'Some Dangers of "Dangerous" Climate Change', *Climate Policy*, 6(5), 527–36.

Robinson, J., Barton, J., Dodwell, C., Heydon, M. and L. Milton (2007), *Climate Change Law – Emissions Trading in the EU and in the UK*, London, Cameron May.

Sánchez Rydelski, M. (2006), *The EC State Aid Regime: Distortive Effects of State Aid on Competition and Trade*, London, Cameron May.

Sanz Rubiales, I, et al. (2007), 'El Mercado de Derechos a Contaminar', Valladolid, Lex Nova.

Sijm, J. (2006), 'EU ETS Allocation: Evaluation of Present System and Options Beyond 2012', ZfE, 30(4), 285–92.

Sijm, J., K. Neuhoff and Y. Chen (2006), 'Cost Pass Through and Windfall Profits in the Power Sector', *Climate Policy*, 6, 49–72.

Smale, R., M. Hartley, C. Hepburn, J. Ward and M. Grubb (2006), 'Emissions, Firm Profits and Market Prices: the Consequences from Emissions Trading', *Climate Policy*, 6, 31–48.

Tridimas, T. (2006), *The General Principles of Community Law*, Oxford, Oxford University Press.

Van den Bergh, R. (1996), 'Economic Criteria for Applying the Subsidiarity Principle in the European Community: The Case of Competition Policy', *International Review of Law and Economics*, 16, 363–83.

Van den Bergh, R. (2000), 'Economic Criteria for Applying the Subsidiarity Principle in European Environmental Law', in R. Revesz, P. Sands and R. Stewart (eds.), *Environmental Law, the Economy, and Sustainable Development*, Cambridge, Cambridge University Press, pp. 80–95.

Vis, P. (2006a), 'Basic Design Options for Emissions Trading', in J. Delbeke (ed.), *EU Environmental Law*, Claeys & Casteels, Leuven, pp. 39–61.

Vis, P. (2006), 'The First Allocation Round: a Brief History', in J. Delbeke (ed.), *EU Environmental Law*, Claeys & Casteels, Leuven, pp. 187–212.

Weishaar, S. (2007), 'The European Emissions Trading System and State Aid: an Assessment of the Grandfathering Allocation Method and the Performance Standard Rate System', *European Competition Law Review*, 28(6), 371–81.

Winter, G. (2004), 'The Legal Nature of Environmental Principles in International, EC and German Law', in R. Macrory, *Principles of European Environmental Law*, Groningen, Europa Law Publishing.

Woerdman, E. (2003), 'Developing Trading in Europe: Does Grandfathering Distort Competition and Lead to State Aid?' in M. Faure, J. Gupta and A. Nentjes (eds.), *Climate Change and the Kyoto Protocol, The Role of Institutions and Instruments to Control Global Change,* Edward Elgar, Cheltenham, UK and Northampton, MA, USA, pp. 108–27.

Zapfel, P., (2005), 'Greenhouse Gas Emissions Trading in the EU: Building the World's Largest Cap-and-Trade Scheme', in B. Hansjürgens (ed.) (2005), *Emissions Trading for Climate Policy: US and European Perspectives*, Cambridge, Cambridge University Press, pp. 162–74.

Zapfel, P. (2007), 'A Brief but Lively Chapter in EU Climate Policy: the Commission's Perspective', in D. Ellerman, K.B.Buchner and Carlo Carraro (eds.), *Allocation in the European Emissions Trading Scheme*, Cambridge, Cambridge University Press, pp. 13–38.

4. The 'Emissions Trading Scheme' case-law: some new paths for a better European environmental protection?[1]

Nicolas Van Aken[2] under the direction of Michel Pâques

1. INTRODUCTION

To assist its Member States in complying with their respective targets, the European Union has established the Emissions Trading Scheme (ETS) within the Community.[3] The scheme, which started to operate from 1 January 2005, 'is a system under which thousands of CO_2-emitting installations in Europe receive a limited amount of allowances in function of their needs'.[4] That ETS directive introduced new concepts and mechanisms in European environmental law.[5] Concerning the gases covered, the practical application is limited. In fact, the ETS Directive covers only the CO_2 emissions. The Directive covers only specific sectors, determined in Annex 1 (representing about 11 500 companies). The emissions taken into account by the Directive represent 40% of all EU-27 greenhouse gases. [6] The creation of tradable EU allowances is the core of the ETS Directive. After each year, the company covered by the ETS Directive has to return an exact number of EU allowances in order to 'justify'

1 This article represents the position as at 10 May 2008.

2 The author would like to thank professor Michel Pâques, Professor at the University of Liège and judge at the Belgian Council of State, for his support and his constructive comments.

3 Directive 2003/87/EC of the European Parliament and of the Council of the 13 October 2003 establishing a scheme for greenhouse gas emission allowance trading within the Community and amending Council Directive 96/61/EC, OJ 2003, L 275/32.

4 Van Hecke and Zgajewski (2007), p. 2.

5 Peeters (2006, pp. 259–80).

6 Commission Staff Working Document, Impact Assessment, Document accompanying the package of Implementation measures for the EU's objectives on climate change and renewable energy for 2020, SEC(2008) 85/3, 23 January 2008, p. 4, b).

its annual emissions. Nevertheless, Member States must also ensure that no installation undertakes any activity producing specified emissions unless its operator holds a permit issued by a competent national authority. That exact prior authorization does not prescribe any threshold or limit to CO_2 emissions. It is only one way to verify whether the operator is able to manage his ETS obligations. Another is to surrender allowances in due time and in such quantity that all emissions during the preceding year are covered.

For each period, Member States have to communicate their national allocation plans to the Commission. The State must develop a plan stating the total quantity of allowances that it intends to allocate for that period and how it proposes to allocate them.[7] The Commission has to control it, but case-law expressly stated that this prior control is not an authorization as such.[8] The plan must be based on objective and transparent criteria including those in Annex III of the Directive. These criteria set the framework for the discretion allowed to the Member States with respect to allocation. The attribution of the EU allowances is made according to the grandfathering method under which 'the operators of covered installations received at least 95% of the EU allowances free of charges in the first allocation period from 2005 to 2007, and at least 90% in the second allocation period 2008–2012'.[9] In other words, the covered installations freely receive a specific number of EU allowances. They will use it to cover their emissions. But the companies can also freely sell them to other operators of covered installations or eventually to other persons.

Some scholars criticized vigorously the Kyoto Protocol and the ETS Directive.[10] In our opinion, regarding the difficulties of finding an agreement in an international range, the Directive and the Kyoto Protocol have the advantage of being a first step against the degradation of the climate. They endeavour to make a compromise between the environmental necessity of bringing

---

7 Article 9 of the Directive 2003/87/EC.

8 Court of First Instance, 30 April 2007, Case T-387/04 EnBW Energie Baden-Württemberg v. Commission, § 124. to be discussed later in this chapter

9 There are three main options for distributing allowances. Next to the 'grandfathering' method, two other options are available. See Barton (2006). For the moment, the system of the Directive is a mix of the 'grandfathering' and 'auctioning' methods.

10 See Prins and Rayner (2007, pp. 973–75). For the authors, 'the Kyoto Protocol was always the wrong tool for the nature of the job' [...] 'Kyoto has failed in several ways, not just in its lack of success in slowing global warming, but also because it has stifled discussion of alternative policy approaches that could both combat climate change and adapt to its unavoidable consequences'. They criticize the Kyoto Protocol because they believe that the global warming struggle needs 'a radical rethink of climate policy'. See also Victor (2001). For the points of view of some Belgian companies about the EU Climate Action 'Energy for a Changing World', see Le Bussy (2001); Comhaire (2001). See also Verhest (2001).

new solutions and paths to the global warming problem in compliance with the economical needs of today.

In this chapter, we will provide an analysis of the case-law concerning the European greenhouse gas emissions trading scheme. This analysis will be presented in eight sections:

- Amendments to the national allocation plans (section 2).
- Access to justice and the 'individually concerned' case-law (section 3).
- The procedure, the delays and the public participation under Directive 2003/87/EC (section 4).
- The connections between the '1st phase' period and the 'Kyoto' period (section 5).
- The interpretation of the Commission's guidelines (section 6).
- Ex post adjustments and the principle of subsidiarity in the ETS Directive's system (section 7).
- The Directive 2003/87/EC and the 'State-aid' matter; the connections with the Articles 87 and 88 of the EC Treaty (section 8).
- Is the scope of the Directive discriminatory or not? (section 9)

2. AMENDMENTS TO THE NATIONAL ALLOCATION PLANS

The case *United Kingdom v. Commission* is the first case pronounced by the Court of First Instance about how the Directive must be interpreted when a Member State wants to modify its national plan.[11] On 30 April 2004, the United Kingdom notified its national allocation plan to the Commission, expressly stating it was provisional. On 7 July 2004, the Commission adopted its decision concerning the United Kingdom national allocation plan. The Commission found the notification incomplete because some important information was missing. The Commission asked the United Kingdom to produce those elements by 30 September 2004. But even then, the United Kingdom was not able to answer the Commission's request. At the time, the comments and the remarks of the public were not yet available. Thus, as soon as the results of the public opinion were made known, the United Kingdom notified the Commission, on 10 November 2004, that it wished to amend its plan to take into account these public inquiries. The Member State proposed, in particular, the increase of the total quantity of allowances to 756.1 million tonnes of carbon dioxide. On 12 April 2005, the Commission adopted Decision

11 Krämer (2006).

C(2005)1081 final concerning the proposed amendment to the NAP notified by the United Kingdom in accordance with the Directive. The Commission concluded that the United Kingdom was not entitled to submit a provisional plan. The Commission added that the United Kingdom is only entitled to amend its NAP in order to address the incompatibilities identified in the decision of 7 July 2004. In other words, the increase of the total number of allowances was not initially rejected by the Commission because it was not proposed, and this is why the Commission did not accept the amended plan. The Member State made an application against this decision.

During the trial, the United Kingdom argued that the Commission had exceeded its legal powers under the Directive. The Commission could not refuse a plan only for the reason that the Member State proposed a new plan modifying some elements that were not initially rejected by the Commission. In order to decide whether the Commission was entitled to reject those amendments as inadmissible, the Court considered that it was necessary to examine the roles and powers allocated to the Commission and the Member States respectively under the Directive.

The Court ruled that the second sentence of Article 9(3) of the Directive does not lay down any limit to the permissible amendments. Therefore [...] any amendments, whether proposed by the Member State of its own initiative or rendered necessary to overcome any incompatibility in the NAP raised by the Commission, must be notified to the latter and accepted by it before the NAP as amended can form a valid basis for the decision taken by the Member State.[12] Finally, 'with regard to the wording of the Directive and from the general structure and objectives of the system which it establishes', the Court considers that the Commission could not restrict a Member State's right to propose amendments, or categories of amendment.[13]

Thus, a Member State can freely propose amendments to its NAP, even those modifying the national measures that were accepted previously by the Commission. However, it still remains that any amendment must be adopted by the Commission in order to become effective. The Commission keeps its power of appreciation of the amendments, under the criteria of Annex III and Article 10 of the European Treaty. As Krämer concluded, 'the lesson to learn from this judgment is thus mainly that the rule of law within the European Union is fundamental and requires a continuous effort to be as precise as any possible in determining rights and obligations of all actors'.[14]

---

12 Court of First Instance, 23 November 2005, Case T-178/05 – United Kingdom/Commission – § 62. (O.J. C 22 of the 28.01.2006, p.14).

13 Court of First Instance, 23 November 2005, Case T-178/05 – United Kingdom/Commission – § 61.

14 Krämer (2006, p. 156).

3. ACCESS TO JUSTICE AND THE 'INDIVIDUALLY CONCERNED' CASE-LAW

3.1 The Cases Related to the 'Individually Concerned' in the Directive 2003/87/EC Case-law

The judges refuse thus far to grant a company the quality of being 'individually concerned' by the Commission's decision about the national allocation plan. The first step of this case-law was taken during the *EnBW Energie Würtemberg v. Commission* case of 30 April 2007.[15] On 7 July 2004, the Commission rejected the German national allocation plan because it contained an 'ex post' possibility, making possible a modification of the EU allowances allocated to an installation after the Commission's approbation of the national allocation plan. The Commission did not accept the plan because of a danger to the effectiveness of the trading mechanism. It is important to emphasize that the Commission rejected it only for that reason.[16] However, the German plan also had another particularity. It contained two specific rules, named the 'transfer rule' and the 'special attribution'.[17] Because old installations pollute more, Germany created those rules to favor their closings. The two rules were an exception to the general rule of the plan. Normally, the allocation of the EU allowances was made, for a 14-year period, on foreseen emissions, calculated pursuant to the 'best available technique' method. This is lower than the previous emissions, before the new system came into force. However, under the 'transfer rule', operators would have, over four years, EU allowances that were calculated on their previous historical emissions, which would have been more favorable for them. Pursuant to the 'special attribution' rule, not so generous as the transfer rule, the nuclear installations would receive temporary, extra EU allowances to cover their closings, which the German law had made mandatory in 2007.[18]

An appeal was brought before the Court of First Instance against the Commission's decision by EnBW Energie Baden-Württemberg, a German nuclear society, submitted to the 'special attribution' rule. In its opinion, the total amount of EU allowances given in accordance with the 'special attribution' rule would be insufficient to cover the loss caused by the closure of its nuclear installation. On the other hand, its concurrent RWE, subject to the 'transfer rule', would receive a greater amount of EU allowances. For that

---

15 Van Aken (2007).

16 See p. 105 of the present book.

17 Ehrmann and Greinache (2006).

18 Gesetz zur geordneten Beendigung der Kernenergienutzung zur gewerblichen Erzeugung von Elektrizität.

reason, that last attribution appeared to be an unfair and unjustified economical advantage.

The Commission, defendant, brought an admissibility argument before the Court. In the Commission's view, the plaintiff lacked sufficient interest to lodge an appeal against its NAP decision, because that latter decision is only addressed to the Member State. The judges took the advantage to make clear the role of the Commission when it examines the NAP of a Member State.[19] The Court rather surprisingly ruled that the evaluation of the Commission is not an 'authorization' *sensu stricto*. Indeed, the Commission can only make its evaluation of the national plan during three months starting from its deposit. If it does not, the NAP will be automatically accepted. Conversely, the Commission's decision is not an 'authorization' either, because it gives an opinion only on some of the specific points of the national allocation plan. However, it cannot be concluded that the Commission has expressed no opinion on the German rules at stake and on their application to the plaintiff in the decision addressed to the German government. Thus the plaintiff is not 'individually concerned' by the decision of the Commission and therefore, in the Court's view, the action is inadmissible.[20]

The operator adduced two other arguments. First, the company was of the view that the Commission's decision about the connection between Articles 87 and 88 of the Treaty and the German provisions was a 'state-aid' decision.[21] Then, the company supported the argument that the decision constituted a grievance, legally speaking.[22] The argument was not followed by the Court because the judges emphasized that the grievance must ensue from the

19 The real reason why the Commission explained this ensued from the Commission's argument before the Court. For the Commission, the plaintiff was not 'individually concerned' by its decision. Of course, the plaintiff disagreed with the opinion of the Commission. For the company, the Commission's evaluation of the national plan is an authorization and, in consequence, it is individually concerned by the Commission's decision.

20 'La décision attaquée ne comporte pas (…) une quelconque autorisation – ni explicite ni implicite – du PNA allemand dans son ensemble, en ce compris la règle de transfert contestée. Dès lors, contrairement à l'annulation d'une décision de compatibilité adoptée en matière d'aides d'État et au but poursuivi par la requérante, l'éventuelle annulation de la décision attaquée ne saurait avoir pour conséquence l'anéantissement de cette autorisation', Court of First Instance, 30 April 2007, Case T-387/04 – EnBW Energie Baden-Württemberg v. Commission – § 124.

21 On that question, see section 8 below.

22 Under an other regular case-law of the Court, see ECJ, 28 January 2004, Case C-164/02 – Netherlands v. Commission – § 21 – Rec. p. I-1177; Court of First Instance, 19 March 2003, T-213/00 – CMA CGM e.a./Commission – § 186 – Rec. p. II-913, both mentioned in Court of First Instance, 25 June 2007, Case T-130/06 – Drax Power e.a./ Commission – § 127.

pronouncement of a decision. In that case, the two contested rules (i.e. the transfer rule and the special attribution rule) were in the grounds of the decision itself and not in the pronouncement. Thus, the Court could not recognize the interest of the litigants because they were not individually concerned because they were not 'damaged' by the pronouncement. However, the Court recognized that the grounds could also contain a grievance, but in that case, it refused to admit the argument because these grounds had no connections with the pronouncement of the decision.

That emerging case-law was confirmed in *Drax Power e.a. and others v. Commission* of 25 June 2007. This case is connected with the above-mentioned case T-178/05, *United Kingdom v. Commission*. Finally, in May 2005, the United Kingdom decided to allocate allowances on the basis of its original plan.[23] In other words, the United Kingdom did not allocate the increase of 2.7% allowances planned for the electricity generation sector in its draft rejected by the Commission. It decided to do that 'subject to and without prejudice to its legal challenge against the Commission' which was still pending before the Court of First Instance. On 22 February 2006, the Commission adopted the decision C(2006) 426 concerning the proposed amendment to the original plan notified by the United Kingdom. In that decision, the Commission rejected the United Kingdom's proposed amendment for the second time. The dismissal was based on the fact that this proposal was notified too late to the Commission. Indeed, the Commission asked the Member State to communicate its amendments by 30 September 2004 at the latest and the Member State sent it to the Commission on 18 February 2005. On 28 April 2006, the United Kingdom announced in a joint statement of the Department for Environment, Food and Rural Affairs and the Department of Trade and Industry, that it had decided not to pursue further Court action against the Commission to procure consideration of its proposed amendment to the original NAP and that it would leave that NAP unchanged. However, several English companies, including Drax Power e.a., made an appeal against the decision. They did not accept the Member State's refusal to allocate them these 'extra' EU allowances.

Once again, the real question was whether the plaintiff was sufficiently concerned by the Commission's decision about the English national allocation plan: 'Given that the contested decision was addressed to the United Kingdom, the Court will examine whether the applicants are directly concerned by that decision'.[24] This time, the judges based their appreciation on *Dreyfus v. Commission* and *DSTV v. Commission* case-law. They recalled

---

23 A total number of 736.3 Mt CO_2.

24 Court of First Instance, 25 June 2007, Case T-130/06 – Drax Power e.a./Commission – § 46. (OJ. C 211 of the 08.09.2007, p. 33).

the two cumulative criteria of its regular case-law in the matter of the 'individually concerned' within the meaning of the Article 230 of the Treaty.[25,26] According to it,

> first, the measure at issue must directly affect the legal situation of the person concerned. Second, that measure must leave no discretion to its addressees who are entrusted with the task of implementing it, such implementation being purely automatic and resulting from Community rules without the application of other intermediate rules [...] The condition required by the second criterion is also satisfied where the possibility for addressees not to give effect to the Community measure is purely theoretical and their intention to act in conformity with it is not in doubt.[27]

So, given that the contested decision was addressed to the Member State, the Court decided to examine whether the applicants were directly concerned by that decision'.[28] The Court of First Instance consecrated this case-law in Drax Power.

The Court examined whether the English decision that did not grant the 'extra' allowances modified the legal situation of the plaintiff or not. For the Court, the possibility that the applicants would receive the additional allowances as envisaged by the proposed amendment to the original NAP was based only on the United Kingdom's declared intention in that regard and cannot be considered as a vested right of the applicants: '(...) the [Commission's] decision did not reduce the total quantity of allowances granted by the United Kingdom by 19.8 Mt CO_2'. 'Furthermore, in order to show that they are directly affected by the contested decision, the applicants are not entitled to rely on future and hypothetical situations which cannot be used as a basis to establish that the contested decision directly affects their legal situation'. [29] So, in other words, 'the direct and definitive determination of the rights and obligations of the operators of those installations can only result from an allocation decision of the Member State. Therefore, the contested decision did not in any way have the effect of depriving the applicants of specific rights

---

[25] ECJ, Case C-386/96 P – Dreyfus v. Commission [1998] ECR I-2309, paragraph 43, mentioned in Court of First Instance, 25 June 2007, Case T-130/06 – Drax Power e.a./ Commission – § 48.

[26] Court of First Instance, Case T-69/99 – DSTV v. Commission [2000] ECR II-4309, paragraph 24, mentioned in Court of First Instance, 25 June 2007, Case T-130/06 – Drax Power e.a./ Commission – § 48.

[27] Court of First Instance, 25 June 2007, Case T-130/06 – Drax Power e.a./ Commission – § 48.

[28] Court of First Instance, 25 June 2007, Case T-130/06 – Drax Power e.a./ Commission – § 47.

[29] Court of First Instance, 25 June 2007, Case T-130/06 – Drax Power e.a./ Commission – §59.

which had vested at the time of its adoption and thus did not result in any change to the applicants' rights or to their legal situation'.[30] 'It follows that a proposed amendment to a NAP may not be seen as definitively fixing the position of the Member State'.[31] In conclusion, the Court said that 'the legal situation of the applicants at the date of the contested decision was that of operators holding specific allocations of allowances for the period from 2005 to 2007 on the basis of the original NAP. The contested decision did not alter that position'.[32]

In the next judgments, *Fels-Werke v. Commission* of 11 September 2007 and the US Steel Kosice cases of 1 October 2007, the Court 'simply' followed its previous interpretation.[33,34]

Saint Gobain Glass Deutschland, Fels-Werke and Spenner Zement made an appeal against the decision of the Court of First Instance. These companies still consider that they are 'individually concerned' by the Commission's decision. However, the European Court of Justice decided, in its order of 8 April 2008, that the Court of First Instance did not make a mistake when it applied Article 230 of the EC Treaty in that case.[35]

3.2 The Lessons from the Court's Decisions

So, if an economic actor, although covered by the Directive's scope, lodges an application against the Commission's decision, the Court will simply reject its application. In other words, the Court accepts only the Member States' appeal.[36] This observation raises a new question: What can operators do if

---

30 Court of First Instance, 25 June 2007, Case T-130/06 – Drax Power e.a./ Commission – §60.

31 Court of First Instance, 25 June 2007, Case T-130/06 – Drax Power e.a./ Commission – §61.

32 Court of First Instance, 25 June 2007, Case T-130/06 – Drax Power e.a./ Commission – § 68.

33 Mentioned below. See section 5.

34 There are two decisions about the US Steel Kosice case, one for the '1st phase' Commission rejection decision (Order of the Court of First Instance of 1 October 2007 – U.S. Steel Kosice v. Commission – Case T-489/04, application. (O.J. C 297 of 08.12.2007; application in O.J. C 82 of 2.4.2005.), and another against the Commission rejection decision for the period from 2008 to 2012 (Order of the Court of First Instance of 1 October 2007, Case T-27/07 – U.S. Steel Kosice v. Commission (O.J. C 297 of 08.12.2007; application in OJ C 69, 24.3.2007). See section 8 of the contribution.

35 E.C.J., 8 April 2008, Case C-503/07 P – Saint-Gobain Glass Deutschland GmbH, Fels-Werke GmbH, Spenner-Zement GmbH & Co. KG v. Commission des Communautés européennes.

36 See also Court of First Instance, 6 November 2007, Case T-13/07 – Cemex UK Cement Ltd v. Commission.

their nationals' authorities refuse to appeal? That specific case-law is well known in the European Law. The Court created it in the *Plaumann* case.[37] The position of the judges was always very restrictive.[38] Many authors criticized the Court's position, specifically under the conformity of those limitations with Article 6, §1, and 13 of the European Convention of Human Rights.[39] A democratic argument was also made. After all, specifically in the environmental matter, are not the physical persons and the companies the first to be concerned by those measures? Even the case-law of the Court was used by the doctrinal authors to criticize that restrictive interpretation.[40] In *Les Verts* case, the Court underlined the necessity that 'the Treaty established a complete system of legal remedies and procedures designed to permit the Court of Justice to review the legality of measures adopted by the institutions'.[41] We can admit that the Court of Justice and the Court of First Instance could not receive and appreciate every single application made by every single company or by a physical person. There is also a well known possibility for judicial abuse by the plaintiffs. However, for us, the 'individually concerned' case-law, which refuses the companies (and the physical persons) the right to access to a Court, can be criticized. We have some doubts about the possible interest of a State in putting an appeal before the Court in the name of the companies located in it.

4. THE PROCEDURE, THE DELAYS AND PUBLIC PARTICIPATION UNDER THE DIRECTIVE 2003/87/EC

4.1 The System of the Directive

The procedure for allocating and issuing allowances consists of three steps. First, the Member State must propose a 'national allocation plan'. Article 9.1. of the 2003/87/EC states that

[37] ECJ, 15 July 1963, Case 25/62 – § 223 – Rec. p. 199; Cf. also ECJ, 24 February 1987, Case 26/86 – Deutz und Geldermann v. Council – §9 – Rec., p. 941; ECJ, 15 February 1996, Case C-209/94 P – Buralux e.a. v. Council – §25 – Rec. p. I-615; ECJ, 2 April 1998, Case C-321/95 P – Stichting Greenpeace Council – Rec. p. I-1651. For an analysis, see Van Raepenbusch (2005), pp. 623–31; Waelbroeck and Waelbroeck (1993).

[38] Some judgments have made more flexible that restrictive case-law: see ECJ, 16 May 1991, Case C-358/89 – Extramet v. Council (Rec. p. I-2501); ECJ, 18 May 1994, Case C-69/89 – Codorniu v. Council – Rec. p. I-1853.

[39] Rasmussen (1980); Waelbroeck and Verheyden (1995); Arnulli (1995); Vandersanden (1995).

[40] About that question, see Van Raepenbusch (2005, p. 627).

[41] ECJ, 23 April 1986 – Case C-194/83 (Rec. p. 1365), §23.

> for each period referred to in Article 11(1) and (2), each Member State shall develop a national plan stating the total quantity of allowances that it intends to allocate for that period and how it proposes to allocate them. The plan shall be based on objective and transparent criteria, including those listed in Annex III, taking due account of comments from the public.

The second stage is that the Commission, which is competent to evaluate the national allocation plan, decides to accept or to refuse it if it seems that the national plan is or is not in accordance with the criteria of Annex III of the Directive. And finally, each year, 'the member States' competent authorities issue a share of the allocated EU allowances to the operators of the covered installations'. As we underlined in section 3, companies can fight the national decision but still then, cannot fight the Commission's decision.

The Member States are not free to notify their national allocations plans at a time of their choosing. The plan has to be published and notified to the Commission and to the other Member States at least 18 months before the beginning of the relevant period. Reasons shall be given for any rejection decision by the Commission.

The 1st phase period was created as a 'test' or 'preparatory' period for the Member States.[42] The majority of the '1st' national allocation plans did not receive any objections from the Commission. The plans had to be published and notified to the Commission by 31 March 2004.

For the 2nd period, known as the 'Kyoto Period', the notification's deadline was fixed as 30 June 2006. The experience of the 1st period acted upon the content of that second round of communications. In these guidelines, the Commission summarized the difficulties, the main ones of which were that in the 2nd period Member States (and the European Community) are submitted to Kyoto's quantum of emissions.[43] This difference explains why the Commission is now stricter in its evaluation of the national allocations plans than during the 1st phase. All the national allocation plans have been rejected by the Commission, without any exceptions, that last requiring each Member State to make some adjustments or some specific modifications of their NAPs' provisions.

---

42 As the Commission said, 'the best preparation for the Community and its member States might be to develop their own emission trading experience'. See the Communication of the Commission (COM(99)230 final).

43 Commission proposal for Directive of the European Parliament and of the Council amending Directive 2003/87/EC so as to improve and extend the EU greenhouse gas emission allowance trading scheme, COM(2008)16 final, 23.01.2008, p. 2. 'However, the environmental outcome of the 1st phase of the EU ETS could have been more significant but was limited due to excessive allocation of allowances in some Member States and some sectors, which must mainly be attributed to reliance on projections and a lack of verified emission data'.

The Directive is particularly vague about how the comments of the public have to be taken into account by the Member States. In our opinion, the Court of First Instance really pushed this question further than the Directive itself. The first time the Court pursued that question was *United Kingdom v. Commission*. The judges discovered a double public consultation system: the first and clearly prescribed public consultation must be made during the elaboration of the national plan, before the notification to the Commission. But, in the Court's view, after the Commission's decision authorizing the allocation and before the national decision of allocation, there is also a second round of public consultation. In its opinion, if the modifications to the national plan were limited to those suggested by the Commission, the second round of public consultation 'would be deprived of its effectiveness and [...] would be rendered purely academic'.[44] In other words, the European judges recognized a kind of 'double procedure' in the matter of the public consultation to grant the Member State a margin of modification of its national allowance plan out of the scope of [and after] the Commission's remarks. A clear double procedure that does not appear clearly in the ETS Directive.

In *Germany v. Commission*, the Court referred to this appreciation of the second public consultation of the *United Kingdom v. Commission* case in order to clarify it:

> Article 9(1) and Article 11(1) of the directive oblige the Member State to '[take] due account of comments from the public', both in the NAP, that is to say following an initial public consultation, and in the allocation decision, adopted following a second public consultation. It follows, first, that, in the absence of an express prohibition in Article 11(1) of subsequent amendment of the individual allocation of allowances, the NAP and the allocation decision may expressly provide for such a possibility of amendment, provided that the criteria for exercise of that power are laid down in an objective and transparent manner.

It must be emphasized that the Court used its own 'two-step public consultation' interpretation to refuse the Commission's point of view. Although the Directive 2003/87/CE specifies that the elaboration of the national plan must take 'due account of the public comments', there is no clear and explicit textual basis about that second round of public consultation. These cases are an example of the influence of the judges, especially when the Directive and the Commission (in its guidelines) are particularly imprecise.

---

44 Court of First Instance, 23 November 2005, Case T-178/05 – United Kingdom/Commission – § 57.

4.2 Lessons from the Court's Case-law: What about the Public Consultation?

Until the start of the trading scheme system, the Commission was aware of that potential possibility of a two-step public consultation. In its first communication, COM(2003)830 final, it expressly provided that

> a Member State should inform the Commission of any intended modifications following public participation subsequent to the publication and notification of the national allocation plan and before taking its final decision pursuant to Article 11. Feedback is to be provided, in a general form, to the public about the decision taken and the main considerations upon which it is based [...] It should be noted that the possibility for the public to comment on the national allocation plan provided for under this criterion constitutes a second round of public consultation.[45]

For each period, the Member State must send its national plan to the Commission. During the elaboration of it, the Member State must take due account of the public comments. Then, the Member State can send its plan to the Commission. If the plan is accepted, the Member State can uphold the distribution of the EU allowances contained in the plan in its so-called 'allocation decision'. Corresponding to the Court's case-law, the national allocation decision has to be adopted following a second public consultation.

5. THE CONNECTIONS BETWEEN THE 1ST PHASE PERIOD AND THE 'KYOTO' PERIOD

5.1 The '*Fels-Werke*' Case-law[46]

The '*Fels-Werke*' case was the opportunity for the European judges to clarify the connections between the '1st phase period' and the 'Kyoto period'. The case derives from a specific approach conducted by Germany in its first national plan.[47] In this '1st phase period' allocation plan, a 'three methods of allocation' mechanism was created. The covered installation received EU allowances upon the basis of its historical emission adapted with one of the three methods respectively in accordance with the moment when the company

---

[45] Communication from the Commission on guidance to assist Member States in the implementation of the criteria listed in Annex III to Directive 2003/87/EC establishing a scheme for greenhouse gas emission allowance trading within the Community and amending Council Directive 96/61/EC, and on the circumstances under which force majeure is demonstrated, COM/2003/0830 final, § 95–96.

[46] Van Aken (2008, p. 1).

[47] See Marr and Schafhausen (2004).

began its exploitation. In consequence, under that mechanism, the younger companies received EU allowances without the intervention of an 'execution factor'. Although this has the effect that they receive more EU allowances than the older installations. When the Commission examined the 1st German national plan, it did not make any objections about the mechanism of the 'three methods of allocations'. The allocations of allowances, without the intervention of the 'execution factor', were made for a period between 12 and 14 years, depending the year of the beginning of the exploitation. As a consequence, allowances can be granted that can exceed their own validity period and cover emissions during the second 'Kyoto' period. The Commission rejected the German plan in the capacity of the ex post mechanism contained in it, and not because of these methods of allocation.[48] Following this, it is interesting to underline the fact that Germany reproduced the same method of allocation for the second plan, with the same duration of validity. However, this time, the Commission rejected the plan because these methods of allocation, favorable to the new installations, would constitute an unfair and unjustified treatment against Articles 87 and 88 of the Treaty. In other words, the Commission emphasized that discrimination occurred between the operators. In regard to the Commission's rejection, the Member State made an application to the European Court of First Instance. To answer that question, it seems that the Court based its interpretation on Article 11.1. That article states 'allowances shall be valid for emissions during the period referred to in Article 11(1) or (2) for which they are issued'. For the Court, the Directive 2003/87/EC clearly discerns the different period of allocations. The allowances allocated are only valid for the period for which they were granted. It implies that Member States must adopt distinct allocation decisions for each period.[49] In consequence, the judges have confirmed that the allowances given for one specific period must be given only for this period.[50] In other words, there is no possibility of being able to use during the 'Kyoto' period an allowance allocated for the 1st period.[51]

---

48 See section 7 below.

49 The Court stated that 'l'article 11, paragraphe 1 et 2, lu conjointement avec l'article 9, paragraphe 1, et avec l'article 13, paragraphe 1, de la directive 2003/87 distingue clairement entre la première et la seconde période d'allocation et restreint la validité des quotas d'émission alloués à une seule période d'allocation, ce qui implique la nécessité pour les Etats membres d'adopter des décisions d'allocations distinctes pour chaque période'. Court of First Instance, 11 September 2007, Case T-28/07 – Fels-Werke GmbH v. Commission – §67 (O.J. C 283 of 24.11.2007, p. 27).

50 The Court had taken an unsurprisingly decision about the application of the company. It said that the company was not 'individually concerned' by the decision of the Commission, and 'the application is dismissed as inadmissible'.

51 Court of First Instance, 11 September 2007, Case T-28/07 – Fels-Werke GmbH v. Commission – §67.

In conclusion, the EU allowances allocated for one specific period are allocated only for this period and cannot be used for the next period (in the case, the 'Kyoto' period). What about the possibility of 'banking'? That 'banking' can be delimited in two different ways. First, the 'annual banking' is a possibility 'of using units in years following the year of their issuance. Banking adds a level of flexibility for participants and may reduce the impact of emission and price fluctuations on the markets. Banking can be unlimited or can be limited to a certain percentage'.[52] Then, next to the 'annual banking', there is also the 'real banking'. It permits the use of allowances allocated for a period of 3 or 5 years prior to the next (5-year) period. In regard to the Court's case-law, it seems that the question is now settled. As there is no formal interdiction of it in the Directive, the Member State can offer the 'annual banking' possibility to the companies. It is a simple use of the subsidiarity principle. However, the Court expressly stated in *Fels-Werke* that 'real banking' is forbidden. Like the Commission, the judges saw in that possibility the major risk of reducing the practical utility of the ETS system. The question of the validity of the allowances allocated for the 'Kyoto' period and possible other periods of five years still has no answer. Maybe, the next future international agreement on the global warming problem will find a solution to that question.

6. THE INTERPRETATION OF THE COMMISSION'S GUIDELINES

Under Article 211 of the EC Treaty, the Commission is charged with ensuring that the provisions of the Treaty and the measures taken by the institutions pursuant to the Treaty are applied. 'This means the Commission has been given primary responsibility for monitoring the application of European law in the Member States.'[53] The most important instrument at the Commission's disposal is the procedure laid down in Article 226. This provides that the Commission may bring a matter before the Court of Justice if it considers that a Member State has failed to fulfill an obligation under the Treaty.[54] The Commission made two applications based on that Article, against the non-transposition of the ETS Directive by Italy and Finland.[55,56] In *Commission v. Finland*, the Court ruled that Finland,

52 Upston-Hooper and Mehling (2007, p. 308).

53 Jans (2000).

54 Ibid.

55 ECJ, Judgment of 18 May 2006, Case C-122/05 – Commission v. Italy: 'By failing to adopt, within the prescribed period, all the laws, regulations and administrative provisions necessary to comply with Directive 2003/87/EC of the European

> by failing, with regard to the province of Åland, to adopt the laws regulations and administrative provisions necessary to comply with Directive 2003/87/EEC of the European Parliament and of the Council of 13 October 2003 establishing a scheme for greenhouse gas emission allowance trading within the Community and amending Council Directive 96/61/EC, the [Member State] has failed to fulfill its obligations under that directive.[57]

The decision was the same in *Italy v. Commission*.

Similarly with Article 211 of the EC Treaty, the European Commission was chosen to be the central organ of control of the Directive's application. In order to fulfill this task, the European Commission made some guidelines to help Member States during their national allocation plan elaboration process.[58,59] The question about the legal validity of these guidelines appeared

---

Parliament and of the Council of 13 October 2003 establishing a scheme for greenhouse gas emission allowance trading within the Community and amending Council Directive 96/61/EC, the Italian Republic has failed to fulfill its obligations under that directive'.

[56] In the same idea, see ECJ, Judgment of 18 July 2007, Case C-61/07 – Grand Duchy of Luxembourg v. Commission of the European Communities. 'The Court declares that, by failing to communicate the information required under Article 3(2) of Decision n° 280/2004/EC of the European Parliament and of the Council of 11 February 2004 concerning a mechanism for monitoring Community greenhouse gas emissions and for implementing the Kyoto protocol, the Grand Duchy of Luxembourg has failed to fulfill its obligation under that provision'.

[57] ECJ, Judgment of 12 January 2006, Case C-107/05 – Commission v. Finland.

[58] Communication from the Commission on guidance to assist Member States in the implementation of the criteria listed in Annex III to Directive 2003/87/EC establishing a scheme for greenhouse gas emission allowance trading within the Community and amending Council Directive 96/61/EC, and on the circumstances under which force majeure is demonstrated, COM/2003/0830 final; Communication from the Commission to the Council and to the European Parliament on Commission Decisions of 7 July 2004 concerning national allocation plans for the allocation of greenhouse gas emission allowances of Austria, Denmark, Germany, Ireland, the Netherlands, Slovenia, Sweden, and the United Kingdom in accordance with Directive 2003/87/EC, COM/2004/0500 final; Communication from the Commission to the Council and to the European Parliament on Commission Decisions of 20 October 2004 concerning national allocation plans for the allocation of greenhouse gas emission allowances of Belgium, Estonia, Finland, France, Latvia, Luxembourg, Portugal, and the Slovak Republic in accordance with Directive 2003/87/EC, COM(2004)0681; Communication from the Commission, 'Further guidance on allocation plans for the 2008 to 2012 trading period of the EU Emission Trading Scheme', COM(2005)0703; Communication from the Commission to the Council and to the European Parliament on the assessment of national allocation plans for the allocation of greenhouse gas emission allowances in the second period of the EU Emissions Trading Scheme accompanying Commission Decisions of 29 November 2006 on the national allocation plans of Germany, Greece, Ireland, Latvia, Lithuania, Luxembourg, Malta, Slovakia, Sweden and the United Kingdom in accordance with Directive 2003/87/EC, COM(2006)0725.

[59] Van Raepenbusch (2005). 'Le caractère non contraignant des recommanda-

immediately. In *Germany v. Commission*, the Court of First Instance underlined that '[the guidelines] do […] not correspond to any of the measures of secondary Community law that are provided for in Article 249 EC'.[60]

> Accordingly, the guidance falls within the category of rules which, as such, do not, in principle, have independent binding effect vis-à-vis third parties and of which the Commission makes extensive use in its administrative practice in order to structure, and increase the transparency of, the exercise of its discretion and supervisory power.
>
> [The Commission] imposes a limit on the exercise of its own discretion and cannot depart from those rules, if it is not to be found, in some circumstances, to be in breach of general principles of law, such as the principles of equal treatment, of legal certainty or of the protection of legitimate expectations.

In consequence, 'the Commission may not depart from them in an individual case without giving reasons that are compatible with the principle of equal treatment'.[61]

So, in the opinion of the Court, the Commission's communications have no real lawful effect, except for the Commission itself. If the Commission wants to depart from its previous declarations, it has to give the reasons why it wants to adopt a new interpretation. As the Court emphasized, these reasons have to be in conformity with the principle of equal treatment. It seems that the Court has based its interpretation on the general principles of legal certainty and the protection of legitimate expectations. In other words, the applicants (the Member States) must take into consideration the Commission's interpretation contained in its declarations during the NAP's process. But, in order to provide legal certainty, they must be sure, during its appreciation of the NAPs, that the Commission cannot change without any justified reasons its interpretation of such and such particular criterion of Annex III in a specific case. In consequence, if the Court does not change its case-law, Member States can use that decision in order to limit any unjustified breach of the Commission's declarations made by the Commission itself.

---

tions empêche de les considérer en tant que telles comme directement applicables. Toutefois, ces actes ne sont pas dépourvues de tout effet direct juridique dès lors qu'ils doivent être pris en considération par les juges nationaux lorsqu'ils interprètent les dispositions nationales dans le but d'assurer la pleine mise en œuvre ou de compléter des dispositions communautaires ayant un caractère contraignant (cf. Aff. C-322/88, Grimaldi - Rec., 1989, p.4407)'.

60 Court of First Instance, 7 November 2007, Case T-384/04 – Germany v. Commission – § 110.

61 Court of First Instance, 7 November 2007, Case T-384/04 – Germany v. Commission – § 111.

7. EX POST ADJUSTMENT AND THE PRINCIPLE OF SUBSIDIARITY IN THE ETS DIRECTIVE'S SYSTEM

7.1 The Problematic of the 'Ex Post' Adjustment in the German National Allocation Plan

This case was an application for partial annulment of Commission Decision of 7 July 2004 concerning the rejection of the national allocation plan for the allocation of greenhouse gas emission allowances notified by the Federal Republic of Germany for the 1st period. The appeal was made because Germany thought that its downward 'ex post' mechanism was not a mechanism against the system created by the Directive. After all, not even one Article of the Directive reads that an ex post mechanism is forbidden or against it. With this mechanism, the Member State can 'take back', after the allocation decision pursuant to Article 11 of the Directive, a number of EU allowances allocated to a covered installation if, in fact, the company does not need so many allowances. It is important to emphasize that the allowances freed up are not cancelled immediately but transferred to the reserve to remain available to new entrants. The guidelines of the Commission are imprecise on that 'ex post' possibility. Like we underlined in section 6, the Commission cannot freely turn aside from its previous declarations. If the European institution does so, its evaluation would be against the fundamental principles of equal treatment, legal certainty and legitimate expectations.[62] So, in view of Germany, the Commission has exceeded its power granted to it by the Directive 2003/87/EC when it refused the 'ex post' mechanism, which was not previously rejected by the Commission. On the other hand, according to the Commission, the German possibility of reviewing the number of allowances allocated to a covered installation was strictly against the Directive and its objectives.[63] The Commission held the view that in an NAP the amount of

[62] Court of First Instance, 7 November 2007, Case T-384/04 – Germany v. Commission – §111.

[63] See the Commission's decision about the German allocation plan, C(2004)2515/ 2 final, 07.07.2004, § 4–7; 'The intention of Germany to potentially make ex-post adjustments to the allocation of allowances to new entrants contravenes criterion 5 of the Annex III, because the application of such ex-post adjustment would unduly favour new entrants compared to the operators of installations that are already listed in the national allocation plan in respect of which no ex-post adjustments to the allocations is permitted during the period 2005–2007 [...] The intention of Germany to adjust the allocation of allowances to an installation listed in the national allocation plan and operating in its territory in the event that an installation whose operation is related to that installation closes within its national territory contravenes with criterion 10 in Annex III to Directive 2003/87/EC which requires the quantity of allowances to

allowances to be allocated to each installation must be determined in advance for the first allocation period and, in any event, can no longer be altered after the Member State has adopted the decision to allocate them.[64] Principally, the Commission considered that 'ex-post adjustments create uncertainty, indeed deter operators from investing, with the consequence that improvements in production technologies and reductions in production are less substantial than they would be in the absence of adjustments'.[65]

The Court recalled the consequences and meanings of the words 'directive' and 'transposition', in order to appreciate the application of the principle of subsidiarity in the environmental matter.[66] When a transposition of a directive in the environmental field is at issue, especially when the directive in question does not prescribe the form and methods for achieving a particular result, the Member States are required to choose the most appropriate forms and methods to ensure the effectiveness of directives. It also follows that, where there is no Community rule prescribing clearly and precisely the form and methods that must be employed by the Member State (like the 'ex post' mechanism), the Commission has the task, when exercising its supervisory power, pursuant in particular to Articles 211 EC and 226 EC, of proving to the required legal standard that the instruments used by the Member State in that respect are contrary to Community law.[67]

---

be allocated to each installation to be stated ex-ante in the national allocation plan covering the period referred to in Article 11(1) of that Directive [...] It is furthermore not compatible with the criterion 10 in Annex III to Directive 2003/87/EC to adjust the allocation of allowances set out in the national allocation plan after the adoption of the decision referred to in article 11(1) of that Directive for the reason that an installation listed in the national allocation plan and operating in its territory experiences lower capacity utilization than foreseen, or the installation's annual emissions are less than 40% of its base periods emissions, or the installation is benefiting from an additional allocation for combined heat and power generates a lower amount of power production from combined heat and power than in the base period [...] In order to bring the national allocation plan in conformity with the criteria listed in Annex III to Directive 2003/87/EC the plan should be amended'.

64 Court of First Instance, 7 November 2007, Case T-384/04 – Germany v. Commission – § 91.

65 Court of First Instance, 7 November 2007, Case T-384/04 – Germany v. Commission – § 127.

66 'According to that principle, in areas which do not fall within its exclusive competence the Community is to take action only if and in so far as the objectives of the proposed action cannot be sufficiently achieved by the Member States and can therefore, by reason of the scale or effects of the proposed action, be better achieved by the Community'. See Court of First Instance, 7 November 2007, Case T-374/04 – Germany v. Commission – § 79.

67 Court of First Instance, 7 November 2007 - Case T-384/04 – Germany v. Commission – § 78.

In order to appreciate whether the German 'ex post' mechanism was, indeed, against the rationale of the directive, the Court adopted a 'four-part' analysis.[68]

First, a literal interpretation: The Court recognized that a literal reading of criterion 10 does not confirm the argument of the Commission before the Court, according to which the NAP and the allocation decision must contain the definitive amount of allowances to be allocated in respect of each of the installations listed. However, the Directive 2003/87/EC, on the basis of the principle of subsidiarity, does 'not preclude that the Community legislature sought to grant some flexibility, indeed some discretion for the Member State, by allowing it the possibility of altering the amount of allowances, as envisaged in the list of installations annexed to the NAP, in a subsequent phase of the implementation of Directive 2003/87'.[69]

Second, an historical interpretation, which did not supply additional factors.

Third, a contextual interpretation: first, the Court, similarly to its *United Kingdom v. Commission* case-law, indicated that an absolute prohibition on amending the individual allocations laid down in the NAP would compromise the practical effect of the second public consultation. The judges decided that the Directive does not expressly prohibit a subsequent amendment of the amount of allowances allocated individually according to the list annexed to the NAP and according to the allocation decision.

> [...] The NAP and the allocation decision may expressly provide for such a possibility of amendment, provided that the criteria for exercise of that power are laid down in an objective and transparent manner [...] It should be added that any subsequent amendment of the individual allocations of allowances [...] does not result in the Commission losing all possibility of review, given the permanent supervision which it exercises as a result of the instruments for management and verification that are provided for by Regulation No 2216/2004, and the general supervisory power with which it is vested under Articles 211 EC and 226 EC and which permits it to act at any time if Community law is infringed.[70]

The Court asserts that the Directive expressly forbids, under its Article 29, a subsequent increase in the amount of individually allocated allowances.

---

[68] The Court also adopted the same method of reasoning about the analysis of article 9 of the Directive 2003/87/EC in EnBW Energie Baden-Württemberg AG v. Commission, o.c., §99-119.

[69] Court of First Instance, 7 November 2007, Case T-384/04 – Germany v. Commission – § 96.

[70] Court of First Instance, 7 November 2007, Case T-384/04 – Germany v. Commission – §106.

> However, the directive contains no express provision limiting the Member State's freedom of action in managing the individual allocation of allowances when that does not result in such an increase but only in downward ex-post corrections. In the latter case, there is no risk of an allocation exceeding the total amount of allowances that is provided for in the NAP, an allocation which would be contrary to the obligation owed by the Member State to reduce emissions.[71]

Finally, from the angle of a teleological interpretation: The Court has confronted the 'ex post' mechanism with the objectives of the ETS Directive. The main objective of reducing emissions is completed with some sub-objectives, as the objective of maintaining cost-effective and economically efficient conditions, the objective of preserving the integrity of the internal market and maintaining conditions of competition, the objective of reducing emissions through improvements in technologies. Under that interpretation, the Court accepted the Commission's argument. In the judges' point of view,

> as the Commission submits, when the operator is aware that any fall in production diverging from his own forecasts will be penalized by the application of ex-post adjustments, his incentive to reduce production in order to free up allowances is affected, not to say removed, even where there is an increase in demand on the trading market from other operators wishing to obtain additional allowances.

In other words, the stimulant able to lead companies to reduce their emissions can no longer exist under the German ex post mechanism. Indeed, companies cannot hold or sell their excessive allowances allocated.

> Consequently, the Commission has demonstrated that certain of the ex-post adjustments at issue, inasmuch as they deter operators from reducing their installations' production volume, are liable to compromise achievement of the objective of efficient functioning of the trading market in accordance with Article 1 and recital 5 of Directive 2003/87. However, the Commission has not put forward evidence or arguments capable of establishing that those adjustments harm the principal objective of Directive 2003/87, namely the reduction of greenhouse gas emissions.

For the Court, as the allowances freed up are included in a reserve to remain available to new entrants [like we underlined above], the Commission's position was no longer relevant, because the reserve has 'the consequence that the total amount of available allowances remains unchanged'.[72] Moreover, the

---

[71] Court of First Instance, 7 November 2007, Case T-384/04 – Germany v. Commission – § 107.

[72] Court of First Instance, 7 November 2007, Case T-384/04 – Germany v. Commission – § 141.

Court underlined the paradoxical position of the Commission. Indeed, 'the Commission appears to contradict its own statements in this regard set out in the Commission Communication of 7 July 2004 (pp. 7 and 8), where it is stated that downward ex-post adjustments "might be argued" to have "a beneficial environmental effect" '. So, in conclusion,

> the mere fact that the ex-post adjustments at issue are liable to deter operators from reducing their production volume [...] is not sufficient to call into question the adjustments' legality in light of the directive's objectives as a whole. Furthermore, it follows from the self-limiting effect created by the Commission guidance that the Commission must accept having the applicant raise against it the lack of clarity and precision of that guidance as regards any prohibition of the ex-post adjustments at issue in light of the directive's objectives.[73]

All those interpretations together came to the same conclusion: the ex post mechanism is not against the Directive's rationale. The Member State keeps a large discretion when it transposes the Directive into its own national legal order and the Commission did not prove that the national measures were against the ETS Directive. In conclusion, the German ex post mechanism is not a mechanism opposed to the Directive's rationale.

Moreover, the Court also recognized that the Commission did not respect the obligation to motivate. The Commission did not explain why the ex post mechanism was a 'risk'. 'The Court [held] that the Commission breached its duty under Article 253 EC to state reasons by failing to provide the slightest explanation regarding the application of the principle of equal treatment in the contested decision, in the Commission Communication of 7 July 2004 or in the context of the adoption of those measures'.[74]

7.2 The Contribution of the *Germany v. Commission* Case to the Emissions Trading Scheme Case-law

This case is probably the most interesting one in the context of the EU global warming policy case-law. It shows a real environmental consideration in the judges' interpretation. After all, the Commission refused a downward modification of the EU allowances only because it was an *a posteriori* modification. The Court clearly rejected that position. They confirmed that an increase of EU allowances is not possible, but it found paradoxical the fact that a downward modification cannot be accepted. The judges used the principle of

---

[73] Court of First Instance, 7 November 2007, Case T-384/04 – Germany v. Commission – § 148.

[74] Court of First Instance, 7 November 2007, Case T-384/04 – Germany v. Commission – § 170.

subsidiarity and the silence both of the Directive and of the Commission's guidelines to consecrate a more environmental perspective. This interpretation calls for another question: Is the system created by the Directive a useful tool to bring new solutions to the global warming problem? Or is it rather a simplistic method to exchange EU allowances as an economic value, without real environmental effect? Maybe, we can consider EU allowances as being merely a new 'possession' called to be exchanged between a limited number of polluting companies or eventually by some persons sensible of the environmental problems.[75] The conclusion that the EU allowance is simply a possession without environmental consequences, especially after the surplus of allowances during the '1st phase', is logical. However, after *Germany v. Commission* which was a decision promoting more protective national measures of the environment, and the stricter behaviour of the Commission during its appreciation of the NAPs of the 'Kyoto' period, that conclusion no longer pertains. The validity of the ETS system is based upon two fundamental conditions: the scarcity of EU allowances allocated and the sanction of any violation of the ETS system. With the scarcity of EU allowances, the operators will really try to limit their polluting emissions, in order to avoid any economical sanctions. That environmental objective, supported by the reduction of the allowances, is reinforced by the Court's case-law.

8. THE DIRECTIVE 2003/87/EC AND THE 'STATE-AID' MATTER

8.1 The Connections with the Articles 87 and 88 of the EC Treaty

The Commission had always recognized the potential effects of EU allowances on the European common market. The European Parliament and the Council are also acquainted with that economical effect. Article 1 states that the 'Directive establishes a scheme for greenhouse gas emission allowance trading within the Community [...] in order to promote reductions of greenhouse gas emissions in a cost-effective and economically efficient manner'. There was a risk that, directly or indirectly, States would try to give an advantage to their national installations through the ETS Directive, whereas state aid is expressly forbidden by the EC Treaty. After all, EU allowances are an economic value for the covered installation.

---

75 On the legal status of an EU allowance, see Pâques (2005); Sepulchre (2005); Mace (2005); Moliner (2003); Peylet (2005).

In *EnBw Energie Baden-Würtemberg AG v. Commission*, the Court has considered for the first time the question of state aid in connection with the ETS Directive. The argument of the plaintiff was that the Commission would have issued a 'state aid' decision through its refusal of the German national plan, and thus, the plaintiff would be sufficiently involved. Again, the Court dismissed the argument 'simply' because the Commission cannot take that kind of decision during the process of the Directive. On the other hand, if it does so, the decision is only a prima facie appreciation of a possible state aid, not a real state aid decision under the Article 88 of the European Treaty.

The *US Steel Kosice* cases are also interesting. A Slovakian society made an application against the Commission's decision about the Slovakian national plan because the European institution wanted to reduce the total amount of EU allowances available for the Slovakian industry. However, the Court used the plaintiff's submission to underline the links between the 'state-aid' matter and the Directive. For the Court,

> a decision based solely on [the] Directive [...] and not [based] on Articles 87 EC and 88 [of the] EC [Treaty], as is the case with the [Commission's] decision [...], allows the Commission to conduct only a prima facie assessment of the State aid aspects of the NAP in the light of the law on State aid, without prejudice to the eventual adoption of a formal decision for the purposes of the third sentence of Article 88(3) EC.[76] In consequence, 'it follows that the contested decision does not have the effect of placing the applicant in the same situation as a recipient of State aid which has been declared to be incompatible with the common market pursuant to a formal decision for the purposes of Article 88 EC. The applicant therefore cannot successfully rely on the case-law which has held that actions for annulment brought by such recipients are admissible.[77]

Thus, for the Court, when the Commission issues a state aid decision, it must be a formal decision under Articles 87 and 88 of the EC Treaty. The Court's interpretation was used to deny the plaintiffs the quality of being 'individually concerned' by the NAP decision as being not a state aid decision as such.[78]

---

[76] Court of First Instance, 1st October 2007, Case T-27/07 – US Steel Kosice v. Commission – § 72.

[77] Court of First Instance, 1st October 2007, Case T-27/07 – US Steel Kosice v. Commission – § 73.

[78] See also Weishaar, 'Verlenen emissierechten "om niet" geen staatssteun'. That publication analysed case T-233/04, Netherlands v. Commission of the 10 April 2008, concerns a state aid decision regarding the Dutch NO_x emissions trading program.

9. IS THE SCOPE OF THE ETS DIRECTIVE DISCRIMINATORY?

9.1 The Limited Scope of the Directive

Annex I of the ETS directive determines the coverage of installations. Included are those installations performing specified activities (energy, ferrous metals, mineral industry, as well as pulp, paper and board) above certain capacity thresholds, which generally cause high CO_2 emissions. Combustion processes involving crackers, carbon black, flaring, furnaces and integrated steelworks, which are typically carried out in larger installations causing considerable emissions, also fall under Annex I.[79]

In case T-183/07, Poland made an application against the Commission's decision to reject its national allocation plan because it allowed too many EU allowances during the 'Kyoto' period. However, Poland did not accept this view because it considered that a further reduction would harm the Polish economy. The Polish authorities estimated a cost of 180 million Euros.[80] The *Poland v. Commission* case followed a different procedure from the other cases. The application was a request to suspend the execution of the Commission's decision, under the Articles 225, 242 and 243 of the EC Treaty. That means that the Member State, to obtain an order from the President of the Court of First Instance, must prove the 'urgency' of the situation and the existence of a 'prejudice'. The Court examined specifically the first condition in this case.

The Court scrutinized the elements brought to complete the condition of urgency. Three reasons were invoked by the Court in rejecting the submissions of the plaintiff. For the President of the Court, Poland had not proved that the companies could not be indemnified according to Article 288 of the EC Treaty of the financial prejudice in case of annulment of the Commission's decision. Besides, the companies could have partially increased the prices of their products to cover these extra-environmental charges. Finally, the President underlined that, if the Commission's decision is annulled, the Polish companies would receive more allowances than they had, with the consequence that the companies could sell those allowances they had bought previously on the market and therefore benefit substantially.[81]

---

79 http://europa.eu/rapid/pressReleasesAction.do?reference=MEMO/06/2&format=HTML&aged=0&language.

80 Court of First instance, 9 November 2007, Order of the Court's President – Case T-183/07 R – Poland v. Commission – § 28.

81 Court of First Instance, 9 November 2007, Order of the Court's President – Case T-183/07 R – Poland v. Commission – § 44.

Poland also asked a very interesting question of the Court. Is the Directive, with its narrow scope, capable of discriminating between the covered installations and the uncovered other polluting installations in the aluminum or plastic sector? The President of the Court of First Instance could not answer this question because the application was made against the Commission's decision about the national plan and not against the ETS Directive itself.

The Arcelor's application puts its finger on the very same question. The steel giant made an application before the highest administrative court in France. The company wanted the annulment of a French decree adopted in 2004 and transposing EU emissions trading rules into French law because the decree would breach French constitutional principles. That administrative court decided to ask the European judges this prejudicial question on 5 March 2007: 'Is Directive 2003/87/EC of the 13 October 2003 valid in the light of principle of equal treatment, in so far as that Directive makes the greenhouse gas emission allowance trading scheme applicable to installations in the steel sector without including in its scope the aluminum and plastic industries?'[82] Lastly, the European Court of Justice has suspended a previous legal challenge to the EU's greenhouse emission trading scheme launched in 2004 by the same company.[83] For the steel giant, the Directive and its limited scope breaches four fundamental principles guaranteed by EU Law: freedom of establishment, proportionality, legal certainty and equality.

The Belgian Constitutional Court took a very interesting decision in this matter. It is important to underline that the Walloon Decree 'simply' transposed the Directive 2003/87/EC without modifications. Basing its reasoning on Article 176 of the EC Treaty, the Belgian judges considered that the Walloon Region could lawfully extend the scope of the Decree. Considering that this limited scope was the choice of the Walloon legislator, the Court affirmed its jurisdiction on the merits of the alleged discrimination and refused to ask a prejudicial question of the European Court of Justice. Concerning the argument of a violation of the equal treatment by the Decree, the Court judged that distinction was based on an objective criterion.[84]

---

82 E.C.J., Case C-127/07 – Reference for a preliminary ruling from the Conseil d'État (France) lodged on 5 March 2007, Société Arcelor Atlantique et Lorraine, Société Sollac Méditerranée, Société Arcelor Packaging International, Société Ugine & Alz France, Société Industeel Loire, Société Creusot Métal, Société Imphy Alloys and Société Arcelor v. Premier ministre, Ministre de l'Économie, des Finances et de l'Industrie, Ministre de l'Écologie et du Développement durable (O.J., C 117 of 25.05.2007, p. 8).

83 Case T-16/04 (suspended) – Action brought on 15 January 2004, Arcelor S.A. v. the European Parliament and the Council of the European Community, O.J. C 71, 20.03.2004, p. 36.

84 See Cour d'arbitrage/Arbitragehof, decision n° 92/2006 of 07.06.2006. For an analysis of the case and a critic of the ruling, see Pâques (2006, pp. 181–90).

9.2 Are the Proposals Including New Sectors in the Scope of the Directive a Solution?

The ETS Directive provides for a possibility for Member States to extend unilaterally its scope. Article 24 states that

> from 2008, Member States may apply emission allowance trading in accordance with this Directive to activities, installations and greenhouse gases which are not listed in Annex I, provided that inclusion of such activities, installations and greenhouse gases is approved by the Commission in accordance with the procedure referred to in Article 23(2), taking into account all relevant criteria, in particular effects on the internal market, potential distortions of competition, the environmental integrity of the scheme and reliability of the planned monitoring and reporting system.[85]

That enlargement needs however two conditions: a temporal condition (from 2008) and an authorization of the Commission.[86,87] However, Article 24 did not put an end to the disputing.

The question of including some others sectors came into play and the European institutions made some propositions for the enlargement of the scope of the Directive.[88,89] This topic is discussed in another chapter of this book.

---

85 Article 24 states also that 'From 2005 Member States may under the same conditions apply emissions allowance trading to installations carrying out activities listed in Annex I below the capacity limits referred to in that Annex'.

86 Pâques (2006, p. 184).

87 Article 24, 2: 'Allocations made to installations carrying out such activities shall be specified in the national allocation plan referred to in Article 9'.

88 See the communication from the Commission to the Council, the European Parliament, the European Economic and Social Committed and the Committee of the Regions: Building a global carbon market – Report pursuant to Article 30 of Directive 2003/87/EC, COM(2006)676 final of 13.11.2006.

89 Proposal for a Directive of the European Parliament and of the Council amending Directive 2003/87/EC so as to include aviation activities in the scheme for greenhouse gas emission trading within the Community, December 2006, COM(2006)818 final. 'This proposal is designed to include aviation activities in the greenhouse gas emissions trading scheme, and is to apply to all flights arriving at or departing from Community airports from 1 January 2012 (2011 for flights between EU airports). Aircraft operators will be responsible for complying with the obligations imposed by the scheme. It is also suggested that the process for allocating allowances should be harmonised across the EU, and that each aircraft operator, including operators from third countries, should be administered by one Member State only'; Communication from the Commission to the Council, the European Parliament, the European Economic and Social Committee and the Committee of the Regions of 27 September 2005: 'Reducing the Climate Change Impact of Aviation' [COM(2005)459

However, we can see that more and more proposals are being made to enlarge the Directive's scope. That is the best solution for an answer to the question of limited scope and for ending the discrimination argument ensuing from it. However, discrimination can also follow from allocation rules. Expansion of the coverage is in our view the best way to improve the environmental effect of the Directive. The enlargement is also a good opportunity to ensure the economical viability of the ETS system. As M.M. Upston-Hooper and Mehling have said, 'the larger the number of market participants [are], the greater the liquidity of the market [is]'.[90] However, the ETS system must be practicable and manageable and thus, every enlargement of its scope needs a serious analysis. As we underlined previously, the ETS Directive is not the only environmental instrument available to improve the quality of the climate. Others options must also be developed.

10. THE ROLE OF THE COURTS IN THE GLOBAL WARMING STRUGGLE

In order to achieve the purpose of a 20% reduction of emissions by 2020, the European authorities must modify the ETS Directive in a manageable way, principally on four fundamental points: the enlargement of the Directive's scope; the scarcity of the EU allowances; the sanction of any illegal emission; and a wider access to justice in environmental matters especially also with regard to community decisions concerning greenhouse gas allowance trading. It seems that the European institutions are on their way to completing these fundamental requirements, except perhaps for a better access to justice. The

– not published in the Official Journal]. 'The air transport sector currently accounts for 3% of all greenhouse gas emissions. However, the rapid growth of this sector means that aviation could eventually become the main source of greenhouse gas emissions, despite improvements in aircraft energy efficiency. Between 1990 and 2003, greenhouse gas emissions from international air transport increased by 73% in the EU. If the sector continues to grow at the current rate, by 2012 emissions will have increased by 150% since 1990'; Commission Staff Working Document of 27 September 2005 – Annex to the Communication from the Commission 'Reducing the Climate Change Impact of Aviation' – Impact Assessment [SEC(2005)1184]. See http://europa.eu/scadplus/leg/en/lvb/l28160.htm. See recently the Common Position (EC) No 13/2008 of 18 April 2008 adopted by the Council, acting in accordance with the procedure referred to in Article 251 of the Treaty establishing the European Community, with a view to the adoption of a Directive of the European Parliament and of the Council amending Directive 2003/87/EC so as to include aviation activities in the scheme for greenhouse gas emission allowance trading within the Community. (O.J.E.U., C 122 E, Volume 51, 20 May 2008, p.19).

[90] Upston-Hooper and Mehling (2007, p. 308). COM (2008) 16 final, p. 9.

industries should be able to be admissible in their claims against a decision of the Commission. The ETS system would probably be better if the subjects of it, the companies, could explain their points of view before a Court that could correctly appreciate their arguments. Of course, the European judges do have not to agree automatically with the companies' arguments, but at least they can listen and give an answer to these claims. In our point of view, the majority of the actual difficulties about the methods of allocations ensue from the 'grandfathering' method, applied by the Member States in accordance with the already mentioned principle of subsidiarity. That principle has the practical effect of allowing a wider margin of appreciation to the Member States when they apply the clauses of the Directive. However, the actual proposals to amend the Directive could bring a (limited) solution to that problem. Under these proposals, the auctioning method would become the main method for allocations.[91] If these amendments are passed, the total number of companies' claims would presumably decrease, because the allowances will be allocated by means of an economic logic, which is much more difficult to contest before a Court. In facts, Member States would lose a sizeable part of their margin of appreciation, and the difficulties ensuing from the national allocation plans elaborated on the grandfathering method would, logically, also disappear (or, at least, be reduced). In other words, there will be even fewer opportunities for industries to go to European and national courts.

Judges also contribute to environmental protection. The real contribution of the Court is that the judges permitted a real and protective environmental dimension to be brought to the ETS system. With *United Kingdom v. Commission* and *Germany v. Commission*, the Court consecrated the importance of the public participation during the NAP process, following a rationale of democratic efficiency of the Directive. When the Court forbade the possibility of 'real banking' between each three- or five-year period in *Fels Werke v. Commission*, the judges also underlined the necessity to ensure that the ETS market is useful and has real environmental integrity. It seems that they authorized Member States to insert the 'annual banking' option in their NAPs, under a subsidiarity perspective. The same logic was followed in *Germany v. Commission*, when the Court of First Instance accepted the German downward ex post mechanisms, sentencing the Commission's position based on its imprecise developments of its guidelines. From a rather institutional perspective, the Court also affirmed the principle that Member States are free to modify their NAPs. That possibility could not be limited by the Commission's

91 See especially the article 10, article 10a and article 10b of the proposal for a Directive of the European Parliament and the Council amending Directive 2003/87/EC so as to improve and extend the greenhouse gas emission allowance trading system of the Community. COM(2008)16 final, 23.01.2008.

suggestions. Then, the European judges expressly underlined the difference existing between the 'state aid' matter, based on Articles 87 and 88 of the EC Treaty, and the Commission 'state aid' appreciation about a provision of a national plan, which is only 'a prima facie assessment of the State aid aspects of the NAP in the light of the law on State aid'. Like we mentioned, the Court has also 'imported' the limited access to justice into the ETS matter, not without some remarks. Those considerations show how the judges can play a role in the environmental matter.

More widely, it is not the first time that judges take the initiative to promote a better environment. In the United States, for some political reason, the federal Government interfered many times with the Environmental Protection Agency's expertise. Under the Clean Air Act, a threshold judgment of that Agency is necessary to trigger regulation of some sectors or pollutants. Political interferences always denigrated the quality of pollutant in the greenhouse gases. The Agency, under some political pressure, decided to decline, in connection with these gases, that threshold judgment necessary to trigger regulation of vehicles' emissions, with the consequence of rejecting it from the Act's scope. That decision had the result of leaving 'the regulatory status quo ante in place'.[92] In the *Massachusetts v. Environmental Protection Agency* case, a 5–4 majority decided to overrule the administrative decision.[93] The judges of the Supreme Court recognized the statutory authority of the EPA to regulate greenhouse gases as pollutants. The real question was to know if, 'on the merits of the relevant statutes, the agency is failing to do something it is legally obliged to do'.[94] For the Court, the Agency 'had failed to justify adequately its denial' because 'EPA may decline to make a statutory judgment only on technocratic and scientific grounds, not political ones'.[95, 96] In fact, this decision can be interpreted as a political decision of the Court. The judges did not want to submit to such an important influence by the American central administration in administrative decisions, especially when this influence pursues a political purpose.[97]

---

[92] Freeman and Vermeule, 'Massachusetts v. EPA: From Politics to Expertise', p. 23. Text to be published in *The Supreme Court Review*, 2007, available on http://www.law.harvard.edu/faculty/freeman/.

[93] 127 S. Ct. 1438 (2007).

[94] Freeman and Vermeule (2007, p. 21).

[95] Ibid., p. 1.

[96] Freeman and Vermeule (2007, p. 21).

[97] Recently, the Regional Greenhouse Gas Initiative was taken. It is an American interstate agreement 'by ten northeastern states to reduce carbon dioxide emissions from power plants in the Region'. To an analysis of the connections between this agreement and the Compact clause, see 'The Compact Clause and the Regional Greenhouse Gas Initiative', in *Harvard Law Review*, Vol. 120, May 2007, Number 7, pp. 1958–1979. See also Driesen (2007).

Next to the European jurisprudence, this decision is one of the most interesting examples to show the importance and the influence of the judges in the environmental law of today.[98] Is that a new example of the judges' activism? However, the role of the European judges in the development of this environmental interpretation has to be alleviated in the capacity of the existence of the subsidiarity principle. When Member States can freely adopt more protective environmental measures, the environmental influence of the judges is less directly marked. On the other hand, it is not a reason to minimize their role within the development of the ETS system.

REFERENCES

Anderson, J. and I. Skinner (2005), 'The European Union's Approach to Reducing Greenhouse Gas Emissions', *Journal for European Environmental and Planning Law* (*JEEPL*) p. 92.

Arnulli, A. (1995), 'Private applicants and the action for annulment under Art.173 of the EC Treaty', *CMLR*, p. 7.

Barton, J. (2006), 'Tackling Aviation Emissions: the Challenges Ahead', *JEEPL*, Berlin, The Lexxion Publisher, Volume 3, Number 4, p. 322.

Comhaire, G. (2001), 'En route pour l'industrie verte', La Libre Belgique, 23 January.

Delnoy, M. (2007), *La participation du public en droit de l'urbanisme et de l'environnement*, Bruxelles, Larcier, p. 937.

De Mulder, J. (2004), 'Richtlijn 2003/87/EG tot vasstelling van een regeling voor de handel in broiekasgasemissierechten binnen de Gemeenschap en tot wijziging van Richtlijn 96/61/EG. Een toelichting', in Tijdschrift voor Milieurecht 2004, Kluwer uitgevers, pp. 138–84.

Driesen, M.D. (2007), 'The Changing Climate for United States Law', in *Carbon & Climate Law Review*, The Legal Publisher, Berlin, Volume 1, Number 1, pp. 35–44.

Dross, M. (2005), 'Access to Justice in EU Member States', *JEEPL*, Berlin, Lexxion Publisher, Volume 2, Number 1, January, pp. 22–30.

Ehrmann, M. and D. Greinache (2006), 'The Transfer Rule in the German Emissions Trading Scheme', *JEEPL*, Volume 3, Number 2, pp. 142–52.

[98] Haritz (2007). See also some other interesting decisions: Center for Biological Diversity et al. v. Brennan et al., U.S. District Court for the Northern District of California (2005–2007); Citizens for Responsibility and Ethics in Washington v. Council on Environmental Quality, Denver District Court (2007); Center for Biological Diversity v. Kempthorne, U.S. District Court for the Northern District of California, moved to Alaska (2007); Natural Resources Defense Council v. Reclamation Board, California Superior Court of Sacramento County (2006); California et al. v. National Highway Traffic Safety Administration, Ninth Circuit Court of Appeals; Green Mountain Chrysler v. Crombie/Dalmasse, District Court of Vermont (2007); New York v. EPA & Coke Oven Environmental Task Force v. EPA, U.S. Circuit Court of Appeals for the District of Columbia (2006).

Ellinghaus, U., P. Ebsen and H. Schloemann (2004), 'The EU Emissions Trading Scheme (EU ETS): a status report' , *JEEPL*, The Legal Publisher Lexxion, Berlin, July, Volume 1, Number 1, p. 4.

Faure, M., D. Grimeaud, H. Kremers, R. Lubbers, P. Nijkamp, J. Koorevaar, H. Verbruggen and S. Wang (2001), 'Climate change policies and international trade', in Dutch National Research Programme on Global Air Pollution and Climate Change, Report no. 410 200 098 (2001), Vrije Universiteit Amsterdam, December.

Freeman, J. and A. Vermeule (2007), 'Massachusetts v. EPA: From Politics to Expertise', p. 10. Text to be published in *The Supreme Court Review*, 2007, available on http://www.law.harvard.edu/faculty/freeman/.

Freestone, D. and C. Streck (2005), *Legal Aspects of Implementing the Kyoto Protocol Mechanisms: Making Kyoto Work*, Oxford, Oxford University Press.

Görgen, R. and U. Lambrecht (2007), 'Particulate matter in Ambient Air', in *JEEPL*, The Legal Publisher Lexxion, Berlin, Volume 4, Number 4, pp. 278–88.

Grimeaud, D. (2003), 'The compatibility of an EU-wide emissions trading scheme and national implementing provisions with Community law', in Grimeaud, M. and Peeters, D., Emissionhandel and concurrentieposities, January, pp. 46–89.

Hadrousek, D. (2004), 'Trading in Greenhouse Gas Emission Allowances: the Czech Approach', in *JEEPL*, The Legal Publisher Lexxion, Berlin, Volume 1, Number 1, July, pp. 32–40.

Haritz, M. (2007), 'Judicial Activism to the Rescue? Case Law in the U.S. with Regard to Climate Change Matter', Ius Commune Congress on 29 November, Liège, pp. 1–10.

Hobley, A. and C. Rowe (2004), 'Transposition of the Emissions Trading Scheme Directive into UK Law and Associated Issues', *JEEPL*, The Legal Publisher Lexxion, Berlin, Volume 1, Number 1, July, pp. 10–22.

Jans, H. (2000), *European Environmental Law*, Oxford, European Law Publishing, pp. 164–65.

Jendroska, J. (2005), 'Aarhus Convention and Community Law: the Interplay', *JEEPL*, Berlin, Lexxion Publisher, Volume 2, Number 1, January, pp. 12–21.

Krämer, L. (2004), 'Data on Environmental Judgments', *JEEPL*, Volume 1, Number 2, September, p. 135.

Krämer, L. (2006), 'Recent Case-law of the European Court of Justice and the Court of First Instance', *JEEPL*, Berlin, The Legal Publisher Lexxion, Volume 3, Number 2, pp. 155–56.

Larssen, C. (2005), L'accès à la justice en matière d'environnement, Actes du colloque organisé le 12 mars 2003 à Bruxelles par l'Association Belge pour le Droit de l'Environnement, Bruylant, Bruxelles, p. 318.

Le Bussy, O. (2001), 'C'est une proposition qui ne tient pas la route', La Libre Belgique, 23 January.

Long, S. and G. Kaminskaite-Salters (2007), 'The EU ETS- Latest Developments and the Way Forward', in *Carbon & Climate Law Review*, Volume 1, Number 1, p. 71.

Mace, M.J. (2005), 'The Legal Status of Emission Reductions and EU Allowances: Issues addressed in an International Workshop', *JEEPL*, p. 123.

Marr, S. and F. Schafhausen (2004), 'Emissions Allowance Trading in Germany: A New Environmental Law Regime is Taking Shape', *JEEPL*, The Legal Publisher Lexxion, Berlin, Volume 1, Number 1, July, pp. 23–31.

Mehlnig, M. and L. Massai (2007), 'The European Union and Climate Change: Leading the Way towards a Post-2012 Regime?', *Carbon & Climate Law Review*, The Legal Publisher Lexxion, Berlin, Volume 1, Number 1, pp. 45–52.

Moliner, M. (2003), 'L'intégration en droit interne du mécanisme des permis d'émission négociables', *CJEG*, p. 489.

Ott, H.E., B. Brouns, W. Sterk and B. Wittneben (2005), 'It Takes Two to Tango – Climate Policy at COP 10 in Buenos Aires and Beyond', *JEEPL*, Volume 2, Number 2, March, pp. 84–91.

Pâques, M. and S. Charneux (2004), 'Le droit européen de l'environnement à l'heure de l'emission trading', Les Cahiers du Juriste, Brussel, Bruylant, 2004/2.

Pâques, M. (2005), 'La nature juridique du quota d'émissions de gaz à effet de serre', in Verhandelbare emmissierechten als klimaatbeleidsintrument – L'échange des droits de pollution comme instrument de gestion du climat, under the direction of Maes, F., Brussel, La Charte, pp. 43–69.

Pâques, M. (2006), 'L'Emission Trading à la Cour d'arbitrage', in Am-Env., Kluwer, Bruxelles, 4, pp. 181–90.

Pâques, M. (2007), 'De l'autorisation administrative', in *Mélanges en l'honneur de Michel Prieur, Pour un droit commun de l'environnement*, Dalloz, Paris, p. 647.

Peeters, M. (2006), Broeikasgasemissiehandel in de EU, in Faure, M. and Peeters, M. (eds), Grensoverschrijdend recht, METRO, Intersentia Antwerpen-Oxford, p. 346.

Peeters, Marjan and Kurt Deketelaere (eds) (2006), 'EU climate change policy. The challenge of new regulatory initiatives', Northampton: Edward Elgar, p. 334.

Peylet, R. (2005), 'Un marché de nouveaux biens meubles les quotas d'émission de gaz à effet de serre', *RJEP/CJEG*, p. 213.

Pfromm, R. and K. Svikis (2004), 'EU-Emissions Trading: The Latvian Allocation Plan', *JEEPL*, The Legal Publisher Lexxion, Berlin, Volume 1, Number 1, July, pp. 41–9.

Prins, G. and S. Rayner (2007), 'Time to ditch Kyoto', *Nature*, **449**(25).

Purdy, R., I. Havercroft, 'Carbon Capture and Storage: Developments under the European Union and International Law', *JEEPL*, Berlin, The Lexxion Publisher, Volume 4, Number 5, pp. 353–66.

Rasmussen, H. (1980), 'Why is Article 173 Interpreted against Private Plaintiffs?', E.L.R., 1980, p. 112.

Schmitt-Rady, B. (2004), 'Allocation of greenhouse gas allowances in Europe: State of affairs and critical legal issues', in International conference 'The challenge of implementing new regulatory initiatives: state of affairs and critical issues of EU climate change', Faculty of Law, University of Leuven, Belgium, 17–18 September, p. 18–20.

Sepulchre, V. (2005), 'Le Protocole de Kyoto et la directive 2003/87: aspects juridiques et comptables des gaz à effet de serre', in Comptabilité et fiscalité pratique, 8, Oct., p. 241.

Streck, C. and B.T. Chagas (2007), 'The future of the CDM in a Post-Kyoto World', *Carbon & Climate Law*, Berlin, The Legal Publisher Lexxion, Volume 1, Number 1, pp. 53–63.

The Compact Clause and the Regional Greenhouse Gas Initiative, in *Harvard Law Review*, Vol. 120, May 2007, Number 7, pp. 1958–1979.

'Time to Ditch Kyoto', *Nature*, Nature Publishing Group, Vol. 449, 25 October 2007.

Upston-Hooper, K. and M.A. Mehling (2007), 'A Nutrient Quota Trading Scheme to Reduce the Eutrophication of the Baltic Sea', *JEEPL*, Volume 4, Number 4, pp. 305–6.

Van Aken, N. (2007), 'Jurisprudence en bref, Cour de Justice et Tribunal de première instance des Communautés européennes – Affaire T-387/04', in Am.-Env., Bruxelles, Kluwer, 4, pp. 222–3.

Van Aken, N. (2008a), 'Jurisprudence en bref, Cour de Justice et Tribunal de première instance des Communautés européennes – Affaire T-28/07', in Am.-Env., Bruxelles, Kluwer, 2008/1, pp. 45–46.
Van Aken, N. (2008b), 'Jurisprudence en bref, Cour de Justice et Tribunal de première instance des Communautés européennes, Affaire T-374/07, in Am.-Env., Bruxelles, Kluwer, 2008/2, pp. 124–28.
Vandersanden, G. (1995), 'Pour un élargissement du droit des particuliers d'agir en annulation contre des actes autres que les décisions qui leur sont adressées', C.D.E., p. 535.
Van Hecke, K. and T. Zgajewski (2007), 'Belgium's Kyoto Policy', pp. 1–2.
Van Raepenbusch, S. (2005), Droit institutionnel de l'Union européenne, Collection de la Faculté de droit de l'Université de Liège, Brussel, Larcier, p. 627.
Verhest, S. (2001), 'Une revolution industrielle', La Libre Belgique, 23 January.
Victor, D. (2001), *The Collapse of the Kyoto Protocol and the Struggle to Slow Global Warming*, Princeton, NJ: Princeton University Press.
Waelbroeck, D. and A.M. Verheyden, (1995), 'Les conditions de recevabilité des recours en annulation des particuliers contre les actes normatifs communautaires à la lumière du droit comparé et de la Convention des droits de l'homme', C.D.E., p. 399.
Waelbroeck, M. and D. Waelbroeck (1993), 'Articles 172, 173 et 174', in Commentaire Mégret, Vol. 10, Bruxelles, p. 89.
Wilson, J.E. and J.A. Gibbons (2007), 'Deploying Carbon Capture and Storage in Europe and the United States: a comparative analysis', *JEEPL*, The Legal Publisher Lexxion, Berlin, Volume 4, Number 5, p. 344.
Wittneben, B., W. Sterk, H.E. Ott and B. Brouns (2006), 'The Montreal Climate Summit: Starting the Kyoto Business and Preparing for post-2012', *JEEPL*, Volume 3, Number 2, p. 90–100.
Ziehm, C. (2005), 'Legal Standing for NGOs in Environmental Matters under the Aarhus Convention and under Community and National Law', *JEEPL*, Berlin, Lexxion Publisher, Volume 2, Number 4, July, pp. 287–300.

APPENDIX

Questions asked to the Court of First Instance

After the approval decision of the Commission, can a Member State freely modify all the propositions of its national allocation plan (NAP) or only the propositions previously rejected by the Commission? (Case T-178/05 – United Kingdom v. Commission – 23 November 2005).

Can a competitor, under a national allocation plan, receive an advantage that an comparable company cannot receive? Is the Commission's decision, concerning the economical validity of the NAP's measures, a decision identical to a decision taken under the Article 87 and 88 of the Treaty? (Case T-387/04 – EnBW Energie Baden-Würtemberg AG v. Commission – 30 April 2007).

A proposition of a NAP's modification, increasing the number of EU allowances, is (or not) a binding decision for the Member State ? (Case T-130/06 – Drax Power e.a. and others v. Commission – 25 June 2007; Case T-489/04 – US Steel Kosice – 1st October 2007; Case T-27/07 – US Steel Kosice – 1st October 2007).

If, due to their validity (12 or 14 years), the allowances allocated to the covered installations for the first period exceed that period, can the societies use them for the emissions made during the 'Kyoto period'? (Case T-28/07 – Fels-Werke GmbH v. Commission – 11 September 2007).

Is an downward ex post adjustment measure, contained in the national plan, compatible with the Directive under the principle of subsidiarity? (Case T-374/04 – Germany v. Commission – 7 November 2007).

The scope of the Directive, as it covers only some specific sectors, is discriminatory or not? (Case T-16/04 – Action brought on 15 January 2004 - Arcelor S.A. v. the European Parliament and the Council of the European, O.J. C 71, 20.03.2004, p. 36 – still pending; Case T-183/07 – Poland v. Commission – 9 November 2007).

Decisions of the Court of First Instance

C.O.F.I.E.C., 23 November 2005, Case T-178/05 – United Kingdom / Commission, O.J. C 22 of 28.01.2006, p. 14.

C.O.F.I.E.C., 30 April 2007, T-387/04 case – EnBW Energie Baden-Württemberg/ Commission, O.J. C 140 of 23.06.2007, p. 27.

C.O.F.I.E.C., 25 June 2007, Case T-130/06 – Drax Power e.a./ Commission, O.J. C 211, 08.09.2007, p. 33.

C.O.F.I.E.C., 11 September 2007, Case T-28/07 – Fels-Werke GmbH v. Commission, O.J. C 283 of 24.11.2007, p. 27.

C.O.F.I.E.C., 1st October 2007, Case T-489/04 case – US Steel Kosice, O.J. C 297 of 08.12.2007, p. 41.

C.O.F.I.E.C., 1st October, Case T-27/07 – US Steel Kosice, O.J. C 297 of 08.12.2007, p. 42.

C.O.F.I.E.C., 6 November 2007, Case T-13/07 – Cemex UK Cement Ltd v. Commission (application O.J. C 56, 10.03.2007, p. 37).

C.O.F.I.E.C., 7 November 2007, Case T-374/04 – Germany v. Commission (application O.J. C 284 of 20.11.2004, p. 25).

C.O.F.I.E.C., 9 November 2007, Case T-183/07 R – Order of the Court's President – Poland v. Commission (application O.J. C 155 of 07.07.2007, p. 41) (available only in French).

Decisions of the Court of Justice

ECJ, Judgment of 12 January 2006, Case C-107/05 – Commission v. Finland. 'By failing, with regard to the province of Åland, to adopt the laws regulations and administrative provisions necessary to comply with Directive 2003/87/EEC of the European Parliament and of the Council of 13 October 2003 establishing a scheme for greenhouse gas emission allowance trading within the Community and amending Council Directive 96/61/EC, the Finnish Republic has failed to fulfill its obligations under that directive'.

ECJ, Judgment of 18 May 2006, Case C-122/05 – Commission v. Italy: 'By failing to adopt, within the prescribed period, all the laws, regulations and administrative provisions necessary to comply with Directive 2003/87/EC of the European Parliament and of the Council of 13 October 2003 establishing a scheme for greenhouse gas emission allowance trading within the Community and amending Council Directive 96/61/EC, the Italian Republic has failed to fulfill its obligations under that directive'.

ECJ, Judgment of 18 July 2007, Case C-61/07 – Grand Duchy of Luxembourg v. Commission of the European Communities. 'The Court declares that, by failing to communicate the information required under Article 3(2) of Decision n° 280/2004/EC of the European Parliament and of the Council of 11 February 2004 concerning a mechanism for monitoring Community greenhouse gas emissions and for implementing the Kyoto protocol, the Grand Duchy of Luxembourg has failed to fulfill its obligation under that provision'.

ECJ, Order of 8 April 2008, Case C-503/07 P – Saint-Gobain Glass Deutschland GmbH, Fels-Werke GmbH, Spenner-Zement GmbH & Co. KG v. Commission.

Cases pending before the Court of First Instance of the European Community

Case T-16/04 (suspended) – Action brought on 15 January 2004 – Arcelor S.A. v. the European Parliament and the Council of the European Community, O.J. C 71, 20.03.2004, p. 36.

Case T-32/07 – Action brought on 7 February 2007, Slovakia v. Commission, O.J. C 69, 24/03/2007, p. 29.

Case T-194/07 – Action brought on 4 June 2007, Czech Republic v. Commission, O.J. C 199, 25.08.2007, p. 41.

Case T-199/07 – Action brought on 5 June 2007, Cementownia 'Odra' v. Commission, O.J. C 170, 21.07.2007, p. 39.

Case T-208/07 – Action brought on 5 June 2007, BOT Elektrownia Belchatow and others v. Commission, O.J. C 184, 04.08.2007, p. 37.

Case T-198/07 – Action brought on 5 June 2007, Cememtownia 'Warta' v. Commission, O.J. C 170, 21.07.2007, p. 39.

Case T-203/07 – Action brought on 5 June 2007, Cemex Polska v. Commission, O.J. C 170, 21.07.2007, p. 40.

Case T-196/07 – Action brought on 5 June 2007, Dyckerhoff Polska v. Commission, O.J. C 170, 21.07.2007, p. 38.

Case T-197/07 – Action brought on 5 June 2007, Grupa Ozarow v. Commission, O.J. C 170, 21.07.2007, p. 38.

Case T-195/07 – Action brought on 5 June 2007, Lafarge Cement SA v. Commission, O.J. C 170, 21.07.2007, p. 37.

Case T-193/07 – Action brought on 5 June 2007, Gorazdze Cement S.A v. Commission, O.J. C 170, 21.07.2007, p. 36.

Case T-263/07 – Action brought on 16 July 2007, Estonia v. Commission, O.J. C 223, 22/09/2007, p. 12.

Case T-368/07 – Action brought on 26 September 2007, Lithuania v. Commission, O.J. C 283, 24.11.2007, p. 35.

Case T-369/07 – Action brought on 26 September 2007, Latvia v. Commission, O.J. C 269, 10.11.2007, p.66.

Case T-127/07 – Reference for a preliminary ruling from the Conseil d'État (France) lodged on 5 March 2007, Société Arcelor Atlantique et Lorraine, Société Sollac Méditerranée, Société Arcelor Packaging International, Société Ugine & Alz France, Société Industeel Loire, Société Creusot Métal, Société Imphy Alloys and Société Arcelor v. Premier ministre, Ministre de l'Économie, des Finances et de l'Industrie, Ministre de l'Écologie et du Développement durable, OJ C 117 of 29.05.2007, p. 8.

Directives and Decisions Relating to the 'Emissions Trading Scheme'

Directive 2003/87/EC of the European Parliament and of the Council of the 13 October 2003 establishing a scheme for greenhouse gas emission allowance trading within the Community and amending Council Directive 96/61/EC, OJ 2003, L 275/32.

Decision No 280/2004/EC of the European Parliament and of the Council of 11 February 2004 concerning a mechanism for monitoring Community greenhouse gas emissions and for implementing the Kyoto Protocol, OJ L 49, 19.2.2004, pp. 1–8.

Commission Decision of 10 February 2005 laying down rules implementing Decision No 280/2004/EC of the European Parliament and of the Council concerning a mechanism for monitoring Community greenhouse gas emissions and for implementing the Kyoto Protocol (notified under document number C(2005) 247), OJ L 55, 1.3.2005, pp. 57–91.

Proposal for a Directive of the European Parliament and of the Council amending Directive 2003/87/EC so as to include aviation activities in the scheme for greenhouse gas emission trading within the Community, December 2006, COM(2006)818 final.

See the Commission Decision of 14 December 2006 determining the respective emission levels allocated to the Community and each of its Member States under the Kyoto Protocol pursuant to Council Decision 2002/358/EC (notified under document number C(2006) 6468), O.J. L 358, 16.12.2006, pp. 87–89, modified by Corrigendum to Commission Decision 2006/944/EC of 14 December 2006 determining the respective emission levels allocated to the Community and each of its Member States under the Kyoto Protocol pursuant to Council Decision 2002/358/EC, O.J. L 367 of 22.12.2006, p.80; European

Community | 19682555325 |, Belgium | 679368682 |, Denmark | 273827177 |, Germany | 4868520955 |, Greece | 694087947 |, Spain | 1663967412 |, France | 2819626640 |, Ireland | 315158338 |, Italy | 2428495710 |, Luxembourg | 45677304 |, Netherlands | 1008565720 |, Austria | 343473407 |, Portugal | 386956503 |, Finland | 355480975 |, Sweden | 375864317 |, United Kingdom | 3412080630 |, Cyprus | not applicable |, Czech Republic | 902890649 |, Estonia | 197902558 |, Latvia | 119113402 |, Lithuania | 221275934 |, Hungary | 578260222 |, Malta | not applicable |, Poland | 2673496300 |, Slovenia | 92934961, Slovakia | 337456459 |.

Commission staff working document, Impact Assessment, Document accompanying the Package of Implementation measures for the EU's objectives on climate change and renewable energy for 2020, SEC(2008)85/3, 23.01.2008.

Commission proposal for Directive of the European Parliament and of the Council amending Directive 2003/87/EC so as to improve and extend the EU greenhouse gas emission allowance trading scheme, COM(2008)16 final, 23.01.2008.

Commission's proposal for Decision of the European Parliament and of the Coucil on the effort of Member States to reduce their greenhouse gas emissions to meet the Community's greenhouse gas emission reduction commitment, COM(2008)17 final, 23.01.2008.

Commission's proposal for a Directive of the European Parliament and of the Council on the promotion of the use of the energy from renewable sources, COM(2008)19 final, 23.01.2008.

Commission's Guidelines

Thematic strategy on air pollution: COM(2005)446 of 21 September 2005.

Communication from the Commission on guidance to assist Member States in the implementation of the criteria listed in Annex III to Directive 2003/87/EC establishing a scheme for greenhouse gas emission allowance trading within the Community and amending Council Directive 96/61/EC, and on the circumstances under which force majeure is demonstrated, COM/2003/0830 final.

Communication from the Commission to the Council and to the European Parliament on Commission Decisions of 7 July 2004 concerning national allocation plans for the allocation of greenhouse gas emission allowances of

Austria, Denmark, Germany, Ireland, the Netherlands, Slovenia, Sweden, and the United Kingdom in accordance with Directive 2003/87/EC, COM/2004/0500 final.

Communication from the Commission to the Council and to the European Parliament on Commission Decisions of 20 October 2004 concerning national allocation plans for the allocation of greenhouse gas emission allowances of Belgium, Estonia, Finland, France, Latvia, Luxembourg, Portugal, and the Slovak Republic in accordance with Directive 2003/87/EC, COM(2004)0681.

Communication from the Commission, 'Further guidance on allocation plans for the 2008 to 2012 trading period of the EU Emission Trading Scheme', COM(2005)0703.

Communication from the Commission to the Council and to the European Parliament on the assessment of national allocation plans for the allocation of greenhouse gas emission allowances in the second period of the EU Emissions Trading Scheme accompanying Commission Decisions of 29 November 2006 on the national allocation plans of Germany, Greece, Ireland, Latvia, Lithuania, Luxembourg, Malta, Slovakia, Sweden and the United Kingdom in accordance with Directive 2003/87/EC, COM(2006)0725.

Communication from the Commission to the Council, the European Parliament, the European Economic and Social Committee and the Committee of the Regions of 27 September 2005: 'Reducing the Climate Change Impact of Aviation' [COM(2005)459 – not published in the Official Journal].

Communication from the Commission to the Council, the European Parliament, The European Economic and Social Committed and the Committee of the Regions: Building a global carbon market – Report pursuant to Article 30 of Directive 2003/87/EC, COM(2006)676 final of 13.11.2006.

5. European emissions trading and the polluter-pays principle: assessing grandfathering and over-allocation

Edwin Woerdman, Stefano Clò and Alessandra Arcuri*

1. INTRODUCTION

The European Union (EU) holds an Emissions Trading Scheme (ETS) for carbon dioxide (CO_2) and other greenhouse gases. This market has been up and running since 2005, based on Directive 2003/87/EC. It is a 'cap-and-trade' scheme that allocates emission caps to polluters. This means that their emission targets are based on absolute standards that define emission ceilings. When a polluter manages to keep emissions below his ceiling, he can sell this surplus in the form of emission rights, called 'allowances', to a polluter that wishes to increase emissions (e.g. Woerdman, 2005).

To create political acceptability, 'grandfathering' has been used as the primary method of allocating the allowances. This means that polluters received most emission rights free of charge primarily based on their historical emissions, so that they did not have to buy rights in an auction. As stated in Article 10 of Directive 2003/87/EC, every EU Member State was required to allocate at least 95% of the allowances free of charge for the three-year period 2005–2007 and at least 90% of the allowances free of charge for the five-year period 2008–2012.

However, a popular perception in the economic and legal literature is that grandfathering is inconsistent with the polluter-pays principle. 'Free allocation

* Corresponding author: Dr. E. Woerdman (Associate Professor of Law and Economics), University of Groningen, Faculty of Law, Department of Law and Economics, P.O. Box 716, 9700 AS Groningen, The Netherlands, Telephone + 31 50 363 5736, e-mail: e.woerdman@rug.nl. Co-authors: Mr. S. Clò (PhD Candidate), European Doctorate in Law and Economics (EDLE), University of Bologna, Italy, and Erasmus University Rotterdam, The Netherlands, and Dr. A. Arcuri (Assistant Professor of Law and Economics), Erasmus University Rotterdam, School of Law, Rotterdam Institute of Law and Economics (RILE), Rotterdam, The Netherlands.

violates the polluter-pays principle (...)', according to Sorrell and Sijm (2003: 427). Also Nash (2000: 13), based on a thorough analysis of the issue, concludes that '(...) grandfathering (...) runs contrary to the polluter pays principle's core (...)'. This suggests that the EU ETS is inconsistent with an important principle of environmental law.

Another problematic feature of the EU ETS is the amount of allowances that have been allocated, generally considered excessive, a phenomenon known as over-allocation (e.g. Ellerman and Buchner, 2006). To give an idea, in 2006 the European Commission published official data showing that the overall CO_2 emissions released by the regulated sectors in the first year of the EU ETS were about 4% or 80 million tonnes lower than the number of allowances distributed to installations for 2005.[1] This 80 million tonnes gap between emissions and allowances raises the question of whether the ETS cap was stringent enough and, subsequently, whether over-allocation violates the polluter-pays principle.

Interestingly, despite the fact that they seem to contravene the polluter-pays principle, both grandfathering and over-allocation are allowed in legal practice. Therefore, the central question of this chapter is two-fold: (1) do polluters pay when allowances have been handed out free of charge (grandfathering) and (2) do polluters pay when too many allowances have been allocated (over-allocation)? We answer these questions by extending an earlier analysis (Woerdman et al., 2008), in which we have primarily studied the grandfathering issue. In addition, we conduct an assessment of over-allocation, both from a theoretical and empirical perspective.

These questions are interesting objects of study for researchers of law and economics, because the polluter-pays principle, by mandating cost internalization in most of its versions, is an eminently economic principle (e.g. Faure and Grimeaud, 2003). In addition, pollution markets have always received considerable attention from some of the founding fathers of law and economics (e.g. Coase, 1960). Aware of the complexities inherent in the interpretation of principles, we distinguish an economic from an equity interpretation of the polluter-pays principle, in order to analyse efficiency aspects without disregarding other goals of law, including distributive justice (Calabresi and Melamed, 1972).

The chapter is structured as follows. In the second section, we describe the economic origin and legal nature of the polluter-pays principle and present a taxonomy of possible interpretations of this principle ranging from efficiency to equity. In the third section, we test whether grandfathering is compatible

[1] For national reports on verified emissions and surrendered allowances for 2005 see http://ec.europa.eu/environment/climat/emission/citl_2005_en.htm.

with these interpretations of the polluter-pays principle by focusing on the concepts of opportunity cost, lump sum-subsidy and capital gift. In the fourth section, we determine a criterion to assess empirically whether allowances have been over-allocated during the first phase of the ETS. We subsequently assess whether over-allocation is consistent with the various interpretations of the polluter-pays principle. In the final section, we present our conclusions.

2. INTERPRETATIONS OF THE POLLUTER-PAYS PRINCIPLE

Let us begin our analysis by emphasizing that principles are not rules; they are characterized by relatively vague formulations. For this reason, understanding the polluter-pays principle is more complex than its wording may suggest. A principle states '(…) a reason that argues in one direction, but does not necessitate a particular decision' (Dworkin, 1977: 26). Therefore, principles work as guidelines: different outcomes might result from the application of a principle since it does not dictate any specific decision. Principles aim at circumscribing the discretion of decision makers and/or judges when they have to shape, apply or interpret the law. 'Discretion, like the hole in a doughnut, does not exist except as an area left open by a surrounding belt of restriction' (Dworkin, 1977: 31). Drawing on Dworkin's analysis, one can conceive the polluter-pays principle as a belt of restriction. Therefore, our challenge is to understand what general goals the polluter-pays principle aims to achieve and how the principle constrains the discretion of the decision maker or judge.

The polluter-pays principle first appeared in 1972 in the *Recommendation of the OECD Council on Guiding Principles Concerning International Economic Aspects of Environmental Policies* (reprinted in OECD, 1975: 11–14). This principle basically means that polluters should pay for pollution prevention and control measures as well as for the environmental damage they cause and that the government should not subsidize pollution. Although the OECD document itself is not binding in international law since it was never ratified by any government, the polluter-pays principle can now be found in an increasing number of international treaties and instruments. For instance, Principle 16 of the 1992 *Rio Declaration on Environment and Development*, a soft law document, reads as follows: 'National authorities should endeavour to promote the internalization of environmental costs and the use of economic instruments, taking into account the approach that the polluter should, in principle, bear the costs of pollution, with due regard to the public interest and without distorting international trade and investment'. Under European Community (EC) law, the polluter-pays principle is laid down in Article 174 of the *EC Treaty*. In this legal document, the principle is mentioned but not defined.[2]

A precise and generally accepted legal definition of the polluter-pays principle is still lacking. As put by Verhoef: '(...) the question of whether the polluter should pay (...) may often lead to different outcomes in terms of both allocative efficiency and equity. (...) This ambiguity in the interpretation of the polluter pays principle is, unfortunately, often overlooked' (Verhoef, 1999: 206–7). To shed light on this issue, we identify two fundamental versions of the polluter-pays principle: an efficiency interpretation, and an equity interpretation.

This distinction warrants further explanation. The efficiency interpretation reflects the idea that pollution costs should be internalized with the aim of achieving an efficient allocation of resources, irrespective of distributive issues. Equity has a wide variety of meanings, but in this context we consider it to be a notion of a fair distribution of costs. We consider the efficiency interpretation to be the core of the polluter-pays principle. Therefore, we frame the equity criterion as an extension of the basic form of this principle, which does not depart from but includes the efficiency dimension.

As emphasized by Faure and Grimeaud: 'one can say that the polluter pays principle is probably the most "economic" of all environmental principles' (Faure and Grimeaud, 2003: 33). Conceptualizing the polluter-pays principle as an eminently 'economic' principle is in line both with its origin (OECD, 1975) and with some of its most representative definitions that explicitly endorse the criterion of cost internalization, such as the above-mentioned Principle 16 of the Rio Declaration. Also legal scholars concede that 'it remains an economic principle that was turned into a legal principle and helps justifying policy decisions – whatever the decisions are' (Krämer, 2005).

Yet, it is clear that next to efficiency also equity has been used as a criterion to impart meaning to the polluter-pays principle. In this context, Bugge (1996) distinguishes between the polluter-pays principle, on the one hand, as an 'economic principle (a principle of efficiency)', and on the other hand, as a 'legal principle of (just) distribution of costs'. In Bugge's view, the efficiency principle is independent from the distributive principle. Alternatively, it is possible to conceive of the polluter-pays principle as a principle endowed with both efficiency and equity dimensions. This view is supported by several authors who have observed that the polluter-pays principle is a principle that allocates the costs on the polluter not only for efficiency but also for equity

2 'Community policy on the environment (...) shall be based on the precautionary principle and on the principles that preventive action should be taken, that environmental damage should as a priority be rectified at source and that the polluter should pay.' (EC Treaty, Title XIX Environment, Article 174 (2)).

reasons (Pearson, 1994: 563; Parikh, 1993). The OECD's 1975 analysis of the principle confirms this viewpoint: 'It should be noted that the problem of cost sharing calls for equity as well as efficiency (…) The question is now whether there is a principle permitting the dual requirements of efficiency and equity to be satisfied together (…)' (OECD, 1975: 25). Our equity interpretation of the principle, by subsuming the efficiency dimension, would satisfy this double requirement.

In relation to the efficiency dimension of the polluter-pays principle, it is possible to further distinguish a weak form (no subsidization) from a strong form (cost internalization). This distinction has been devised by Jonathan Remy Nash, building upon Wirth (1995), in the context of an extensive study on the potential conflict between tradable allowances and the polluter-pays principle (Nash, 2000). The weak form prohibits governmental subsidies for pollution control equipment to ensure that product prices reflect the costs of pollution abatement. The strong form calls for governments to assure the internalization of environmental costs (and not just to refrain from subsidizing pollution control equipment). This means that the strong form subsumes the weak form: both versions require that companies internalize pollution costs (Nash, 2000: 31 (footnote 31)). Therefore, both the weak and the strong form are manifestations of an efficiency interpretation of the polluter-pays principle.

In addition, our equity interpretation means that equity is used as a criterion on top of (and not instead of) efficiency. Therefore, we speak of an extended form of the polluter-pays principle. To be more precise, equity refers to the distributive implications of institutional arrangements, in a world where wealth transfers matter. In the specific case of the polluter-pays principle, this means that regulatory measures complying with the principle allocate wealth-burdens on the polluter.

The taxonomy outlined above allows us to sharpen our initial research question as follows: are grandfathering and over-allocation consistent (i) with a weak and a strong efficiency interpretation of the polluter-pays principle and (ii) with an extended equity interpretation of the polluter-pays principle?

3. DO POLLUTERS PAY UNDER GRANDFATHERING?

Rather than examining the law, we use economic theory to answer the question of whether polluters pay under grandfathering. Under grandfathering, polluters receive their emission rights free of charge, whereas under auctioning, polluters have to purchase the allowances. Nash (2000: 13) finds that '(…) grandfathering (…) runs contrary to the polluter pays principle's core, violating even the principle's weak form'. He states: 'The core of the polluter pays principle argues that neither the government nor society-at-large should subsidize pollution and polluters and that polluters should internalize the costs of pollution abatement' (Nash, 2000: 3). Nash defends his claim by arguing: 'Grandfathering of allowances creates a government subsidy of polluters (...) The recipients are at liberty to sell the allowances, which they received at no cost, on the market for cash payments' (Nash, 2000: 13).[3] He then concludes that grandfathering is *inconsistent* with the polluter-pays principle.

Contrary to his views, we demonstrate the *consistency* of grandfathering with the polluter-pays principle by arguing that grandfathered allowances internalize pollution costs because of their opportunity costs and that grandfathered allowances constitute lump sum-subsidies that do not distort competition.

3.1 Efficiency Interpretation

Grandfathered allowances internalize pollution costs because of their opportunity costs. Everyone understands that a firm must pay for its emission allowances at an auction and that it saves these costs when those rights are allocated free of charge. But this does not mean, as Nash (2000: 3) states, that grandfathering distributes the allowances 'at no cost' to existing polluters. We emphasize that allowances allocated free of charge also involve costs for firms. Grandfathered allowances used for covering the emissions of the allowance owner have an 'opportunity cost' (e.g. Sijm et al., 2006; Grafton and Devlin, 1996; Nentjes et al., 1995). The opportunity cost is the revenue forgone by refraining from selling the allowances and by employing them in producing output. This opportunity cost, which is equal to the price at which the allowance can be sold, must be included in the product price, despite the fact that allowances have been assigned free of charge. The reason for this is as follows.

In economics, the concept of opportunity cost must be taken into account whenever a resource can be used in alternative ways. In the EU ETS, a firm can decide to produce and to use the allowances to cover its emissions or, alternatively, it can produce less, leading to fewer emissions (or stop producing) and sell the allowances that exceed its emissions. The opportunity cost of grandfathered allowances is the revenue the firm renounces by opting for one use over another. Producing and using allowances to cover emissions generated from production is a first-best option only if the gained profits are at least equal to the profits it could earn by reducing production (at the extreme, closing the

3 Additional arguments by Nash (2000) are summarized, and criticized, in Woerdman et al. (2008).

plant) and selling its exceeding allowances.[4] Therefore, when a firm under the EU ETS decides to continue production, the opportunity cost of the allowances, which is equal to the price at which they can be sold, has to be internalized into the marginal production cost and incorporated in the product price. The implication is that grandfathering does not induce a price deviation from the welfare optimum.

Another example to explain why a producer must pass on the value of allowances as costs to consumers is the analogy between labor and emission rights (Woerdman et al., 2008). An entrepreneur does not have to pay for his own labor (in contrast with the labor of his employees to whom he must pay salaries), but he does employ his labor and he must pass on the value of this in the product price. The same can be said of emission rights. Although the entrepreneur does not have to pay for them, he does employ them to cover the emissions when producing output and therefore he must pass on the value of those rights in the product price. Consequently, if the entrepreneur were not allowed to pass on the opportunity costs of the grandfathered allowances, he would incur an economic loss.

Although polluters should fully pass on the opportunity costs of grandfathered allowances in their product prices, electricity companies in the EU have done this only to a limited extent. Sijm et al. (2005) argue that the main reason for a limited pass-through is the oligopolistic nature of the electricity market. Economic theory learns that any price variation caused by a marginal cost change is greater in perfectly competitive markets than in oligopolistic ones. This result can be explained on the basis of different market equilibrium conditions: marginal costs equal marginal revenues in both perfectly competitive and oligopolistic markets, but the equivalence between price and marginal revenue is guaranteed only under perfect competition.[5] Intuitively, we can say that in oligopolistic markets where prices are already above marginal costs there is little opportunity for a further marginal price increase, but when markets become more competitive, prices tend to be aligned more closely with costs (e.g. Ten Kate and Niels, 2005). The implication is that the less competitive the electricity market is, the lower the pass-through rate will be. Consequently, the opportunity costs of free allowances are only partly incorporated in a higher power price when the electricity market is oligopolistic.

The internalization of pollution costs makes grandfathering consistent with the strong form of the polluter-pays principle ('cost internalization'). Because

---

4 Most EU Member States have determined that a firm loses its allowances after it has shut down an installation. The consequence of this is that the closure of old and inefficient plants is discouraged.

5 A firm in an oligopoly faces a downward sloping marginal revenue curve.

the strong form subsumes the weak form ('no subsidization'), we can deduce full compatibility with the polluter-pays principle. But if we actually check this, instead of making that derivation, we indeed find that grandfathered allowances constitute lump sum-subsidies that do not distort competition. This implies, in our view, that Nash's analysis is incomplete on this point.

Since grandfathering implies a capital gift to the firm, a firm with grandfathered allowances has more financial resources, or own capital, than an identical firm with auctioned allowances. Grandfathering thus implies a transfer of wealth to firms because they receive an input that has a certain market value. This means, as Nash also notices correctly, that grandfathering allowances could be viewed as granting a subsidy to the firm (e.g. Hepburn et al., 2006a; Nash, 2000; Böhringer et al., 1998). However, we emphasize that this subsidy is a capital gift to the firm which has the character of a lump sum-subsidy (e.g. Hepburn et al., 2006a; Hargrave et al., 1999). In other words, there is a subsidy, but it is one that is conceptually different from a subsidy directly linked to the costs of pollution control and prevention measures. If a firm receives its allowances free of charge, it obtains a non-distortionary windfall profit (e.g. Bohm, 1999). In efficiency terms, a lump sum-subsidy is not distorting in the product market, since it does not affect marginal emission reduction costs. The lump sum-subsidy implied by grandfathering does not alter the output and price decisions of firms. Consequently, the incentive to abate is not affected.

Nash's (2000: 13) idea that grandfathered allowances imply a government subsidy for polluters, which they received at no cost and which they can sell for actual cash, is clearly incomplete. We have indicated that a lump sum-subsidy in the form of gratis emission rights does not alter the output and price decisions of firms. Moreover, we have just seen that not only auctioning, but also grandfathering entails costs for firms, namely the opportunity costs when they are used for covering the emissions of the permit owner. These are part of the cost price and must be incorporated in the product price. The implication of all this is that the government does not subsidize pollution when allocating emission rights free of charge, which makes grandfathering also consistent with the weak form of the polluter-pays principle ('no subsidization'). Product prices under an environmental regime of grandfathered allowances will reflect the costs of pollution abatement as the weak form of this principle requires.

3.2 Equity Interpretation

Grandfathering is efficient since it internalizes pollution costs, but its distributive effects are more problematic. It might be argued that grandfathering is inconsistent with the extended form of the polluter-pays principle (efficiency plus equity), since polluters do not actually purchase their allowances and the State does not raise any revenues. Under grandfathering there is a wealth transfer from the public to the polluter. This improves the financial position of the shareholders: the value of a share increases because the polluter has received an asset with a market value free of charge. Even if the polluter pays under grandfathering because of the opportunity costs faced, the polluters receive a capital gift equal to the revenues that the government would have obtained at an auction. Such a capital gift, while not distortive in efficiency terms, does have a redistributive impact that is beneficial for the polluter. Grandfathering may thus be perceived as unfair from a polluter-pays perspective.

The idea that auctioning provides a 'better reflection' of the polluter-pays principle (noted in Egenhofer and Fujiwara, 2006: 25) can only be defended based on an equity interpretation of the principle. Members of the European Parliament seem to have endorsed such an extended interpretation of the polluter-pays principle by pleading in favor of more auctioning. Article 10 of the EU emissions trading Directive requires that every Member State allocate at least 95% of its allowances free of charge in the period 2005–2007 and at least 90% in the period 2008–2012. Before the adoption of this Directive, the European Parliament made the following remark on grandfathering: 'Since it involves no cost, the proposed method (…) does not incorporate the 'polluter pays' principle' (EP, 2002: 52). After the adoption of this Directive, members of the Parliament said that the increasing possibility of auctioning allowances – from 5% in the period 2005–2007 to 10% in the period 2008–2012 – would ensure the 'progressive' application of the polluter-pays principle, insisting that further harmonization should be considered, including auctioning, for the period after 2012 (Worsley and Freedman, 2003: 15). Apparently, in their view, a harmonized scheme that prescribes 100% auctioning would ensure the 'full' application of the polluter-pays principle, which reflects the equity view.

More recently, the European Commission presented a proposal to amend the emissions trading Directive 2003/87/EC and indicated that: 'Auctioning best ensures efficiency of the ETS, transparency and simplicity of the system and avoids undesirable distributional effects. Auctioning also best complies with the polluter-pays principle (...)' (COM, 2008: 7). Also the Commission's opinion reflects the equity interpretation of this principle.

From an equity perspective, the polluter does not pay under grandfathering, since polluters receive a capital gift equal to the revenues that the government would have obtained in an auction. This raises an interesting political acceptability trade-off in the ETS, which basically comes down to the observation that auctioning is more acceptable to consumers, while grandfathering is more acceptable to producers.

In the case of auctioning, each producer will have to buy emission rights to cover its emissions. This entails additional costs for polluters. As a consequence, the producers will pass on these costs in their product prices. The costs of purchasing emission rights that producers incorporate in their product prices are probably easier to understand for consumers than the opportunity costs of gratis rights that are passed on to them under grandfathering. In both cases, the cost pass-through is in accordance with economic theory, but consumers are likely to find the price mark-up more acceptable when producers directly purchase the allowances at an auction. Moreover, auctioning implies that the capital gift will shift from the shareholders to the government, so that shareholders will not become any richer from the allocation of emission rights. This suggests that auctioning may be more acceptable to consumers than grandfathering (e.g. Cramton and Kerr, 1998; see also Hepburn et al., 2006b).

However, the financial advantage of grandfathering ensures that the emissions trading scheme becomes more politically acceptable for producers compared to emission taxation or permit auctioning. Moreover, some even argue that gratis allowances compensate the owners of existing plants for the 'stranded costs' they bear as a result of the new requirement to reduce emissions (e.g. Harrison and Radov, 2002). In addition, auctioning could spark a 'secondary allocation debate' by shifting the allocation problem to the issue of how to recycle the auction revenues (Egenhofer and Fujiwara, 2005: 26).

People may have different perceptions of what is equitable. Therefore, we do not draw the conclusion that grandfathering is 'unfair' in absolute terms on a macro-level. Instead, we consider the more nuanced possibility that grandfathering is inconsistent with an equity interpretation of the polluter-pays principle as defined above, because polluters, while bearing opportunity costs, are granted a capital gift.

4. DO POLLUTERS PAY IN CASE OF OVER-ALLOCATION?

Having assessed that grandfathering is an efficient means of allocating allowances, we cannot conclude that the EU ETS is efficient altogether. For instance, politicians may have set an inefficient overall emission target due to imperfect information on the environmental damage function. Even without questioning the efficiency of the emission target itself, there may still be another problem: the government caps emissions inappropriately, assigning too many allowances to polluters. This is referred to as over-allocation. Related to the latter issue, the question investigated in this section is whether an over-allocation of allowances is inconsistent with the polluter-pays principle. But what is over-allocation precisely? In order to answer this question, we develop a theoretical framework to construct a benchmark against which over-allocation can be tested. On the basis of our constructed benchmark, we assess empirically whether allowances have been over-allocated in the EU for the period 2005–2007 (Clò, 2007).

4.1 Empirical Assessment

The EU ETS covers only part of the CO_2 polluting sources. The ETS sectors include mainly energy and energy-intensive industries, while agriculture, households and transport are generally excluded. When we refer to the emission cap we mean the total amount of allowances allocated to the ETS sectors. In other words, this cap indicates the maximum amount of CO_2 the ETS sectors are allowed to produce and, symmetrically, the emission reduction burden imposed on the ETS sectors.

Over-allocation is not a clear concept. It implies that too many allowances have been assigned, but it does not give a precise indication of how many allowances have been given in excess. To assess if and to what extent over-allocation has occurred, we have to define a benchmark, that is, a cap which reflects the ideal amount of allowances that should be allowed. It follows that any amount superior to this ideal quantity would imply over-allocation. The Kyoto emission reduction target as such cannot be used as a benchmark, since it applies to all greenhouse gases in the EU, whereas the European emissions trading scheme covers only a subset of those emissions. Consequently, we need some criterion to define which part of the Kyoto target can be directly compared with the ETS cap.

The first-best candidate is the efficiency criterion: emissions should be abated at the lowest marginal cost. Being produced by both ETS and non-ETS sectors, the European emission reduction burden should be divided among ETS and non-ETS sectors according to their marginal abatement curves (MACs). Although we know that on average marginal abatement costs are higher for non-ETS sectors than for ETS sectors (e.g. Criqui and Kitous, 2003; Böhringer et al., 2005; Peterson, 2006), the ETS and non-ETS aggregated MACs by countries are not publicly known. The implication is that the efficient ETS cap cannot be determined with precision.

A second-best option is the proportionality criterion: the ETS cap should be set in a way to impose on the ETS sectors an emission reduction burden proportional to the percentage of the emissions they produce in the EU, called the ETS share. This theoretical benchmark, called the ETS proportional cap, can be calculated for each Member State by multiplying the allowed emissions (following from the Kyoto emission reduction target) by the percentage of emissions produced by the ETS sector (Clò, 2007) (see Appendix, Table 5.3).

While the emission reduction targets for the EU and its Member States are publicly available (see Appendix, Table 5.1), official data about the pre-2005 ETS share are not, so that this share can only be estimated. Before 2005 emissions were monitored and aggregated at a national level and the amount of emissions released by the ETS sector was unknown. Georgedopolou et al. (2006) estimate the pre-2005 ETS share for each Member State dividing the ETS historical emissions baseline of 2002 assessed in each National Allocation Plan (NAP) by the national GHG emissions produced in the same year as reported in the yearly GHG emission reports of the European Environment Agency (EEA) (see Appendix, Table 5.2). Although this is a reliable estimation of the pre-2005 ETS share, there may still be an upward bias in these data. The reason is that historical emission data in the NAPs were collected by voluntary, self-reported submissions on behalf of the ETS installations, so that firms had an incentive to signal higher emissions in order to receive more grandfathered allowances (e.g. Ellerman and Buchner, 2006).

In 2007 the EEA published official data of the GHG emissions produced in 2005, making it possible to calculate the exact ETS share (see Appendix, Table 2). However, this ratio cannot take into account the possibility that in 2005 the ETS share varies from 2004 because of the establishment of the ETS itself.

Given the potential problems of both pre-2005 and 2005 ETS shares, referring to only one of them might lead to an imprecise assessment of the actual over-allocation. Instead of choosing between the pre-2005 and the 2005 ETS share, we consider both of them obtaining two different benchmarks:

Benchmark 1: *the pre-2005 ETS proportional target*, derived from multiplying the Kyoto target by the pre-2005 ETS share estimated by Georgedopolou et al. (2006);
Benchmark 2: *the 2005 ETS proportional target*, derived from multiplying the Kyoto target by the 2005 ETS share.

These two benchmarks define the ETS proportional target range. Assessing over-allocation in relation to a range rather than to a single benchmark allows us to derive more robust conclusions. When allowances are over-allocated in respect of this range the same would be true if over-allocation was assessed using only one of the two benchmarks – but not vice versa. Moreover, the unrealistic assumption of a constant ETS share (e.g. Georgedopoulou et al., 2006; Betz et al., 2006) can be relaxed (Clò, 2007).

The EU ratified the Kyoto Protocol in 2002, well before the EU enlargement of 2004. Thus, the 8% emission reduction target refers only to the former EU-15 Member States. The EU-15 will comply with its Kyoto commitment if it succeeds in reducing emissions to 3925 Mton CO_2-eq. in 2012 (see Appendix, Table 3). For the period 2005–2007, the EU-15 Member States allocated an amount of allowances equivalent to 42% of the EU-15 target. This percentage is higher than both the pre-2005 and 2005 EU-15 ETS shares of respectively 41% and 38%. We can therefore conclude that, accordingly to the proportionality criterion, allowances have been on average over-allocated to the ETS sectors in the EU-15 Member States.

A similar analysis can detect which Member State over-allocated allowances to its national installations (see Appendix, Table 5.3). The graph, Figure 5.1, compares the amount of allocated allowances during the first period, 2005–2007, with this range.[6]

We classify the Member States into three categories:

Those whose 2005–2007 ETS cap is above the ETS Kyoto proportional range (Austria, Denmark, Finland, France, Germany, Greece, Ireland, Italy, Luxemburg, The Netherlands, Portugal, Spain and Sweden);

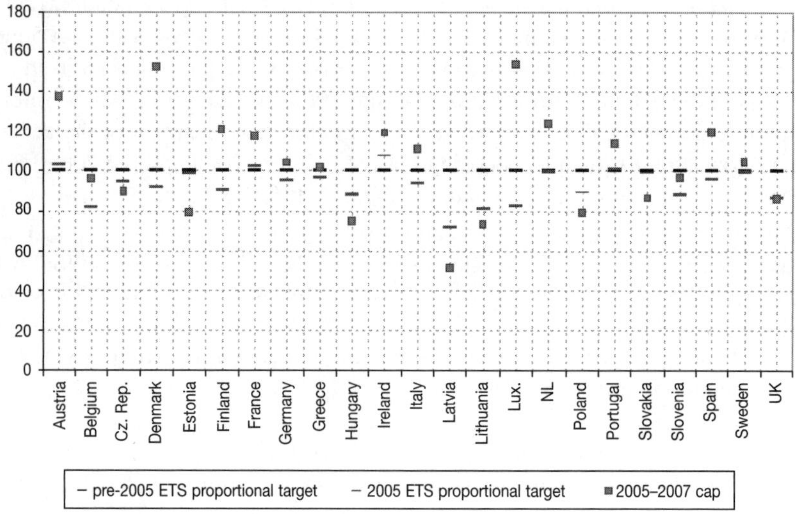

Source: Clò (2007)

Figure 5.1 ETS proportional target range and ETS 2005–2007 real cap (normalized values)

[6] For any Member State the theoretical pre-2005 ETS proportional Kyoto target is normalized to 100. Then the 2005 ETS proportional Kyoto target and the amount of ETS allowances allocated during the two phases are expressed in relation to the normalized theoretical pre-2005 ETS target. All values used to build this graph are reported in Appendix 1.

Those whose 2005–2007 amount of allocated allowances is included in the range (Belgium and Slovenia);

Those whose 2005–2007 amount of allocated allowances is below the range (Czech Republic, Estonia, Hungary, Lithuania, Latvia, Slovakia and the United Kingdom).

States belonging to the first category (all original EU-15 Member States, except the United Kingdom and Belgium) over-allocated allowances to their national ETS sectors. These sectors received an amount of allowances that is larger than their proportional ETS share, independently from which ETS share we consider (either pre-2005 or 2005). The same conclusion holds for Belgium and Slovenia if we consider the range's upper benchmark, but the opposite would be true when the amount of allocated allowances is compared to the range's lower limit. Given this ambiguity, we abstain from drawing a firm conclusion about the stringency of their caps. Finally, the amount of allocated allowances is actually lower than their proportional pre-2005 and 2005 ETS emissions share for all new Member States that joined the EU in 2004. The United Kingdom also did not over-allocate allowances.

4.2 Efficiency Interpretation

Having concluded that over-allocation occurred in a significant number of cases, we now turn to our main question: Is the efficiency version of the polluter-pays principle (weak and/or strong) violated when allowances are over-allocated? If allowances are over-allocated on an aggregate level, thus for the EU as a whole, there will be no demand, leading to an allowance price of zero. Although one might be tempted to believe that *de facto* no subsidy is given (since the allowances are literally worthless), over-allocation is a form of cross subsidization because the environmental costs are shifted to the non-ETS sectors. Those sectors, which are not part of the emissions trading system, will have to pay for the emission reductions of the ETS sectors if the country wants to comply with its emission target. This means that over-allocation violates the weak form of the polluter-pays principle. The cross-subsidy, which has the character of an environmental cost shifted from ETS to non-ETS sectors, is not efficient because marginal abatement costs are higher for non-ETS sectors than for the ETS sectors (e.g. Criqui and Kitous 2003, Böhringer et al. 2005, Peterson, 2006). Consequently, the ETS sectors should actually bear a higher emission reduction burden than the non-ETS sectors, and not vice versa.

The national government could reduce emissions on behalf of the ETS sectors, for instance by acquiring credits from Joint Implementation and Clean

Development Mechanism projects. If allowances are over-allocated to ETS sectors, Ministries of Finance as well as tax-payers will have to pay for these credits, transforming the international Kyoto mechanisms into largely public-funded markets (Neuhoff et al., 2006). Also in this case the weak form of the polluter-pays principle is violated: over-allocation constitutes a subsidy to the ETS sectors as governments would abate on behalf of the ETS sectors to comply with the Kyoto targets.

Moreover, if over-allocation leads to a zero price of allowances, there is no cost internalization, leading to a violation of the strong form of the polluter-pays principle as well. When the European Commission reported officially in May 2006 that in 2005 ETS verified emissions were about 4% lower than the number of allowances distributed to installations for 2005, CO_2 prices initially dropped from 30 to 10 Euros and then to even below 1 Euro during 2007. This surplus of allowance supply implies that an emissions trading market had been created in the EU without scarcity, failing to give polluters an incentive to reduce emissions.

Without over-allocation, the environmental cost to be internalized by the ETS sectors would be proportional to the percentage of pollution they generate. With over-allocation, the environmental costs are only partly internalized by the polluters in the emissions trading scheme and the remaining costs are shifted (or 'externalized') to those outside the scheme. Thus we conclude that over-allocation violates both the weak and the strong form of the polluter-pays principle.

4.3 Equity Interpretation

Over-allocation is not only inefficient, but it also violates the extended form of the polluter-pays principle. In fact, the environmental costs that the ETS sectors should bear in order to contribute proportionally to achieving the Kyoto target are shifted to the non-ETS sectors which are not legally required to abate. The implication is that those outside the ETS system actually have to abate on behalf of those sectors that are regulated under the ETS. While the ETS sectors will be bearing an emission reduction burden that is less proportional than their ETS share, the non-ETS will have to abate an amount of emissions that is excessive compared to the percentage of emissions they generate. This burden transfer is both inefficient and unfair from a proportionality point of view.

5. CONCLUSION

The EU ETS has been up and running since 2005 and it is now the largest emissions trading scheme in the world. It is therefore important to understand how the scheme works in practice and how it operates in relation to general principles of environmental law.

In this chapter we have assessed the compatibility of some of the scheme's features, notably grandfathering and over-allocation, with the polluter-pays principle. Until 2012, the EU ETS primarily uses grandfathering to allocate the allowances free of charge based on historical emissions. Moreover, in 2005 there was an 80 million tonnes gap between emissions and allowances of the ETS sectors, suggesting that polluters received more allowances than they needed. Are grandfathering and over-allocation consistent with the polluter-pays principle?

To set out clear boundaries of our research question, we have built a taxonomy of possible interpretations of the polluter-pays principle where we distinguished an efficiency from an equity version. In relation to the equity interpretation, we speak of an 'extended' form of the polluter-pays principle, because equity is used as a criterion on top of (and not instead of) efficiency. Within the efficiency interpretation, we have identified a 'weak' form (no subsidization) and a 'strong' form (cost internalization).

We have concluded that grandfathering is consistent with the efficiency interpretation of the polluter-pays principle. Because grandfathered allowances entail opportunity costs, pollution costs are internalized, which makes grandfathering consistent with the strong form of the efficiency version of the polluter-pays principle. Moreover, grandfathered allowances constitute lump sum-subsidies that do not distort competition, which makes grandfathering consistent with the weak form as well. However, the claim that grandfathering does not violate the polluter-pays principle can only be defended from an efficiency perspective. Grandfathering improves the financial position of the shareholders, since polluters receive an asset with a market value for free. Such a capital gift, while not distortive in efficiency terms, has a redistributive impact which is beneficial for the polluter. Accordingly, grandfathering is unfair from an extended polluter-pays perspective. Only auctioning is consistent with this form of the polluter-pays principle, because it internalizes pollution costs and forces polluters to purchase their allowances.

To assess over-allocation, we have constructed a benchmark against which over-allocation can be tested and on this basis we have shown that over-allocation has indeed occurred to a considerable extent in the first phase of the EU ETS. We then explained why over-allocation is *inconsistent* with all aforementioned interpretations of the polluter-pays principle. Indeed, over-allocation entails a form of cross-subsidization, since environmental costs are shifted from the ETS to the non-ETS sectors. Those sectors outside the scheme will have to pay for the emission reductions of the ETS sectors if the country wants to comply with its emission target. This means that over-allocation violates the weak form of the polluter-pays principle. The cross-subsidy is inefficient because marginal abatement costs are higher for non-ETS sectors than for the ETS sectors. Because over-allocation leads to an allowance price of (almost) zero, there is no cost internalization, leading to a violation of the strong form of the polluter-pays principle as well. Finally, over-allocation violates the extended form of that principle, because of the burden transfer from ETS to non-ETS sectors.

From our analysis, it emerges that the polluter-pays principle is frustrated in a number of ways by the early design of the EU ETS. Some rethinking of the scheme is required if this principle is deemed important for European environmental law. Auctioning the allowances and strengthening the emission caps is necessary to bring the scheme more in line with the polluter-pays principle.

REFERENCES

Betz R. and M. Stato (2006), 'Emissions Trading: Lessons Learnt from the 1st Phase of the EU ETS and Prospects for the 2nd Phase', 6 *Climate Policy* 351–9.

Betz R., K. Rogge and J. Schelich (2006), *EU Emissions Trading: An Early Analysis of National Allocation Plans for 2008–2012*, Working Paper, Fraunhofer Institute.

Bohm, P. (1999), *International Greenhouse Gas Emission Trading – With Special Reference to the Kyoto Protocol*. TemaNord 1999: 506. Stockholm: Department of Economics.

Böhringer, C., T.F. Rutherford and A. Voß (1998), 'Global CO_2 Emissions and Unilateral Action: Policy Implications of Induced Trade Effects', 11(1–4) *International Journal of Global Energy Issues* 18–22.

Böhringer, C., T. Hoffmann, A. Lange, A. Löschel and U. Moslener (2005), 'Assessing Emission Regulation in Europe: An Interactive Simulation Approach', 26 *Energy Journal* 1–22.

Bugge, H. C. (1996), 'The Principles of "Polluter-Pays" in Economics and Law', in E. Eide and R. Van den Bergh, eds., *Law and Economics of the Environment* 53, Oslo: Juridisk Forlag.

Calabresi, G. and A.D. Melamed (1972), 'Property Rules, Liability Rules, and Inalienability: One View of the Cathedral', 85 (6) *Harvard Law Review* 1089–1128.

Clò, S. (2007), *Assessing the European Emissions Trading Scheme Effectiveness in Reaching the Kyoto Target: An Analysis of the ETS 1st and 2nd Phase Cap Stringency*, Working paper presented at the Annual Conference of the European Association of Law and Economics 2007, available at: www.cbs.dk/content/download/67304/930289/file/Stefano%20Clň.pdf.

Coase, R.H. (1960), 'The Problem of Social Cost', 3 *Journal of Law and Economics* 1–44.

COM (2008), *Proposal for a Directive of the European Parliament and of the the Council Amending Directive 2003/87/EC so as to Improve and Extend the Greenhouse Gas Emission Allowance Trading System of the Community*. 2008/0013 (COD). Brussels: European Commission.

Cramton, P. and S. Kerr (1998), *Tradable Carbon Permit Auctions: How and Why to Auction Not Grandfather*. Discussion Paper 98-34, Washington DC: Resources for the Future (RFF).

Criqui, P. and A. Kitous (2003), *Kyoto Protocol Implementation (KPI) Technical Report: Impacts of Linking JI and CDM Credits to the European Emissions Allowance Trading Scheme – A Report for DG Environment.* CNRS-IEPE and ENERDATA S.A.

Dworkin, R. (1977), *Taking Rights Seriously*. Cambridge: Harvard University Press.

Egenhofer, C. and N. Fujiwara (2005), *Reviewing the EU Emissions Trading Scheme: Priorities for Short-Term Implementation of the Second Round of Allocation (Part I)*. Brussels: Centre for European Policy Studies (CEPS).

Egenhofer, C. and N. Fujiwara (2006), *Reviewing the EU Emissions Trading Scheme: Priorities for Short-Term Implementation (Part II)*. Brussels: Centre for European Policy Studies (CEPS).

Ellerman D. and B. Buchner (2006), *Over-Allocation or Abatement? A Preliminary Analysis of the EU ETS Based on the 2005 Emissions Data*. Nota di Lavoro 139.2006. Milan: Fondazione Eni Enrico Mattei (FEEM).

EEA (2006), *Greenhouse Gas Emission Trends and Projections in Europe 2006*. European Environment Agency (EEA) Report No. 9/2006.

EEA (2007), *Annual European Community Greenhouse Gas Inventory 1990–2005 and Inventory Report 2007*. European Environment Agency (EEA) Technical Report No. 7/2007.

EP (2002), *Report on the Proposal for a European Parliament and Council Directive Establishing a Scheme for Greenhouse Gas Emission Allowance Trading within the Community and Amending Council Directive 96/61/EC*. Final A5-0303/2002 Par1 (RR/477361EN.doc), 13 September, Brussels: European Parliament.

Faure, M. and D. Grimeaud (2003), 'Financial Assurance Issues of Environmental Liability', in M. Faure, ed., *Deterrence, Insurability and Compensation in Environmental Liability*. ECTIL, Vienna. Springer.

Georgedopoulou, E., Y. Sarafidis, S. Mirasgedis and D.P. Lalas (2006), *Next Allocation Phase of the EU Emissions Trading Scheme: How Tough Will the Future Be?*, 34 *Energy Policy* 4002–23.

Grafton, R.Q. and R.A. Devlin (1996), 'Paying for Pollution: Permits and Charges', 98(2) *Scandinavian Journal of Economics* 275–88.

Hargrave, T., N. Helme, T. Denne, S. Kerr and J. Lefevere (1999), *Design of a Practical Approach to Greenhouse Gas Emissions Trading Combined with Policies and Measures in the EC*. Washington: Center for Clean Air Policy (CCAP).

Harrison, D. and D.B. Radov (2002), *Evaluation of Alternative Initial Allocation Mechanisms in a European Union Greenhouse Gas Emissions Allowance Trading Scheme*. National Economic Research Associates (NERA).

Hepburn, C., J.K.-H. Quah and R.A. Ritz (2006a), *Emissions Trading and Profit-Neutral Grandfathering*. Department of Economics Discussion Paper Series Nr. 295. Oxford: Oxford University.

Hepburn C., M. Grubb, K. Neuhoff, F. Matthes and M. Tse (2006b), 'Auctioning of EU ETS Phase II Allowances: How and Why?', 6 *Climate Policy* 137–60.

Ten Kate, A. and G. Niels (2005), 'To What Extent are Cost Savings Passed on to Consumers? An Oligopoly Approach', 20 *European Journal of Law and Economics* 323–37.

Krämer, L. (2005), 'Directive 2004/35 on Environmental Liability and Environmental Principles', 4 *Tijdschrift voor Milieuaansprakelijkheid* 131–4.

Nash, J.R. (2000), 'Too Much Market? Conflict Between Tradable Pollution Allowances and the "Polluter Pays" Principle,' 24(2) *Harvard Environmental Law Review* 1–59.

Nentjes, A., P. Koutstaal and G. Klaassen (1995), *Tradeable Carbon Permits: Feasibility, Experiences, Bottlenecks*. Dutch National Research Programme on Global Air Pollution and Climate Change (NRP), NRP Report no. 410 100 114, Groningen/Bilthoven: RuG / NRP.

OECD (1972), *Environment and Economics: Guiding Principles Concerning International Economic Aspects of Environmental Policies*, 26 May 1972, Annex Par. 1, Doc. No. C(72)128, 1972 WL 24710. Paris: Organization for Economic Co-operation and Development (OECD).

OECD (1975), *The Polluter Pays Principle: Definition, Analysis, Implementation*. Paris: Organization for Economic Co-operation and Development (OECD).

Parikh, K.S. (1993), 'The Polluter-Pays and User-Pays Principles for Developing Countries: Merits, Drawbacks and Feasibility', in E. Dommen, ed., *Fair Principles for Sustainable Development: Essays on Environmental Policy and Developing Countries*. Aldershot: Edward Elgar, pp. 81–91.

Pearson, C.S. (1994), 'Testing the System: GATT + PPP = ?', 27 *Cornell International Law Journal* 553–75.

Peterson, S. (2006), *Efficient Abatement in Separated Carbon Markets: A Theoretical and Quantitative Analysis of the EU Emissions Trading Scheme*, Kiel Working Paper 1271.

Sijm, J.P.M., S.J.A. Bakker, Y. Chen, H.W. Harmsen and W. Lise (2005), *CO_2 Price Dynamics: The Implications of EU Emissions Trading for the Price of Electricity*. ECN-C-05-081. September 2005. Petten: Energieonderzoek Centrum Nederland (ECN).

Sijm, J., K. Neuhoff and Y. Chen (2006), 'CO_2 Cost Pass-Through and Windfall Profits in the Power Sector', 6 *Climate Policy* 49–72.

Sorrell, S. and J. Sijm (2003), 'Carbon Trading in the Policy Mix', 19(3) *Oxford Review of Economic Policy* 420–37.

Verhoef, E.T. (1999), 'Externalities', in J.C.J.M. van den Bergh, ed., *Handbook of Environmental and Resource Economics*. Cheltenham: Edward Elgar, pp. 197–214.

Wirth, D.A. (1995), 'The Rio Declaration on Environment and Development: Two Steps Forward and One Back, or Vice Versa?', 29 *Georgia Law Review* 599–653.

Woerdman, E. (2005), 'Tradable Emission Rights', in J.G. Backhaus, ed., *Elgar Companion to Law and Economics*. Cheltenham: Edward Elgar, pp. 364–80.

Woerdman, E., A. Arcuri and S. Clò (2008), 'Emissions Trading and the Polluter-Pays Principle: Do Polluters Pay under Grandfathering?', 4 (2) *Review of Law and Economics* forthcoming.

Worsley, R. and R. Freedman (2003), *Europarl Daily Notebook: 02-07-2003*. Brussels: European Parliament.

APPENDIX

MS Kyoto Targets

The Kyoto target for the EU has been redistributed among Member States (MS) via the Burden Sharing Agreement. This agreement assigns to each country a specific target varying from a minimum of –28% (Luxembourg) to a maximum of +27% (Portugal).

Table 5.1 MS Burden Sharing and Distance to the Kyoto target (Mt CO_2)

	Base year emissions	Burden sharing agreement	Kyoto target	2005 GHG emissions	Distance to the target	Distance to the target
Austria	78.9	–13%	68.68	93.3	–24.62	35.8%
Belgium	146.9	–7.5%	135.87	143.8	–7.93	5.8%
Czech Rep.	196.3	–8%	180.58	145.6	34.98	–19.4%
Denmark	69.3	–21%	54.77	63.9	–9.13	16.7%
Estonia	42.6	–8%	39.23	20.7	18.53	–47.2%
Finland	71.1	0%	71.10	69.3	1.8	–2.5%
France	567.1	0%	567.09	553.4	13.69	–2.4%
Germany	1230	–21%	971.67	1001.5	–29.83	3.1%
Greece	111.1	+25%	138.82	139.2	-0.38	0.3%
Hungary	122.2	–6%	114.89	80.5	34.39	–29.9%
Ireland	55.8	+13%	63.03	69.9	–6.87	10.9%
Italy	519.6	–6.5%	485.83	582.2	–96.37	19.8%
Latvia	25.9	–8%	23.82	10.9	12.92	–54.2%
Lithuania	50.9	–8%	46.86	22.6	24.26	–51.8%
Lux.	12.7	–28%	9.14	12.7	–3.56	38.9%
NL.	214.3	–6%	201.45	212.1	–10.65	5.3%
Poland	565.3	–6%	531.34	399	132.34	–24.9%
Portugal	60.0	+27%	76.15	85.5	–9.35	12.3%
Slovakia	73.2	–8%	67.36	48.7	18.66	–27.7%
Slovenia	20.2	–8%	18.60	20.3	–1.7	9.1%
Spain	289.4	+15%	332.79	440.6	–107.81	32.4%
Sweden	72.5	+4%	75.35	67	8.35	–11.1%
UK	767.9	–12.5%	671.90	657.4	14.5	2.2%
EU-15	4266.4	–8%	3925.11	4192	–266.89	6.8%
EU-23	5363.2	No target	4946.3	4940.1	6.22	–0.1%

Source: EEA (2006: 61) and EEA (2007).

MS ETS Share

Article 14 of the Directive 2003/87/EC establishes that MS have the duty to monitor ETS emissions in accordance with specific monitoring and reporting guidelines (Commission Decision 2004/156/EC). Every year (t) MS have to report the amount of emissions produced in the previous year (t-1) by each ETS installation. These data are aggregated at a national level and officially published by the European Commission. Publicly available data of the ETS emissions and the national GHG emissions produced in the same year were finally available. Moreover, according to Council Decision 280/2004/EC, each year (t) EU MS have to report the national GHG emissions for year (t-2) to the European Commission (assisted by the EEA). The EEA inventory reports cover all the trading and non-trading sectors and all GHG emissions. Thus, in 2007 it finally became possible to calculate the exact 2005 ETS share and its marginal annual change.

Table 5.2 Assessing the ETS share

	2005 ETS verified emissions (Mt) (1)	2004 total GHG emissions (mt) (2)	2002 ETS emissions (Mt) (3)	2002 total GHG emissions (Mt) (4)	Pre-2005 ETS share (5) = (3)/(4)	2005 total GHG emissions (Mt) (6)	2005 ETS share (%) (7) = (1)/(6)
Austria	33.4	91.3	30.2	86.4	35%	93.3	36%
Belgium	55.4	147.9	63	145.3	47.5%	143.8	39%
Cz Rep	82.5	147.1	89	142.9	60.3%	145.6	57%
Denmark	26.5	68.1	30.9	69	44.8%	63.9	41%
Estonia	12.6	21.3	12	19.5	61.5%	20.7	61%
Finland	33.1	81.4	40.9	77.2	53%	69.3	48%
France	131.3	562.6	132.4	554.1	23.5%	553.4	24%
Germany	474	1015.3	501	1015.2	49.3%	1001.5	47%
Greece	71.3	137.6	71	133.6	52.8%	139.2	51%
Hungary	26	83.1	29.4	80.8	36.4%	80.5	32%
Ireland	22.4	68.5	20.6	69.4	29.7%	69.9	32%
Italy	225.3	582.5	228.1	555	41.4%	582.2	39%
Latvia	2.9	10.7	3.7	10.6	37.6%	10.9	27%
Lithuania	6.6	20.3	8.5	19.6	35.7%	22.6	29%
Lux	2.6	12.7	2.6	10.8	24.2%	12.7	20%
NL	80.4	217.8	81.7	213.5	38.2%	212.1	38%
Poland	205.4	386.4	219.8	370.2	57.1%	399	51%
Portugal	36.4	84.5	36.6	86.1	42.5%	85.5	43%
Slovakia	25.2	51	26.7	50.9	52.4%	48.7	52%
Slovenia	8.7	20.1	20.6	69.5	48.9%	20.3	43%
Spain	182.9	427.9	174.5	398.6	43.8%	440.6	42%
Sweden	19.3	69.9	9.8	20.1	29.1%	67	29%
UK	242.5	659.3	276.7	643.7	42.5%	657.4	37%
EU-15	1636.8	4227.3	1663.8	4078	41%	4192	38%
EU-23	2006.7	4967.3	2073.5	4842	43%	4940.1	41%

Source: Clò (2007)

The following table calculates the *ETS Proportional target* for each MS as it has been defined in section 4.

Table 5.3 Range Benchmarks for the 2005–2007 ETS Cap

	Kyoto target (Mt CO_2) (1)	Pre-2005 ETS share (2)	2005 ETS share (3)	Pre-2005 ETS proportional target (4) = (1) x (2)	2005 ETS proportional target (5) = (1) x (3)	2005–2007 ETS real cap	Normalized values		
							Pre-2005 ETS proportional target	2005 ETS proportional target	2005–2007 cap
Austria	68.68	35%	36%	24.0	24.7	33	100	102.9	137.3
Belgium	135.87	47.5%	39%	64.5	53.0	62.1	100	82.1	96.2
Czech Rep.	180.58	60.3%	57%	108.9	102.9	97.6	100	94.5	89.6
Denmark	54.77	44.8%	41%	24.5	22.5	37.3	100	91.5	152.0
Estonia	39.23	61.5%	61%	24.1	23.9	19	100	99.2	78.8
Finland	71.1	53%	48%	37.7	34.1	45.5	100	90.6	120.7
France	567.09	23.5%	24%	133.3	136.1	156.5	100	102.1	117.4
Germany	971.67	49.3%	47%	479.0	456.7	499	100	95.3	104.2
Greece	138.82	52.8%	51%	73.3	70.8	74.4	100	96.6	101.5
Hungary	114.89	36.4%	32%	41.8	36.8	31.3	100	87.9	74.8
Ireland	63.03	29.7%	32%	18.7	20.2	22.3	100	107.7	119.1
Italy	485.83	41.4%	39%	201.1	189.5	223.1	100	94.2	110.9
Latvia	23.82	37.6%	27%	9.0	6.4	4.6	100	71.8	51.4
Lithuania	46.86	35.7%	29%	16.7	13.6	12.3	100	81.2	73.5
Lux.	9.14	24.2%	20%	2.2	1.8	3.4	100	82.6	153.7
Netherlands	201.45	38.2%	38%	77.0	76.6	95.3	100	99.5	123.8

Table 5.3 Continued

	Kyoto target (Mt CO_2) (1)	Pre-2005 ETS share (2)	2005 ETS share (3)	Pre-2005 ETS proportional target (4) = (1) x (2)	2005 ETS proportional target (5) = (1) x (3)	2005–2007 ETS real cap	Normalized values		
							Pre-2005 ETS proportional target	2005 ETS proportional target	2005–2007 cap
Poland	531.34	57.1%	51%	303.4	271.0	239.1	100	89.3	78.8
Portugal	76.15	42.5%	43%	32.4	32.7	36.9	100	101.2	114.0
Slovakia	67.36	52.4%	52%	35.3	35.0	30.5	100	99.2	86.4
Slovenia	18.6	48.9%	43%	5.4	8.0	8.8	100	87.9	96.8
Spain	332.79	43.8%	42%	145.8	139.8	174.4	100	95.9	119.6
Sweden	75.35	29.1%	29%	36.8	21.9	22.9	100	99.7	104.4
UK	671.9	42.5%	37%	285.6	248.6	245.3	100	87.1	85.9
EU-15	3925.11	41%	38%	1.609	1.491.5	1.657			
EU-23	4946.32	43%	41%	2.127	2.028	2.174			

Source: Clò (2007)

6. EU greenhouse gas emissions trading and competition law

Stefan Weishaar

INTRODUCTION

This chapter addresses Competition law issues that arise under the European Emissions Trading System (EU ETS). Where appropriate, comments on the proposed amendment of Directive 2003/87/EC[1] that emphasizes auctioning and benchmarking are made. Pursuant to the proposal of the European Commission, sectors exposed to strong international competition and that give rise to carbon leakage will benefit from free allocation. Sectors other then electricity – electricity will in principle not benefit from any free allocation[2] – will be subject to ever more auctioning.[3] A transitory rule envisages a partial free allocation of 80% of the average measured emissions during the period 2005–2007. The free allocation is reduced annually by equal amounts so that by 2020 free allocation will have completely faded out in these sectors.[4]

Undue interventions by Member States are largely foreclosed through the application of the four freedoms, while EC Competition law (Articles 81 and 82 EC Treaty) is geared to the prevention of competitive distortions arising in particular from undue behaviour of firms. Not prejudicing the application of the four freedoms, Member States' involvement in practices that distort competition between undertakings is addressed through the application of EU Competition law rules. There are two main alleys in which unduly distorting State measures taken within the EU ETS framework can be contained. Firstly, through the joint application of Articles 3(g), 10(2), 81 and 82 EC Treaty which was developed by the ECJ upon recognition that State measures can

1 COM(2008)16 final of 23.01.2008.

2 See the derogation for cogeneration installations, Article 10 (a) (3), COM(2008)16 final of 23.01.2008.

3 Recital 16–19 of the Explanatory memorandum and Article 10 (a) (3), COM(2008)16 final of 23.01.2008.

4 Recital 17, COM (2008) 16 final of 23.01.2008. Free allocations are also subject to the general reduction in emission allowances available under the EU ETS as provided for in Article 9, COM(2008)16 final of 23.01.2008.

undermine the effectiveness of the EC Treaty, and secondly through the application of Article 87 regarding State aid.

The first section of the paper briefly examines the possibilities and requirements for the joint application of Articles 3(g), 10(2), 81 and 82 to induce Member States to select the least distortive measures when allocating emission allowances. Since the actual situations in which the Court would be able to apply these Articles will still have to arise, this treatment is restricted in scope. The more extensive second section addresses State aid issues. Here, too, it is examined how Competition law could guide Member States to select the least distortive allocation measure.

JOINT APPLICATION OF ARTICLES 3(G), 10(2) AND 81 OR 82 EC TREATY

Government measures taken to allocate EU ETS emission allowances within the framework of the National Allocation Plans (NAPs) can be the source of distortions of competition. NAPs, and to a varying degree allocation mechanisms, can lead to distortions that influence firms' propensity to collude and abuse.

Allocation formats employed by the NAPs can, for example, create 'barriers to entry'. Such barriers will be created if the allocation format affords incumbents an absolute cost advantage over new entries. NAPs granting preferential treatment to incumbents or which do not afford new entrants equal treatment in the presence of resource depletion or transfer rules, can give rise to such barriers. Their existence gives rise to concern from a competition policy perspective since they facilitate cartelization and abuse of dominant positions that work to the detriment of society. The frequently employed grandfathering allocation mechanism[5] appears more likely to give rise to unequal treatment of undertakings than the Performance Standard Rate (PSR) system.[6] Within the so-called Performance Standard Rate system there is no absolute cap identifying the total amount of allowed emissions. Instead, a relative approach is taken by establishing a performance standard, indicating the allowed amount of emissions *per unit of production,* or *per unit of fuel.* If an

---

5 There are three bases which can be used for (historical) grandfathering: input-based (used historic energy input), output-based (e.g. kilowatt-hours of electricity production) or emission-based (direct or indirect i.e. total emission from emitting facilities). Furthermore, the base period for historic data has to be determined, Harrison and Radov (2002), p. 60.

6 The System contains all elements of an Emission Trading System and thus goes beyond an allocation system. See Weishaar (2007a).

operator were to produce fewer emissions than indicated by this relative standard, it could sell these credits to other industries or reserve these credits for future use. If an operator exceeds the relative standard, it is obliged to cover the extra emissions with an emission-credit, bought from another industry, or taken from its own reserve. Since the same rules are applicable to incumbents and new entrants the only source of distortions is thought to stem from depletion of the new entrants' reserve. The same concern is present under the proposed amendment to Directive 2003/87/EC that envisages a benchmarking system with a 5% new entry reserve available to all new entrants that can benefit from free allocation.[7]

Besides barriers to entry, entrenchment of market shares is identified as facilitating collusion. Compensation for losers of market shares under the grandfathering allocation system, for example, renders winning of market shares more difficult and emphasizes the recognition of interdependence between undertakings, which in turn can alter their propensity to collude. Here too, a PSR system does not appear to give rise to such concerns, since undertakings that have historically been polluting more cannot benefit from cross-subsidising present production by selling allowances. A similar finding may be warranted under a Community-wide benchmark system that is envisaged under the proposed amendment to Directive 2003/87/EC.[8]

It should be noticed, however, that such distortions of competition that could stem from allocation mechanisms have been subject to compensatory justification balancing acts under State aid investigations where the European Commission compared environmental benefits to distortions of competition.[9] Upon finding a positive societal effect, State measures were declared to be compatible with the common market.[10] Even though it is unlikely that the ECJ would call into question the substance of the Commission's Decision, a number of interesting academic questions arise.

Legislative interventions[11] by Member States are largely contained through the application of the four freedoms, while EC Competition law is geared to the containment of competitive distortions arising in particular from undue behaviour of firms. The legislator has introduced Article 86 EC Treaty to contain State measures affecting public undertakings and the granting of special or exclusive rights to undertakings by bringing them within the field of

---

[7] See explanatory memorandum recital 18 and Article 10 (a) (2), (3) and (6) COM(2008)16 final of 23.01.2008.

[8] See explanatory memorandum recital 18 and Article 10 (a) (1) COM(2008)16 final of 23.01.2008.

[9] See the second part of this chapter.

[10] See Weishaar (2007a).

[11] As contrasted to subject matters dealt with under State Aid legislation.

application of EC law in general and Competition law in particular. The scope of this Article was and is, however, too narrow to address all possible anti-competitive measures Member States can take. The resulting legal gap in the Treaty placed some State measures inducing cartelization and abuse beyond reach of the four freedoms and Competition law alike. The European Court of Justice (ECJ) has recognized this shortcoming.

Based on Articles 3(g), 10(2), 81 or 82 EC Treaty, the Court in 1977 developed case law establishing that Member States are obliged to abstain from taking measures which could deprive Articles 81 and 82 of their effectiveness (*effet utile*). This jurisprudence was intended to contain undue State interference with the objectives of the EC Treaty. In its case law the ECJ established clear parallels to Article 86(1) EC Treaty and the Articles were applied jointly in a myriad of contexts including price regulations, social security provisions and maximum credit rates.[12]

In the following two sections the current application of these Articles is dealt with in the context of the EU ETS in order to examine if they can be employed to guide Member States towards selecting the least distortive allocation measures. First cartelization is presented and subsequently abuse.

CARTELIZATION

A State measure will only be caught by the joint application of Articles 3(g), 10(2) and 81 EC Treaty if it (i) introduces new infringements, (ii) reinforces infringements or (iii) delegates authority to private entities. Since the third point is not of relevance in the context of the EU ETS only the first two are considered. Yet it should be noted that a *conditio sine qua non* for the application of European Competition law is that trade between Member States must be affected.[13]

---

[12] Since their joint application requires infringements of Article 81 or 82 to be applicable it is not surprising that the Court did not apply it in the context of collective agreements between management and labour, in pursuit of social policy objectives such as the improvement of conditions of work and employment which do not fall within the ambit of either Article. See Joined cases C-115/97 to C-117/97 Brentjens' Handelsonderneming BV v Stichting Bedrijfspensioenfonds voor de Handel in Bouwmaterialen [1999] ECR I – 06025, para. 66; Case 219/97 Maatschappij Drijvende Bokken BV v Stichting Pensioenfonds voor de Vervoer- en Havenbedrijven [1999] ECR I-06121, para. 52.

[13] For a statement of this criterion with regard to State measures see Case 136/86, Bureau national interprofessionnel du cognac v. Yves Aubert, [1987] ECR 4789, para. 16; Case 311/85, ASBL Vereniging van Vlaamse Reisbureaus v. ASBL Sociale Dienst van de Plaatselijke en Gewestelijke Overheidsdiensten, [1987] ECR

This section examines how anticompetitive measures taken within the framework of the EU ETS are to be assessed in light of the foregoing discussion. The same structure as applied above will be used.

i) The measure requires or favours the adoption of agreements, decisions or concerted practices contrary to Article 81[14]

In light of the case law it is clear that obligations created by national measures that are self-sufficient or self-contained cannot be regarded to reinforce or favour cartelization.[15] Since neither Directive 2003/87/EC nor any of the National Allocation Plans submitted by the Member States obliges undertakings to collude, it can be concluded that national measures taken within the EU ETS framework are neither self-sufficient nor self-contained and can therefore not be excluded from joint application of Articles 3(g), 10(2), 81 EC Treaty.

While it is clear that the Court requires the violation of Article 81 EC Treaty before it will condemn any State measure, it is, however, not at all clear how strong the link the ECJ requires between a State measure and undue behaviour of undertakings has to be. Unfortunately the Court failed to clarify how it interprets the terms 'requiring' or 'favouring'. The flagrant violation the ECJ condemned in *CNSD* expressly requiring an association of undertakings to form agreements, granting it relative decision-making powers and providing legal compliance rules is clearly a strict point of reference.[16]

If the *CNSD* judgment was taken as a benchmark, barriers to entry established under allocation mechanisms would not be caught but would escape legal sanctioning. This is based on the finding that they do not contain the express obligation to collude but nevertheless increase benefits from cartelization.

ii) The measures reinforce the effects of a violation of Article 81 EC Treaty[17]

Another branch of the ECJ's approach to State measures under Article 81 EC

3801, para. 18; Case C-60/91, José António Batista Morais, [1992] ECR I 02085, para 12; Case C-35/96, Commission v. Italy (CNSD), [1998] ECR I-03851, para. 48; and Case C-35/99, Manuele Arduino, [2002] ECR I-01529, para. 33.

14 This criterion has been expressed in Case 209–213/84, Lucas Asjcs and others, Andrew Gray and others, Andrey Gray and others, Jacques Maillot and others and Léo Ludwig and others, [1986] ECR 1425, para. 72. Neergaard (1998), p. 72, maintains that it can also be found in Leclerc, Case 229/83, Association des Centres distributeurs Édouard Leclerc and other v. SARL 'Au blé vert' and others, [1985] ECR 1, para. 15.

15 Case 2/91, Wolf W. Meng, [1993] ECR I–05751, para. 15; Case 245/91, Ohra Schadeverzekeringen NV., [1993] ECR I–05851, para. 11.

16 Case C-35/96, Commission v. Italy (CNSD), [1998] ECR I-03851, para. 55.

17 As introduced in Case 209 – 213/84, Lucas Asjes and others, Andrew Gray and others, Andrey Gray and others, Jacques Maillot and others and Léo Ludwig and others, [1986] ECR 1425, para. 72.

Treaty requires the presence of a pre-existing infringement.[18] In its assessment, the Court would at first closely examine the relationship between advising experts and regulated undertakings. The Court held that representatives, in particular when appointed by public authorities upon proposal of the undertakings they are charged to regulate,[19] must be acting upon their own behalf, not be bound by their respective undertakings and take interests of other sectors as well as the public[20] at large into account.[21] The Court has been accepting requirements for impartiality contained in national legislation.[22] With regard to the EU ETS it would thus be examined whether the choice and design of the particular allocation method and the amount of allowances granted would favour the undertakings and sectors of the advising experts. If the Court were to find that advisors were neither impartial nor obliged to take the interests of other sectors and the public at large into account, it would proceed to test a second criterion.

In a second step the Court would examine whether an original contractual prohibition was transformed into legislative provisions[23] with legal remedies and effective sanctions[24] and whether it would either wholly or in part contain the terms of the pre-existing cartel agreement. Since the creation of the EU ETS introduces a new production factor to the market process, it is rather unlikely that there are any prior collusive agreements between undertakings regulating prices or sectoral output based on the overall amount of CO_2 emissions. Consequently, it is quite unlikely that the Court would be able to condemn State measures taken within the framework of the EU ETS.

---

18 Case 2/91, Wolf W. Meng, [1993] ECR I – 05751, para. 19.

19 Case 123/83, Bureau national interprofessionnel du cognac v. Guy Clair, [1985] ECR 391, para. 3, 19–20.

20 In Case C-38/97, Autotrasporti Librandi, [1998] ECR I 05955, para. 38–42, the Court held that the term public interest applied in Case C-96/94, Centro Servizi Spediporto, [1995] ECR I – 02883, para. 42, corresponds to the term general interest as applied in Case C-185/91, Bundesanstalt für Güterfernverkehr v. Gebrüder Reiff GmbH & Co. KG., [1993] ECR I – 05801, paras. 17–19 and Case C-153/93, Delta Schiffahrts- und Speditionsgesellschaft mbH, [1994] ECR I – 02517, paras. 21–22. Public interest criteria are to be determined by national legislation and their application is observed by national courts. See Case C-38/97, Autotrasporti Librandi, [1998] ECR I 05955, para. 47.

21 Case C-185/91, Bundesanstalt für Güterfernverkehr v. Gebrüder Reiff GmbH & Co. KG., [1993] ECR I-05801, paras. 17–19.

22 Case C-185/91, Bundesanstalt für Güterfernverkehr v. Gebrüder Reiff GmbH & Co. KG., [1993] ECR I-05801, para. 4.

23 Case 136/86, Bureau national interprofessionnel du cognac v. Yves Aubert, [1987] ECR 4789, para. 24.

24 Case 311/85, ASBL Vereniging van Vlaamse Reisbureaus v. ASBL Sociale Dienst van de Plaatselijke en Gewestelijke Overheidsdiensten, [1987] ECR 3801, para. 23.

Yet irrespective of the fact that CO_2 emission allowances constitute new products in most Member States, the second criterion employed by the Court is too legalistic to be of practical relevance in most cartel cases. Illegal cartel agreements are rarely drafted in contract form and, unless revealed by whistle blowers lured by effective leniency policies, not available to the Court. It is therefore unlikely that this criterion – if applied strictly by the Court – would permit the joint application of Articles 3(g), 10(2), 81 EC Treaty towards the EU Emissions Trading System.

This section has reviewed the joint application of Articles 3(g), 10(2) and Article 81 EC Treaty to State measures that induce undertakings' propensity to collude. It found little guidance as to how State measures taken within the framework of the EU ETS are to be assessed in the absence of pre-existing cartels. The Court is called upon to take due consideration of economic insights to assess the link between State measures and collusion. The benchmark employed by the Court, that violations of Article 81 can only be reinforced if they do include part of a cartel agreement, is criticized as very legalistic and as an element that will constitute a prohibitively high burden of proof in practice and appears to differ from the Court's case law under Article 81 procedures.[25] Little support has therefore been found that allocation of emission allowances will fall within the ambit of the joint application of these Articles as it stands today.

ABUSE

Whether national measures taken within the framework of the EU ETS system that impact the propensity to abuse could be declared incompatible with the common market on the basis of a joined application of Articles 3(g), 10(2) and 82 EC Treaty constitutes a far from trivial question. This is particularly so in light of the considerable gaps existing in the Courts interpretations to this date.

A national measure can only be declared incompatible with the Community law if it (1) gives rise to dominance,[26] (2) there is an abuse by one or more undertakings and (3) there is a link between the national measure and the

[25] In Case T-41/96, Bayer AG v. Commission [2000] ECR II – 3383, para. 69, the Court states that the form of an agreement is unimportant but that it represents the full expression of the parties' intention.

[26] Joined Cases C-140/94, C-141/94, C-142/94, DIP, [1995] ECR I – 03257, paras. 20–27; Case C-96/94, Centro Servizi Spediporto, [1995] ECR I – 02883, paras. 31–35, Case C–85/76, Hoffmann-La Roche, [1979], ECR I – 00461, para. 38, Case C-70/95, Sodemare SA, [1997] ECR I – 03395, paras. 44 and 47–48, Case C-38/97, Autotransporti Librandi Snc di Librandi F. & C., [1998] ECR I – 05955, para. 27.

infringement of Community law. With regard to the first two elements, it is expected that the ECJ applies the same standards as under Article 82 EC Treaty. Therefore NAPs or allocation mechanisms can only be subject to judicial scrutiny under the joint application of these Articles if undertakings are abusing market power. The condemnation of national measures will crucially depend on the link between legislation and abuse.

The link between national legislation and an infringement is, however, not addressed by existing case law. Drawing parallels between the granting of statutory monopolies and abuse under the Court's case law regarding Article 86 is not only insightful but also warranted given the object and purpose of the joint application doctrine with regard to Article 82 EC Treaty.

In proceedings jointly applying Article 82 and 86, four approaches to State measures creating legal monopolies – epitomes of dominant firms – can be recognized.[27] Both the 'absolute-' and the 'limited sovereignty approach' view monopolies generally as favourable. The first appears to suggest that monopolies are compatible with EC legislation unless the Commission can prove the contrary,[28] while the limited sovereignty approach condemns monopolies only when they cannot avoid abusing their dominance when executing their normal operations.[29] By contrast, both the 'limited-' and the 'absolute competition approach' view legal monopolies rather critically. While in the first case, the ECJ appears to view the granting of exclusive rights creating a dominant position as impermissible unless a Member State can prove its necessity,[30] the Court's interpretation under the absolute competition approach[31] goes much further. Here the Court holds that the granting of rights is not permissible if it creates a situation where an undertaking is led to or induced to infringe Article 86. This comes very close to stating that State measures creating dominance are illegal per se.[32]

In particular, the developments under the absolute competition approach are noteworthy with respect to the assessment of the link between NAPs and allocation schemes and abuse. Here the Court recognizes the interaction between legislation placing undertakings artificially into a dominant position

27 For the following see in particular Edward and Hoskins (1995).

28 Case C-202/88, French Republic v. Commission, [1991] ECR I – 01223.

29 Case C-323/93, Crespelle, [1994] ECR I-05077, para. 18; Case C-41/90, Höfner v. Macrotron, [1991] ECR I-01979, para. 29.

30 Case C-155/73, Giuseppe Sacchi, [1974] ECR 00409; Case C-320/91, Corbeau, [1993] ECR I – 02533.

31 See Case C-179/90, Merci convenzionali porto di Genova Spa v. Siderurgica Gabrielli SpA, [1991] ECR I – 05889, para. 19; Case C-18/93, Corsica Ferries Italia Srl. v. Corpo dei Piloti del Porto di Genova, [1994] ECR I – 01783; Case C-260/89, Elliniki Radiophonia Tiléorassi AE, [1991] ECR I – 02925, para. 38.

32 See Craig and De Búrca (2003), p. 1129.

and the creation of incentives for undertakings to engage in abusive behaviour. This is also the case with regard to State measures taken within the framework of the European Emissions Trading System. It appears that such State measures could fall within the ambit of Competition law if the Court were to follow the footsteps of its absolute competition approach and if it were extending it from the realm of Article 86 to Article 82 EC Treaty. In this way, undue distortions of competition arising from Member State interference with the market could be addressed.

From a legal point of view an important observation has to be made. The condemnation of dominance applied by the Court contrasts sharply with the traditional approach it takes under Article 82 EC Treaty. Here it holds that dominance as such is not against Community law. Despite the seeming paradox in legal terms, it reflects good economic judgement and deserves praiseworthy recognition. In economic terms, dominance can be a natural result of competition on the merits – as recognized by the ECJ under Article 82 EC Treaty. Yet State interventions creating dominance in markets where it would not have been able to evolve nor survive is clearly distorting free competition and is consequently to be rejected on economic grounds – this is recognized by the Court through application of the absolute competition approach under Article 86 EC Treaty. Thus from a conceptual point of view the proposed legal paradox is resolved by reference to State intervention.

This section has shown that even in the absence of case law guiding the joint application of Articles 3(g), 10(2) and 82, some important findings can be derived through the drawing of legal parallels. If the Court were to extend its absolute competition approach from Article 86 to Article 82, it would theoretically be able to close the existing legal gap and bring State measures distorting competition – such as barriers to entry that are present under the EU ETS – into the realm of the core provisions of Competition law. Thereby it would be enabled to safeguard competition on the merits and mitigate social welfare reducing distortions stemming from abusive behaviour.

Whether the ECJ is indeed willing to extend the scope of the joint application of these Articles will depend on the impairment of the *effet utile* of Community law. This effectiveness is expected to depend on the perceived gravity of the anticompetitive effects and the possibility that the infringements can satisfactorily be addressed through application of the existing provisions directed against undertakings. In the absence of empirical studies for abuse being directly attributable to the change in the propensity to abuse originating in State measures taken within the EU ETS framework, it is to be expected that the Court will not broaden the joint application of these Articles. It is therefore not believed that the joint application of Articles 3(g), 10(2) and 81 or 82 will be effective in guiding Member States towards selecting the least distortive allocation measure.

STATE AID

This section examines to what extent the State aid provision of EC Competition law guides Member States to select the least distortive allocation format. As in the previous section, the two free allocation mechanisms examined are grandfathering and the Performance Standard Rate System. In order to address this question, first it has to be established if both allocation mechanisms fall within the ambit of the State aid rules and which one of them gives rise to less anticompetitive concern. Given that the current legislative proposal for amending the EU ETS is not very specific, it is assumed here that the findings will be similar to a PSR system that also uses a benchmarking approach.[33]

According to Article 87(1) EC Treaty State aid is incompatible with the common market – unless exempted by derogations – if the criteria below cited are fulfilled. Even though criteria v and vi are assessed jointly from a legal point of view, for didactical reasons they are treated separately here.

(i) Transfer of a benefit or an advantage (notion of aid);
(ii) Aid favouring a certain undertaking over others (selectivity principle);
(iii) Granted by the State or through State resources;
(iv) It should be an undertaking or … production;
(v) Distorts or threatens to distort competition;
(vi) Community dimension: aid capable of affecting trade between Member States.

Each will be treated in turn.[34]

i) Transfer of a benefit or an advantage (notion of aid)
In order to avoid distortions of competition,[35] the Commission has the power to interpret the concept of aid.[36] The ECJ does not distinguish between measures of State intervention by reference to their causes or aims but determines aid solely based on their effects.[37] Transferral of an advantage is the

---

33 See explanatory memorandum recital 18 and Article 10 (a) COM(2008)16 final of 23.01.2008. Though it should be noticed that the proposed Directive envisages a uniform benchmarks at EU level and may therefore affect trade between Member States to a lesser degree.

34 For a more extensive examination of State aid and the EU ETS see Weishaar (2007b).

35 Evans (1997), p. 27.

36 Case T-459/93 Siemens SA v. Commission [1995] ECR II-1675, para. 52.

37 Case 173/73 Italy v. Commission [1974] ECR 709, para. 13; Case C-241/94 France v. Commission [1996] ECR I-4551, paras. 19–20; joined Cases T-228/99 and T-233/99, Westdeutsche Landesbank Girozentrale and Land Nordrehein-Westfalen v. Commission, [2003] ECR II-435, para. 180.

objective test employed to determine its existence.[38] Given its broad scope, direct benefits and cost reductions constitute aid.[39]

Both examined allocation formats satisfy this first criterion. Grandfathering constitutes a windfall profit in the form of a lump sum transfer of emission allowances with a market value. Under a PSR system an undertaking's CO_2 savings are accredited in the form of intangible assets with a market value. Consequently, undertakings that are more CO_2 efficient than the benchmark fixed by the government receive benefits that are satisfying this State aid requirement.

ii) Aid favouring a certain undertaking over others (selectivity principle)

This criterion differentiates between aid favouring certain undertakings or the production of goods[40] and general State interventions regarding fiscal rules, social security measures, etc. The ECJ looks at the effects in order to distinguish between a selective measure directed to aid certain undertakings and general economic measures.[41] The selectivity principle requires that emission allowances are allocated in an objective, non-discriminatory and non-discretionary way.[42]

EU ETS grandfathering is potentially selective and liable to constitute State aid. Selectivity stems from a Member State's discretion to allocate allowances to particular entities. Furthermore, abatement cost differentials and the setting of historical standards generate discriminatory effects across undertakings and sectors. The underlying differentiation between trading and non-trading sectors that was crucial in the Danish system[43] could be extended and brought into the realm of the discussion of covered and uncovered sectors and incumbents and new entrants. How the PSR system is assessed with regard to the

38 Case T-64/94 Ladbroke Racing Ltd. v. Commission [1998] ECR II-1, para. 52; Case T-46/97 SIC v. Commission [2000] ECR II-2125, para. 83; joined Cases T-228/99 and T-233/99, Westdeutsche Landesbank Girozentrale and Land Nordrhein Westfalen v. Commission, [2003] ECR II-435, para. 180.

39 Case 30/59 De Gezamenlijke Steenkolenmijnen Limburg v. High Authority [1961] ECR I, para. 19.

40 Case T-55/99 CETM v. Commission [2000] ECR II-3207, para. 39; joined Cases T-92/00 and T-103/00 Territorio Histórico de Álava v. Commission [2002] ECR II 1385, para. 48, Case C-143/99 Adria-Wien Pipeline GmbH and Wietersdorfer & Peggauer Zementwerke GmbH v. Finanzlandesdirektion fuer Kaernten [2001] ECR I-8365, para. 41.

41 Case 173/73 Italy v. Commission [1974] ECR 709, para. 13; Case C-241/94 France v. Commission [1996] ECR I-4551, paras. 19–20; Case 310/85 Deufil GmbH & Co. KG v. Commission [1987] ECR 901, para. 8.

42 For a more extensive discussion on this point see De Sepibus (2007 pp. 13 ff.).

43 European Commission, (2000), Statsstøttesag Nr. N 653/1999 – CO_2-kvoter, SG(2000) D/, 12.04.2000, pp. 5–6.

selectivity principle is not a trivial question. The answer depends upon the objectiveness of the employed criteria and the scope of the governmental benchmarks. The higher and broader they are, the greater the likelihood that they are judged as being of a general economic nature. Yet, to the extent that covered sectors have heterogeneous abatement cost structures and distinct demand curves, the more likely that a measure, though not directly discriminatory, has such an effect. If, for example, large undertakings were favoured by a PSR, the measure would be selective[44] unless justified by the nature or generality of the scheme.[45] Therefore the result is ambiguous. In a recent case the CFI has ruled that the Dutch NOx system, the epitome of a PSR system, did not constitute state aid on the basis of the selectivity criterion.[46]

iii) Granted by the State or through State resources

Aid granted through an institution associated with the State[47] or directly or indirectly affecting public accounts falls within the meaning of Article 87(1)[48] if the aid is a result of an unilateral and autonomous decision[49] of the State and not motivated by its obligations under the EC Treaty. Yet even if there is an advantage granted through State action that goes beyond its obligations aris-

---

44 Case C-143/99 Adria-Wien Pipeline GmbH and Wietersdorfer & Peggauer Zementwerke GmbH v. Finanzlandesdirektion fuer Kaernten [2001] ECR I-8365, para. 41.

45 Case C-143/99 Adria-Wien Pipeline GmbH and Wietersdorfer & Peggauer Zementwerke GmbH v. Finanzlandesdirektion fuer Kaernten [2001] ECR I-8365, para. 42.

46 Case T-233/04 The Netherlands v. Commission [2004] nyr. para. 96 following.

47 Case 78/76 Steinike v. Bundesamt für Ernährung und Forstwirtschaft [1977] ECR 595, para. 21; joint Cases 67, 68 and 70/85, Van der Kooy BV v. Commission [1988] ECR 219, para. 35; Case C-303/88 Italian Republic v. Commission [1991] ECR I-1433, para. 11; Case C-305/89, Italy v. Commission [1991] ECR I-1603, para. 13; Case C-482/99 France v. Commission [2002] ECR I-4397, para. 48; Case C-200/97 Ecotrade Srl v. AFS [1998] ECR I-7907, para. 35.

48 Case 82/77, Opebaar Ministerie v. Van Tiggele [1978] ECR 25, paras. 23–25; joined Cases C-72/91 and C-73/91 Sloman Neptun v. Bodo Ziesemer [1993] ECR I-887, paras. 19 and 21; Case C-189/91 Kirsammer-Hack [1993] ECR I-6185, para. 16; Cases 213-215/81 Norddeutsches Vieh- und Fleischkontor v. BALM [1982] ECR 3583, paras. 22 and 23; joined Cases C-52/97, C-53/97 and C-57/94 Viscido, Scandella, Terragnolo and Others v. Ente Poste Italiane [1998] ECR I-2629, para. 13, C-295/97, Industrie Aeronautiche e Meccaniche Rinaldo Piaggio SpA v. International Factors Italia SpA [1999] ECR I-3735, para. 35, Cases T-204/97 and T-270/97 EPAC v. Commission [2000] ECR II-2267, para. 80; Case C-200/97 Ecotrade Srl v. AFS [1998] ECR I-7907, para. 43; joint Cases 67, 68 and 70/85, Van der Kooy BV v. Commission [1988] ECR 219, para. 28. See also Quigley, C. and Collins, A. (2003), p. 26.

49 Case T-351/02 Deutsche Bahn v. Commission [2006] nyr., para. 100.

ing from the Treaty, as long as no extra financial burden is placed upon public authorities, there is no aid[50] within the meaning of the Article.

Due to its broad concept any grandfathering institution meets the 'State' element. State resources are affected if more than the obliged 90% allowances for the three-year period are allocated free of charge.[51] Since in practice it is not discernable whether allowances derive from the permissible amount, all proposals containing over-allocations affect State resources.

Furthermore, in the present trading period Member States are subject to Kyoto obligations to meet particular national targets. Some Member States are also bound to emission targets within the framework of the EU Burden Sharing Agreement. They do, however, enjoy discretion to allocate relatively more or fewer allowances to the trading sector as a whole. Favouring trading sectors at the expense of, for instance, transportation influences the available allowances under the Directive and impacts State resources. It can be concluded that grandfathering mechanisms are liable to constitute State aid.

Regarding PSR systems, the Commission has taken the view that the Dutch NOx system satisfies the criterion of aid granted by the State or through State resources and that it consequently constitutes State aid. This assessment has been upheld by the CFI.[52]

The Dutch government argues that allowances under a PSR system are proof of compliance with administrative efficiency benchmarks[53] and are thus not being distributed.[54] In *Preussen Elektra* support was granted through legislation and in the absence of direct State involvement, the Court was not concerned with the position of the State but with the financial burden placed upon it,[55] thus emphasising the question of a financial burden.[56] In its recent

---

50 Case C-379/98 Preussen Elektra v. Schleswag AG [2001] ECR I-2099, para. 58, 63–65, Case T-613/97 Ufex v. Commission [2000] ECR II-4055, para. 108–10.

51 While in the first trading period 2005–2007 the amount was 95%, it has been reduced to 90% for the second period beginning 1 January 2008, see Article 10, Directive 2003/87/EC.

52 Case T-233/04 The Netherlands v. Commission [2004] nyr. para. 78.

53 European Commission, (2003), Steunmaatregelen van de Staten N35/2003 – Nederland Systeem van verhandelbare emissierechten voor NOx, C(2003) 1761 fin, 24.06.2003, under 3.2.

54 The Dutch government thus follows a similar line of argumentation as used in Belgium Green Certificates. European Commission, (2001), Steunmaatregel nr. N 550/2000 België Groenestroomcertificaten, SG(2001) D/290545, 25.07.2001, pp. 5–6.

55 C-379/98 Preussen Elektra v. Schleswag AG [2001] ECR I-2099, paras. 58–61.

56 See Case 53/00 Ferring SA v. Agence centrale des organismes de sécurité sociale [2001] ECR I-9067, para. 27, for direct sales taxes falling within the ambit of 'State aid'.

ruling the Court stressed that the tradability of allowances as such constituted an advantage.[57]

Regarding transfers of the State, the Commission and the CFI contended in *Dutch NOx*[58] and *Netherlands v Commission* that allowances constitute intangible assets with market value,[59] and that the government was deliberately forgoing revenue while having full discretion to sell allowances to operators[60] and that this constitutes State aid.[61] This reasoning is criticized because the incorporation of payment schemes like auctioning into a PSR system creates unnecessary uncertainty for investments in environmentally friendly technology, and thus runs counter to the system's environmental objective. This may be viewed to be questionable under the protection of confidence considerations because the PSR system is based on the accreditation of additional emission reductions. The Commission's emphasis placed on the 'willingness to pay' of undertakings is particularly questionable in light of *Belgium Green Certificates*[62] where a similar willingness has been present and the measure was not held to constitute State aid.

While the EC Treaty does not compel Member States to levy taxes or to generate profits, the revenue argument extends the scope of Article 87(1) EC Treaty. Yet in *Preussen Elektra* the ECJ did not condemn the employment of statutory provisions to fix minimum prices above real market prices and to oblige private parties to bear the costs.[63] Despite the fact that tax revenues would be lower,[64] it distinguished the cases by reference to emission allowances. This decision therefore contrasts the ECJ's rejection of previous

---

[57] Case T-233/04 The Netherlands v. Commission [2004] nyr. para. 74.

[58] European Commission, (2003), Steunmaatregelen van de Staten N35/2003 – Nederland Systeem van verhandelbare emissierechten voor NOx, C(2003) 1761 fin, 24.06.2003.

[59] European Commission, (2003), Steunmaatregelen van de Staten N35/2003 – Nederland Systeem van verhandelbare emissierechten voor NOx, C(2003) 1761 fin, 24.06.2003, under 3.1, Case T-233/04 The Netherlands v. Commission [2004] nyr. para. 75.

[60] European Commission, (2003), Steunmaatregelen van de Staten N35/2003 – Nederland Systeem van verhandelbare emissierechten voor NOx, C(2003) 1761 fin, 24.06.2003, under 3.2, Case T-233/04 The Netherlands v. Commission [2004] nyr. para. 75.

[61] European Commission, (2001), Steunmaatregel nr. N 550/2000 België Groenestroomcertificaten, SG(2001) D/290545, 25.07.2001, pp. 5–7. Case T-233/04 The Netherlands v. Commission [2004] nyr. para. 78.

[62] European Commission, (2001), Steunmaatregel nr. N 550/2000 België Groenestroomcertificaten, SG(2001) D/290545, 25.07.2001, pp. 5–6.

[63] C-379/98 Preussen Elektra v. Schleswag AG [2001] ECR I-2099, para. 66.

[64] C-379/98 Preussen Elektra v. Schleswag AG [2001] ECR I-2099, para. 62.

Commission attempts to extend the field of application of Article 87(1).[65] If the argument was motivated by the perception that the measures taken by the Dutch government constitute a *fràude a la loi*, it has appropriate tools at its disposal and should employ these.

While the above is relevant for a national PSR system, its application in a multilateral context is complicated by the Burden Sharing Agreement which is binding upon the former EU 15 Member States. PSR systems could be set at European levels according to objective criteria and with the aim to minimize competitive distortions for particular product groups. To the extent that the national allocation under the PSR system would not be identical to the number of allowances granted under the Burden Sharing Agreement, transfers between national governments would be required. Such transfers would then imply direct budgetary consequences.

iv) It should be an undertaking or . . . production

It is believed that the notion of undertaking under Articles 87 and 88 EC Treaty is identical to the one applied under Articles 81 and 82 EC Treaty. In *Höfner*[66] the ECJ held that any entity engaged in an economic activity regardless of its legal status and the way it was financed amounted to an undertaking.

Directive 2003/87/EC restricts the allocation of emission allowances to operators[67] defined as persons operating or controlling installations.[68] Any recipient of allowances fulfils this criterion.

v) Distorts or threatens to distort competition

Under Article 87 EC Treaty aid must distort or threaten to distort competition. In order to determine distortions of competition, the ECJ assesses the direct and immediate effects of aid on the competitive position of the recipient. The market position prior to and after the granting of aid is compared and distortion established if the position of the undertaking is more favourable *ex post*.[69] Any aid granted to an undertaking is held to give an advantage in relation to actual or potential competitors and as affecting competition.[70] Even though

---

[65] In C-290/83 Commission v. France [1985] ECR 439, para. 18 the ECJ ruled that the scope of Articles 87 and 88 EC Treaty did not leave sufficient room for a competing concept of 'measures having equivalent effect' to State aid. See also Quigley, C. and Collins (2003, p. 17).

[66] Case C-41/90 Höfner and Elser v. Macroton GmbH [1991] ECR I-1979, para. 21.

[67] See Article 11(1) Directive 2003/87/EC.

[68] See Article 3(f) Directive 2003/87/EC.

[69] Case 173/73 Italy v. Commission [1974] ECR 709, para. 17.

[70] Evans (1997, p. 77).

actual proof of an anticompetitive distortion is not required[71] and its potential presence is sufficient, the Commission must at least set out those circumstances in the statement of reasons for its decision.[72] Thus, unlike under Articles 81 and 82 EC Treaty, here the Commission is under no obligation to prove distortions of competition on relevant product markets and the Court appears to be reluctant to do so.[73]

The ECJ does not accept arguments that aid lowered the relatively higher costs of a sector or that other States made similar payments.[74] The Court looks at the anticompetitive effects without taking into account any grounds for motivation of the aid.

The Commission has promulgated Regulation (EC) No 1998/2006 on threshold levels below which all cumulated aid not exceeding 200 000 EUR, granted within a period of three years, is adjudged not to be capable of distorting competition – the so-called *de minimis* rule.[75] Therefore only aid exceeding this threshold can fall within the ambit of Article 87(1) EC Treaty.

Legal considerations aside, determining the monetary value of allowances granted under a National Allocation Plan is not a trivial issue. 'Aid' should include all economically quantifiable advantages accruing to an entity. This is complicated by the volatility of market prices of allowances and because the marginal benefit of an allowance to an undertaking depends on its ability to pass on increased production costs to consumers. Sectors unable to do this will experience a larger cut in profits. Both from a firm as well as a social welfare point of view, particular regard of the undertakings' marginal abatement cost structures, including expected technological developments and growth forecasts, have to be taken into account. Thus in summary the actual effects of aid granted to undertakings may vary and are not easily determined.

---

71 Case T-288/97 Regione Friuli Venezia Giulia v. Commission [2001] ECR II-1169, paras. 49–50; Case T-35/99 Keller SpA v. Commission [2002] ECR II-261, para. 85; Case T-214/95, Vlaamse Gewest v. Commission [1998] ECR II-717, para. 67.

72 Joined Cases 296/82 and 318/82, Netherlands and Leeuwarder Papierwarenfabriek v. Commission [1985] ECR 809, para. 24, joined Cases C-329/93, C-62/95 and C-63/95 Germany, Hanseatische Industrie-Beteiligungen GmbH and Bremer Vulkan Verbund AG v. Commission [1996] ECR I-5151, para. 52.

73 Case 730/79 Philip Morris v. Commission [1980] ECR 2671, paras. 9–13; Case 53/00 Ferring SA v. Agence centrale des organismes de sécurité sociale [2001] ECR I-9067, para. 21.

74 Case 78/76 Steinike v. Bundesamt für Ernährung und Forstwirtschaft [1977] ECR 595, para. 24; Case T-214/95, Vlaamse Gewest v. Commission [1998] ECR II-717, para. 54.

75 Article 2(2) of Commission Regulation (EC) No 1998/2006 of 15 December 2006.

Based on the particular constellations between undertakings, one can distinguish four kinds of competitive relationships that can be distorted. These are firstly, relations between incumbents and newly entering firms; secondly, between trading and non-trading sectors; thirdly, relationships between competing firms of the same Member State, in particular regarding covered and non-covered parts of a sector as well as undertakings within a covered sector; and fourthly between trading sectors. Each will be discussed in turn. If any of these relationships demonstrates actual or potential competitive distortions, the measure at hand is liable to fall within the ambit of Article 87(1) EC Treaty, as long as the granted aid exceeds the *de minimis* threshold.

1) Between incumbents and newly entering firms

Competitive distortions arise from the NAPs if they were not to award potential competitors equal treatment. Granting new entrants relatively fewer allowances sets them at a comparative disadvantage vis-à-vis incumbent undertakings by increasing barriers to entry. This can lead to higher consumer prices and to potential x-inefficiency.[76] Thus, competition will be distorted if new entrants were not grandfathered adequate amounts of emission allowances. A full industrial economic analysis is required to determine the substance of such a claim. Yet in practice the Commission is not obliged to prove actual competitive distortions; their mere potential detriment is sufficient.[77]

The Directive 2003/87/EC affords Member States discretion by obliging them to take the need to provide access to allowances for new entrants into account.[78] Equally authentic language versions differ in clarity. While the English version speaks of the obligation to take into account the need of new entrants, the Dutch version of the text speaks of the necessity to keep emission allowances available for new entrants.

This ambiguity translates into disadvantages when new entrants' reserves are depleted and new entrants are not entitled to allowances as in most Member States.[79] Interestingly enough, while the Commission acknowledges that it is crucial that new entrants have access to allowances, it at the same time states that new entrants' interests are sufficiently safeguarded by afford-

---

[76] Frank (1997, p. 412), defines x-inefficiency as a condition in which a firm fails to obtain maximum output from a given combination of inputs.

[77] See for example Case 730/79 Philip Morris v. Commission [1980] ECR 2671, para. 11.

[78] See Article 11(3) and Annex III criteria 6 of Directive 2003/87/EC.

[79] Notable exceptions include Germany (see German NAP, p. 37) and Poland (see Polish NAP, p. 39).

ing them the possibility of buying allowances,[80] even though resulting differences in production costs set new entrants at a comparative disadvantage.

Besides resource depletion and direct expenses, new entrants can suffer from unequal treatment. Under the Dutch NAP, incumbent top performers can receive up to 10% more allowances.[81] Closure combined with transfer rules can constitute barriers to entry. Rules allowing incumbents to retain allowances for less CO_2-efficient plants while shifting production to more efficient installations unilaterally benefits incumbents[82] and may even further set them at a comparative advantage if new entry is made in another EU ETS country.[83] Thus new entrants are not only set at a disadvantage under grandfathering schemes but equally so under the existing NAPs.

By way of construction the PSR system is not prone to such fallacies because all undertakings abide identical benchmarks. Yet also here the obligation to comply with costly environmental standards could deter entrance. To the extent that the new entrant is not credit rationed,[84] investments are not sunk,[85] or the financial burden of lending money is positive, new entrants will require a higher level of profitability to find it attractive to enter a market. To the extent that such effects are not compensated by benefits from new and more efficient technology, barriers to entry could be created.

2) Between trading sectors and non-trading sectors

Distortions of competition arise if competing sectors do not fall under the trading system. Under both allocation formats considered increases in production costs in the sector falling under the EU ETS lead to distortions of relative prices and impact the structure of the economy. To the extent that such price increases reflect the internalization of negative externalities, changes of relative prices can be considered to lead to increases in social welfare. Otherwise society may be worse off.

3) Between competing firms of the same Member State

Distortions between competitors in the same market are related to unequal treatment of undertakings, coverage and entrenched market shares.

---

80 COM(2003)830 final, p. 12.

81 Dutch NAP, p. 27.

82 See Dutch NAP, p. 41.

83 See Dutch NAP p. 41. For an effective prevention of this see German NAP, section C2 and C3.3.

84 This implies that a potential entrant is unable to attain sufficient funds to make the necessary investments to enter the market.

85 Sunk costs refer to irrecoverable costs once invested. They do not have a bearing on a firm's decision to exit a market but constitute a decisive factor for market entry.

Under grandfathering, distortions stem from temporal production downturns that affect the data used to grant emission allowances, different growth rates, and prior investments in abatement technology.[86] Punishment of early movers creates distributive effects impacting undertakings' financial position. Grandfathering is thus directly liable to distort competition on the merits. While the PSR system does not give rise to unequal treatment of covered undertakings, distortions are generated if marginal abatement costs and investment capabilities of firms differ.[87] Hence PSR systems are not liable to distort competition within a sector provided that benchmarks reflect the equilibrium market price.

Grandfathering allowances mitigate the direct financial cost burden and alleviate the competitive pressure experienced vis-à-vis non-coverage undertakings. Since it does not take into account the changing of firms' market shares but only their historic emissions, it has an inherently static focus. Under PSR, differences in coverage can give rise to differential cost burdens and to market distortions. The system must be constructed properly so as to mitigate distortions.

Because producers can cross-subsidize under grandfathering and part of the comparative production cost advantage of more competitive producers is absorbed by the necessity to buy additional emission allowances, gaining market share is difficult.[88] Such entrenchment of market shares distorts competition, creates real welfare losses and contravenes the polluter pays principle. Under a PSR system there is no compensation for losers of market shares, and winners are not burdened with the obligation to buy additional allowances.

4) Between trading sectors

Unlike grandfathering, the PSR system does not excessively burden sectors depicting different growth or technical innovation potentials. Undue burdens are, however, created anew every trading period if production benchmarks do not reflect marginal sector abatement costs. This distorts the market equilibrium

---

86 One example that underlines the self-defeating rationale behind the setting of historic standards are Carbon Capture and Storage projects. Electricity producers that could capture and store part of their CO2 emission have an incentive to first pollute in order to be eligible for the grandfathering of emission allowances before they can actually benefit from their investments in CO2 abatement. For a description of such a project see the Vattenfall's newsletter on the CO2 free power plant project, No. 3, November 2005.

87 From an economic efficiency point of view losses of inefficient operators constitute pecuniary effects and are the result of a competitive selection process which only allows the fittest market participants to stay on the market.

88 Here a market with a stable market size is assumed.

and leads to inefficient resource allocation. The existence of State aid is thus dependent on the particular benchmarks set by the government and cannot be answered at any level of generality.

If aid creates anticompetitive effects in any of the above situations it will be considered *de minimis* if it does not exceed a total of 200 000 Euros in the last three consecutive years.[89] Any aid below this threshold is presumed not capable of distorting competition adjudged to be compatible with the common market.

vi) Community dimension: aid capable of affecting trade between Member States

Member State trade is affected if the position of an undertaking vis-à-vis undertakings competing in intra-community trade is strengthened.[90] Even though limited aid or aid to small recipients also is distortive,[91] the *de minimis* regulation[92] applies. Here too, the potential detriment is sufficient.[93] Though the Commission must at least set out those circumstances in the statement of reasons for its decision.[94]

With regard to national Emissions Trading Systems, the same elements cited under the fifth criterion are to be mentioned. They are not repeated here.

The selection of different reference periods under grandfathering can give rise to competitive distortions leading to differential treatment of comparable undertakings. That historical standards differ can easily be seen by comparing

---

89 Commission Regulation (EC) No 1998/2006 of 15 December 2006.

90 Case 730/79 Philip Morris v. Commission [1980] ECR 2671, para. 11; Case T-214/95, Vlaamse Gewest v. Commission [1998] ECR II-717, para. 50; Case T-288/97 Regione Friuli Venezia Giulia v. Commission [2001] ECR II-1169, para. 41; Case T-152/99 Hijos de Andrés Molina, SA v. Commission [2002] ECR II-3049, para. 220; Cases T-298/97, T-312/97, T-313/97, T-315/97, T-600 to 607/97, T-1/98, T-3/98 to T-6/98, T-23/98, Alzetta Mauro and Others v. Commission [2000] ECR II-2319, para. 81. In joined Cases C-278/92, C-279/92 and C-280/92 Spain v. Commission [1994] ECR I-4103, para. 40, the Court clarified that the beneficiary undertaking itself does not need to engage in exports. See also Case 102/87 France v. Commission [1988] ECR-4067, para. 19; Case C-75/97, Belgium v. Commission [1999] ECR I-3671, para. 47.

91 Case C-142/87 Belgium v. Commission [1990] ECR I-959, para. 43.

92 Commission Regulation (EC) No 1998/2006 of 15 December 2006.

93 Cases T-298/97, T-312/97, T-313/97, T-315/97, T-600 to 607/97, T-1/98, T-3/98 to T-6/98, T-23/98, Alzetta Mauro and Others v. Commission [2000] ECR II-2319, para. 78.

94 Joined Cases 296/82 and 318/82, Netherlands and Leeuwarder Papierwarenfabriek v. Commission [1985] ECR 809, para. 24, joined Cases C-329/93, C-62/95 and C-63/95 Germany, Hanseatische Industrie-Beteiligungen GmbH and Bremer Vulkan Verbund AG v. Commission [1996] ECR I-5151, para. 52.

NAPs. Differences in Member States' closure and transfer rules hinder the free movement of production capacity and competition. 'Barriers to entry' created by NAPs can lead to the reduction of competition and real welfare losses in the form of x-inefficiency and cartelization.[95]

It is also obvious that a PSR system applied only in one Member State cannot ensure equal treatment of similar enterprises in a multilateral environment; the PSR system is liable to constitute State aid with regard to this criterion. If a PSR system were to be introduced on a European level, it would not be liable to generate intra-community competitive distortions stemming from the selection of historical standards nor would it be subject to diverging closure and transfer rules.[96] By the same token the proposed Community-wide benchmarks may not be liable to constitute State aid.[97]

SECTION SUMMARY

The examination of the six State aid criteria has shown that grandfathering systems can be liable to constitute State aid within the meaning of Article 87(1) EC Treaty. Whether this is indeed the case depends on the particular allocation made to each installation under an NAP and the characteristics of the particular market the entity is operating on. It cannot be answered on any level of generality. PSR systems should not constitute State aid within the meaning of Article 87(1) EC Treaty. In principle, State aid rules should work towards the selection of the least distortive allocation system. The ECJ reached the conclusion that a particular PSR system would amount to State aid if it did not pass the selectivity test. Aid could, however, still be compatible with the common market if one of the derogations discussed below were applicable. The following section therefore considers whether State aid provisions guide Member States towards the selection of the least distortive allocation format even in those cases where it selects among mechanisms that fall within the ambit of the Article.

95 See Weishaar (2006).

96 It should be noticed that a grandfathering system with equal rules applying in all Member States would also give rise to less concern. It would, however, still lead to potential distortions regarding the choice of historical standards and create distortions between market share winners and losers. A PSR system could give rise to distortions of the general market equilibrium if the benchmarks were not set on appropriate levels.

97 See Article 10 (a) COM(2008)16 final of 23.01.2008.

STATE AID DEROGATIONS

The Commission's wide discretion to allow aid under Article 87(3)[98] EC Treaty is only subject to the marginal control of the ECJ which ruled in *Italy v Commission* that the Commission had to take all relevant factors into account.[99] The new Environmental guidelines[100] that have been adopted on 23 January 2008 seek to correct market failures, promote sustainable development and raise levels of environmental protection.[101] Besides command and control approaches that Member State governments can follow to enhance environmental protection, they can share the financial burden of private enterprises and thereby set positive incentives for investment in more environmentally friendly production. As was the case for its predecessor, it is the object and purpose of these guidelines to ensure that there will not be badly targeted or excessive State aid that distorts competition while failing to meet its environmental objectives.

In order to assess the compatibility of an aid measure with the common market the Commission balances the positive impact of the aid on the common interest against any negative side effects such as distortions of trade and competition.[102] The Commission employs a three ties test, the so called 'balancing test'. It first examines if the objective belongs to a well-defined common interest and then examines whether the policy instrument is appropriate to address the market failure. In order to satisfy this criterion whether and under what conditions State aid may be regarded as *necessary* to ensure environmental protection, and whether the environmental gain is proportional to the amount of aid granted, are examined. Whether the environmental aid causes *disproportionate* effects on competition and economic growth or whether the overall balance of the aid is positive is evaluated in the last step of the test.

The previous Commission guidelines on State aid for environmental

---

98 Case 730/79 Philip Morris v. Commission [1980] ECR 2671, para. 17; joined Cases 62/87 and 72/87 Exécutif Régional Wallon v. Commission [1988] ECR 1573, para. 21; Case T-152/99 Hijos de Andrés Molina, SA v. Commission [2002] ECR II-3049, para. 48; Case C-142/87 Belgium v. Commission [1990] ECR I-959, para. 56; Case C-39/94 SFEI [1996] ECR I-3547, para. 36; Case 78/76 Steinike v. Bundesamt für Ernährung und Forstwirtschaft [1977] ECR 595, para. 8; Case C-156/98 Germany v. Commission [2000] ECR I-6857, para. 67; Case C-303/88 Italian Republic v. Commission [1991] ECR I-1433, para. 34. See also Woerdman, E. (2004) pp. 175ff.

99 Case C-261/89 Italy v. Commission [1991] ECR I-1437, para. 20.

100 OJ C 37, 03.02.2001, pp. 3–15.

101 European Commission (2008) Environmental aid guidelines, paras. 5 and 6.

102 European Commission (2008) Environmental aid guidelines, para. 16.

protection[103] were silent about tradable permit schemes and offered the Commission the choice between Articles 87(3)(b) and 87(3)(c) for the examination of the NAPs in the first trading period. The new guidelines offer more guidance and legal certainty. Tradable permit schemes fall under Article 87(3)(c) EC Treaty.[104]

The balancing test that is executed to examine whether a measure can benefit from the derogation under Article 87(3)(c)[105] is supplemented by further specifications.[106] The Commission emphasizes[107] the environmental objective of the aid and underlines the importance of transparency and objectivity of the allocation methodology as well as its firm grounding on reliable data. Allocation methodologies may not favour certain undertakings or sectors and may not treat new entrants more favourably or create undue barriers to entry.

The methodology that has been applied for assessing State aid involvement in the NAPs under the EU ETS for the second trading period is described in paragraph 140 of the Commission's guidelines. For the trading period post 2012 it is noticeable that the Commission emphasizes economic analysis to ensure the prevention of passing on cost increases from tradable permit schemes to consumers and seeks to prevent the generation of windfall profits.[108]

In order to examine whether Competition law is able to lead to the selection of the least distortive allocation mechanism, proportionality is of particular importance. Here, whether the applied measure is proportional or excessive in relation to the objective sought is examined. If one of two comparable allocation mechanisms was less distortive to competition, the proportionality principle would in general work towards the selection of the least distortive system. Whether it indeed is effective does essentially depend on the degree of comparability between the allocation formats, and the arguments underlying the Member State's choice. If they are very political in nature, the Court may refrain from criticizing the government's selection.

It therefore follows that the degree of comparability of both PSR and grandfathering has to be examined. The systems are comparable to the extent that neither burdens undertakings with direct financial expenditures and both fall within the category of free allocation mechanisms. They do, however, differ in

103 OJ C 37, 3.2.2001, p.3 have been repealed, see European Commission (2008) Environmental aid guidelines, para 12.

104 European Commission (2008) Environmental aid guidelines, para. 139.

105 European Commission (2008) Environmental aid guidelines, paras. 71–72.

106 European Commission (2008) Environmental aid guidelines, paras. 71–72, 139–141.

107 European Commission (2008) Environmental aid guidelines, para. 141.

108 European Commission (2008) Environmental aid guidelines, para. 141.

their design[109] as in the distortions of competition they create, in particular regarding new entrants, unequal treatment, entrenchment of market shares, closure and transfer rules.[110] While the former does not appear to be decisive from a proportionality point of view, the latter is. Grandfathering allows allocations which are tailor-made to the specific requirements of firms and thus entails different effects on undertakings that impact the level playing field, while the PSR system, in contrast, pursues a sectoral approach that may include several benchmarks for specific product groups. If the Court were to indeed differentiate between both systems, EC Competition law[111] would not encourage the selection of the less distortive allocation format.

In comparison to grandfathering, the level of distortions of competition that are likely to be created by the PSR system is lower and consequently the environmental net benefit required to justify the application of a measure will be lower too. It therefore appears that with regard to the *quid pro quo* examination applied in the last stage of the balancing test a PSR system is preferable to a grandfathering system. It has therefore to be concluded that EC Competition law may not lead to the selection of the allocation mechanism that is least distortive.

CONCLUSION

This chapter has examined competition law issues arising under the European Emissions Trading System. How EC Competition law can limit distortions of competition stemming from Member State measures that are taken pursuant to Directive 2003/87/EC has been examined.

With regard to the first section, the examination of the joint application of Articles 3(g), 10(2) and 81 and 82 EC Treaty, it has been found that the current jurisprudence is unlikely to prevent Member States from taking measures that create incentives for undertakings to engage in anticompetitive practices. Concerning cartelization little guidance has been found as to how the joint application of these Articles is interpreted in the absence of pre-existing agreements. The benchmark employed by the Court that violations of Article 81 EC

109 Important design characteristics are that the State assumes a less central role in the granting of allowances and that the PSR system in its generic form as established under the Dutch NO_x system does not have a legally enforceable cap. Yet with regard to the EC emissions trading system it is clear that a cap is intended.

110 See Weishaar (2007b).

111 The polluter pays principle, may however be considered and support the selection of a system that burdens operators employing less environmentally efficient means of production.

Treaty can only be reinforced by State measures if they include part of the cartel agreement is criticized as very legalistic and as a prohibitively high standard of reference.

Similarly little direct guidance can be inferred from the jurisprudence pertaining to the joint application of Articles 3(g), 10(2) and 82 EC Treaty. It is through the drawing of parallels to its application under Article 86 EC Treaty, in particular to the absolute competition approach, that the Court would theoretically be able to close the perceived legal gap and bring State measures taken under the EU ETS within the realm of Competition law. Whether the Court would be inclined to do so will, however, ultimately depend on the impairment of the *effet utile* of Community law.

Regarding the second section of the chapter, the overall finding is that grandfathering systems are liable to constitute State aid while PSR systems should not. Clearly, the Commission has the power to interpret the concept of aid and held that the Dutch NOx system, as an epitome of a PSR system, constitutes State aid. The Court held that there are State resources involved in a PSR system but that it was not selective and hence did not constitute State aid. In addition, a PSR system established on an EU-wide level would not impede trade between Member States and hence undermine a further criterion for State aid. If a PSR system were to fall within the ambit of Article 87(1) EC Treaty, it may fall under a derogation and be declared compatible with the common market.

It also appears that the required level of environmental benefit as a precondition for the granting of aid is lower for a PSR system than for grandfathering. With regard to the balancing test, a PSR system is therefore preferable to grandfathering. Since the balancing test does not entail an examination of all possible aid formats that could be employed to address the market failure under consideration, it is concluded that State aid rules also may not be effective in guiding Member States towards selecting the least distortive allocation measures.

REFERENCES

Craig, P. and G. De Búrca (2003), *EU Law, Text, Cases and Materials*, third edition, Oxford University Press, p. 1241.

COM (2003) 830 final, 'Communication from the Commission on guidance to assist Member States in the implementation of the criteria listed in Annex III to Directive 2003/87/EC establishing a scheme for greenhouse gas emission allowance trading within the Community and amending Council Directive 96/61/EC, and on the circumstances under which force majeure is demonstrated', pp. 27, Brussels, 7.1.2004.

COM (2008) 16 final, 'Proposal for a Directive of the European Parliament and of the

Council amending Directive 2003/87/EC so as to improve and extend the greenhouse gas emission allowance trading system of the Community', pp. 51, Brussels, 23.01.2008.

Commission Regulation (EC) No 1998/2006 of 15 December 2006 on the application of Articles 87 and 88 of the EC Treaty to de minimis aid, OJ L 379, 28.12.2006, pp. 5–10.

Community Guideline on State aid for environmental protection (2001/C 37/03), OJ C 37 of 03.02.2001, p. 3.

De Sepibus, J. (2007), The European emission trading scheme put to the test of state aid rules, *NCCR trade regulation, swiss national centre of competence in research*, Working paper No 2007/34, September 2007, p. 34.

Edward, D. and M. Hoskins (1995), 'Article 90: Deregulation and EC Law. Reflections Arising from the XVI Fide Conference', *Common Market Law Review*, Vol. 32, pp. 157–86.

European Commission (2000), *Statsstøttesag Nr. N 653/1999 – CO_2-kvoter*, SG (2000) D/, 12.04.2000.

European Commission (2001), *Steunmaatregel nr. N 550/2000 België Groenestroomcertificaten*, SG(2001) D/290545, 25.07.2001.

European Commission (2003), *Steunmaatregelen van de Staten N 35/2003 – Nederland*, C(2003) 1761 fin, 24.06.2003.

European Commission (2008), Community guidelines on state aid for environmental protection, adopted 23.01.2008 http://ec.europa.eu/comm/competition/state_aid/reform/environmental_guidelines_en.pdf. This document is published for information purposes only and without any prejudice to the official text as will be published in the Official Journal.

Evans, A. (1997), *European Community Law of State Aid*, Clarendon Press, Oxford, p. 484.

Frank, R. (1997), *Microeconomics and Behaviour*, 3rd edition, Irwin McGraw-Hill, Boston, p. 744.

Germany (2004), *National Allocation Plan for the Federal Republic of Germany 2005–2007*, Federal Ministry for the Environment, Nature Conservation and Nuclear Safety, Berlin, 31 March 2004, Translation of 07 May 2004, available at http://www.bmu.de/files/pdfs/allgemein/application/pdf/nap_kabi_en.pdf.

Harrison, D. and D. Radov (2002), *Evaluation of Alternative Initial Allocation Mechanisms in a European Union Greenhouse Gas Emissions Allowance Trading Scheme*, NERA, pp. 168.

Neergaard, U. (1998), *Competition & Competences, the Tension between European Competition law and Anti-Competitive Measures by the Member States*, DJØF Publishing Copenhagen, pp. 358.

Poland (2004), *National Allocation Plan for CO_2 Emission Allowances 2005–2007 Trading Period*, Warsaw 2004, unofficial translation, available at http://www.mos.gov.pl/she/prace_nad_kpru/NAP_2005-2007.pdf.

Quigley, C. and A. Collins (2003), *EC State Aid Law and Policy*, Hart Publishing, Oxford and Portland, OR, pp. 394.

The Netherlands (2004), *Allocatieplan CO_2-emissierechten 2005 t/m 2007, Nederlands nationaal toewijzingsplan inzake de toewijzing van broeikasgasemissierechten aan bedrijven, Bijlage E: samenvatting convenanten energie efficiency*, http://www.novem.nl/default.asp?documentId=113926, viewed on 08.02.2005, pp. 1–5.

Vattenfall's newsletter on the CO_2 free power plant project, No. 3, November 2005.

Weishaar S. (2006), 'The European Emission Trading System and competition – anti-competitive measures beyond reach? An assessment of the grandfathering allocation method and the Performance Standard Rate system', written for the contract research: Emissions trading and equal competition, September 2006, METRO/Maastricht University, available at www.rechten.unimaas.nl/metro.

Weishaar, S. (2007a), 'CO2 emission allowance allocation mechanisms, Allocative efficiency and the environment: A Static and dynamic perspective', *European Journal of Law and Economics*, Volume 24, Issue 1, pp. 29–70.

Weishaar, S. (2007b), 'The European CO_2 Emission Trading System and State Aid an assessment of the grandfathering allocation method and the performance standard rate system', *European Competition Law Review*, Volume 28, Issue 6, pp. 371–81.

Woerdman, E. (2004), *The Institutional Economics of Market-Based Climate Policy*, Developments in Environmental Economics, Elsevier, Amsterdam, pp. 340.

7. The underestimated possibility of ex post adjustments: some lessons from the initial greenhouse gas emissions trading scheme

Chris Backes, Kurt Deketelaere, Marjan Peeters and Marijke Schurmans

1. INTRODUCTION

An ex post intervention within an emissions trading scheme is a governmental decision that changes the legal circumstances under which the market and thus the market participants may operate. A specific example of such an ex post intervention, which is the focus of this chapter, is the upwards or downwards adjustment of the amount of tradable rights (also called allowances) to which a company is entitled. From a viewpoint of legal certainty, ex post adjustments concerning allocated tradable allowances need careful attention. Indeed, the withdrawal of issued traditional 'permits', let alone modern tradable allowances, needs to be balanced between, on the one hand, the specific legal position of the permit-holder, and, on the other hand, the specific policy goal that motivates the administration to withdraw the permits. Secondly, particularly for an emissions trading market, stability of the legal conditions under which trade can occur is seen as an important stimulus for letting the market work. Without confidence in the trading system, participants would be reluctant to trade. Moreover, as with all environmental and other legislation, the administrative tasks for implementing the emissions trading instrument should be transparent and should be kept as feasible and simple as possible. Following these considerations, the governmental competence to conduct ex post adjustments with regard to allocated emissions rights needs meticulous consideration. However, these considerations need to be done in view of the specific emissions trading model. There is a difference between, on the one hand, allocation models where the allowances will be distributed ex ante and, on the other hand, the credit and trade model where at the end of a certain period (like a calendar year) a calculation will be made of the total amount of allowances to which an industry is entitled.[1] The choice between ex post adjustments of allocated tradable allowances or ex post allocations of allowances is one of the design options for the legislator that wants to introduce an emissions trading model.

Within the initial European greenhouse emissions trading scheme, the possibility of intervention regarding the allocated amount of tradable rights has been introduced by most of the Member States, specifically with respect to closures of installations. Many Member States also intended to conduct interventions in order to be able to correct allocations in case the predicted production growth, on which the ex ante allocation had been done, did not occur. An ex post adjustment would then bring the allocation in line with the real production. This method, as being introduced by Germany for some specific situations, has been rejected by the European Commission in its decision on the German National Allocation Plan for the period 2005–2007. The Court of First Instance in its ruling of 7 November 2007, however, has emphasized the freedom of Member States to choose form and methods when implementing a directive like the greenhouse gas directive.[2] It concludes that the Commission wrongly rejected the choice of ex post adjustments as being proposed by Germany. This means that the German government is allowed to ask around 700 companies to return 15 million EU allowances for the first phase of the Emissions Trading Scheme.[3]

This chapter discusses some preliminary experiences with ex post adjustments of allocated rights within the initial framework of the EU ETS. As such, it aims to provide an initial start for a systematic analysis of possible ex post interventions within emissions trading schemes. Section 2 first discusses the current legislative framework of the EU Greenhouse Gas Emissions Trading Scheme (EU ETS), and section 3 gives some illustrative examples derived from the implementing legislation of The Netherlands, Germany, and Belgium. Section 4 provides an analysis of the relevant case law, both from national courts and the Court of First instance. In section 5, a conclusion and a look ahead will be presented.

---

1 See for both models chapter 2, and for a more detailed discussion of the credit and trade approach that includes ex post allocations Peeters et al. (2007).

2 CFI November 2007, T-374/04 (which will be discussed in section 4.3).

3 Newsletter New Values http://community.newvalues.net/2007/11/ germanys_expost_corrections_an.html#more, visited 23 January 2008.

2. THE LEGISLATIVE FRAMEWORK ON THE EU LEVEL

2.1 ETS Directive

One could define ex post adjustment as an 'adjustment after the fact'. Neither Directive 2003/87/EC of the European Parliament and of the Council of 13 October 2003 establishing a scheme for greenhouse gas emission allowance trading within the Community and amending Council Directive 96/61/EC[4] (henceforth: 'ETS Directive'), nor the Commission Regulation no. 2216/2004/EC of 21 December 2004 for a standardized and secured system of registries,[5] as amended by the Commission Regulation no. 916/2007 of 31 July 2007[6] (henceforth: 'Registry Regulation'), prohibits explicitly ex post adjustments in general. The ETS Directive prescribes explicitly one situation where the competent authority is authorized to issue additional and non-transferable allowances,[7] namely in case of force majeure.[8]

The allocation of allowances is governed by Articles 9, 10 and 11 of ETS Directive as well as Annex III. Allocations are to be made before the beginning of each period (the first period being 2005–2007, and the second period being 2008–2012). In accordance to Article 11.1 and 11.2. of ETS Directive the allocation and issuance of allowances is based on the National Allocation Plan (henceforth: 'NAP'). Thus the elaboration of a NAP by each Member State is one of the most important tasks to be accomplished prior to the commencement of the allowance issuing.

Member States draw up a NAP, which is a statement of how they intend to allocate allowances to individual operators.[9] The objective in drawing up a NAP is to indicate the amount of greenhouse gas emissions from installations participating in the EU ETS and to ensure a reasonably fair share-out of the task of emission reductions:[10]

- between sectors participating in the trading scheme and the rest of the economy;

---

4 OJ L 275, 25 October 2003, p. 32.

5 OJ L 275, 25 October 2003 p. 32.

6 OJ L 200, 1 August 2007, p. 5.

7 Article 29 of ETS Directive.

8 A definition of force majeure can be found in the guidance of the Commission COM/2003/0830.

9 Non-paper of the EU Commission of 1 April 2003 'The EU Emission Trading Scheme: How to develop a National Allocation Plan', http://ec.europa.eu/environment/climat/pdf/030401nonpaper.pdf.

10 Ibid.

- among sectors participating in the trading scheme, and
- among installations in the participating sectors.

In this context the NAPs have to respect the Commission Decision 2006/944/EC of 14 December 2006 determining the respective emission levels allocated to the Community and each of its Member States under the Kyoto Protocol pursuant to the Council Decision.

The NAP must contain an allowance methodology, which does not favour particular sectors or firms unless it can be justified under the Annex III criteria. The allocation should avoid unjustifiable differences between covered sectors and sectors which are not covered, and between and within covered sectors. The principle of equality and transparency can be found in Article 9(1) of the ETS Directive.

The NAP has to meet at least the (binding) criteria of Annex III, optionally also the non-binding criteria of Annex III.[11] The criteria of Annex III allow Member States to take a variety of approaches to establish absolute quantities (including historical emissions or a 'national benchmarking' approach). The ETS Directive does not lay down how Member States determine the quantities of allowances allocated to each operator.

Once the NAP has been adopted by the Member State, the plan is notified to the Commission.[12] Furthermore it needs to be published at the latest upon notification in order to allow the general public to express comments prior to

---

[11] A Member State has an obligation to apply all elements of criteria (2 – assessments of emission developments), (5 – non-discrimination between companies and sectors), (9 – involvement of the public) and (10 – list of installations), and some elements of the criteria (1 – the Kyoto commitments), (3 – potential to reduce emissions) and (4 – consistency with other legislation). It can, therefore, choose whether it wants to take specific action with respect to some elements of criteria (1 – the Kyoto commitments), (3 – potential to reduce emissions) and (4 – consistency with other legislation), and the criteria (6 – new entrants), (7 – early action), (8 – clean technology) and (11). The Commission will not reject a plan if all mandatory criteria and mandatory elements of criteria are applied in a correct manner. The Commission will not reject a plan if optional criteria or optional elements of criteria are not applied. However, if these optional criteria or optional elements of criteria or additional transparent and objective criteria are applied, the Commission will assess their application. In all cases, the Commission does require information from a Member State with respect to criteria (7 – early action) and (8 – clean technology), even if this is only to state that a criterion has not been applied. In respect of criterion (6 – new entrants) a Member State must state the manner in which new entrants will be able to begin participating in the Community scheme in that Member State (see the Guidance of the European Commission of 7 January 2004).

[12] Non-paper of the EU Commission of 1 April 2003 'The EU Emission Trading Scheme: How to develop a National Allocation Plan', http://ec.europa.eu/environment/climat/pdf/030401nonpaper.pdf.

a decision being taken on allocating of the allowances. The public consultation is required to be in accordance with the Aarhus Convention. The public consultation is also explicitly foreseen in Article 17 of the ETS Directive. Within three months of notification of the NAP by a Member State, the Commission can reject a plan, and ask for changes to be made.[13]

Once the NAP has been adopted and approved by the Commission, the Member State has to decide upon the total quantity of allowances it will allocate for that period and the allocation of those allowances to the operator of each installation. For the first trading period, the decision had to be taken at least three months before the beginning of the period; for the second trading period this decision shall be taken at least twelve months before the beginning of the relevant period.[14] The decision to allocate must be based on its NAP, having taken due account of comments from the public.

Only a proportion of allowances will be issued each year to an operator mentioned in the NAP, the total quantities to be allocated to each operator for the whole period will be known from the outset of the NAP. Further 'adjustments' to an operator's holding will be carried out through buying and selling with other participants in the scheme.

According to the European Commission any allocation discussion *during an ongoing trading period*, can only concern the initial allocation for the *next* period.[15] Any possibility of making revisions of the NAP during the trading period would create uncertainty for businesses. According to the European Commission, ex post adjustments after the conclusion of the national allocation decision are not allowed, except in two cases. Firstly, where an installation is closed during a trading period the Member States may determine that there is no longer an operator to whom allowances will be issued. However, it is not prescribed by the ETS Directive that a Member State should cancel allowances in case of closure. Secondly, for allowances for new entrants, the Member States then need to determine the exact amount of allowances to be allocated to the new entrant.[16] However, the NAP should already contain information about the manner in which new entrants will be able to become participants in the ETS, so that the main criteria for ex post allocation to new entrants are known beforehand.

13 Article 9, 3 of the ETS Directive.

14 Article 11, 1 and 2 of the ETS Directive.

15 See e.g. decision of 16 January 2007 of European Commission regarding second NAP of Belgium.

16 See e.g. decision of 16 January 2007 of European Commission regarding second NAP of Belgium.

2.2 Commission Regulation No. 2216/2004/EC

Registration of the allocation and transfer of allowances is quite a complicated administrative matter. The Community Independent Transaction Log (CITL) records the issuance, transfer, cancellation, retirement and banking of all allowances that take place in the registry. CITL is regulated by Commission Regulation (EC) No. 2216/2004/EC of 21 December 2004 for a standardized and secured system of registries pursuant to Directive 2003/87/EC of the European Parliament and of the Council and Decision No. 280/2004/EC of the European Parliament and of the Council,[17] amended by Commission Regulation (EC) No. 916/2007 of 31 July 2007[18] (henceforth: 'Registry Regulation').

The task of CITL is to connect the Member State registries and to maintain an independent record of the issuance, transfer and cancellation of the allowances. It is mandatory for each Member State to have a national registry. According to Article 15 of the Registry Regulation, an operator of a GHG installation that holds a GHG permit where the installation has not previously been covered by such a permit or the activation of the communication link between the registry can hold an 'operator holding account' in the national registry (next to a personal holding account). According to Article 4 of the ETS Directive those GHG permits are required for each installation listed in Annex I of the ETS Directive. Furthermore, those installations are listed in the NAP, which determines the total quantity of allowances it will allocate for that period and the allocation of those allowances to the operator of each installation. According to Articles 38–40 (first trading period) and 44–47 (second trading period) of the Registry Regulation the allocation of allowances to GHG operators is based on the national allocation plan table, which is based on the decision taken in Article 11 of ETS Directive. As mentioned above, that decision is taken in response to the NAP. Each year, by 28 February at the latest,[19] the registry administrator shall transfer from the Party holding account to the relevant operator holding account the proportion of the total quantity of allowances issued under Article 45 of Registry Regulation that has been allocated to the corresponding installation for that year in accordance with the relevant section of the national allocation plan table. Once the GHG permit has been revoked or surrendered, the operational holding account will be closed.[20]

---

[17] OJ L 386, 29 December 2004.

[18] OJ L 200, 1 August 2007.

[19] According to Article 46, in fine of Registry Regulation where foreseen for an installation in the NAP of the Member State, the registry administrator may transfer that proportion at a later date of each year.

[20] Article 17 of Registry Regulation.

Thus, the Registry Regulation links the holding of an operational holding account to a GHG permit and the issuance of allowances to the national allocation plan table, responding to the NAP.

In the case of a correction having been made to the NAP, and this correction having been notified by a Member State to the Commission, which did not reject it and the correction results from improvements in data, the Commission shall instruct the Central Administrator to enter the corresponding correction into the national allocation plan table held in the Community independent transaction log.[21] The registry administrator shall, subsequent to the correction made, which occurs after allowances have been issued under Article 45 of Registry Regulation and which reduces the total quantity of allowances issued under Article 45 for the 2008–2012 period or subsequent five-year periods, convert the number of allowances specified by the competent authority into AAUs by removing the allowance element from the unique unit identification code of each such AAU comprising the elements set out in Annex VI of Registry Regulation.[22]

To conclude, the Registry Regulation allows ex post adjustments in cases where such adjustments correspond to the NAP. Some other ex post adjustments are also explicitly allowed.[23] The holder can also request, on a voluntary basis, that its allowances be cancelled.[24]

3. IMPLEMENTATION BY MEMBER STATES

3.1 Introduction

The ETS Directive leaves quite an amount of discretion to Member States regarding the possibility of ex post adjustments, and it is interesting to see in what way the Member States sought to use this discretion and established ex post adjustment competences. By doing so, we need to take into account that, because of the strongly held view of the Commission to only allow ex post adjustments in case of closure and new entrants, Member States were reluctant to act against the view of the Commission.[25] The case of Germany is an exemption, and leads to an intriguing outcome, meaning that the Commission

---

21 Article 44.2. of Registry Regulation.
22 Article 44.3. of Registry Regulation.
23 See for instance art. 47 of the Registry Regulation.
24 Article 62 of Registry Regulation.
25 For instance, in light of the difficulties with the Commission, Luxembourg decided not to include ex post adjustments in the second national allocation plan, EEA report 2007, pp. 55–56.

was unable to uphold its view against ex post adjustments. This case will be discussed in the next section. This section provides some examples from the legislative frameworks of The Netherlands, Germany and Belgium regarding the design of ex post adjustments.

3.2 The Netherlands

Contrary to most of the other Member States, the Dutch legislator did not address the possibility of closures for the first period running from 2005–2007. This means that in cases of closure, the allowances could simply be sold by the industry at hand. However, from 2008 onwards a competence is available for the government to withdraw the greenhouse gas permit in case of closure of the greenhouse gas installation, which means that after the withdrawal of the greenhouse gas permit, no greenhouse gas allowances can be transferred any longer to the account of the operator of the installation.[26]

The Act introducing the closure provision entered into force on 1 September 2007.[27] Remarkably, there has been no discussion about the closure provision during the legislative process: the Council of State made no explicit comment on the legislative proposal from the government, and also the Second and First Chamber did not discuss possible comments or problems regarding the legislative proposal.

In fact, the closure provision concerns the situation where an installation cannot qualify any longer as a greenhouse gas installation. In such a case, the competent authority may withdraw the greenhouse gas permit.[28] As soon as this permit has been withdrawn, it is no longer possible anymore to issue allowances to the account of the operator.[29]

The legislative provision does not explicitly deal with a part-withdrawal of the greenhouse gas permit: the competence only refers to the 'withdrawal of a permit'. Also, the explanatory memorandum to the legislative proposal does not elaborate on the possibility of withdrawing the permit specifically in case of closure of only a part of the installation.

As the closure provision is linked to the possibility of withdrawing the greenhouse gas permit, it is thus necessary for the competent authority to determine whether an activity can qualify as a greenhouse gas installation. Following Article 3 of directive 2003/87, the Dutch Environmental Management Act provides a definition of a greenhouse gas installation,

26 Kamerstukken II, 2005–2006, 30 964, and Staatsblad (2006, 611).
27 Staatsblad 2007, 295.
28 Art. 16.20 b Dutch Environmental Management Act.
29 Art. 16.35 Dutch Environmental Management Act.

complemented by a prescription given by an Executing Order.[30] At the least, the following aspects are relevant for applying the competence to withdraw the permit. First, the definition of a greenhouse gas installation as given in the Act needs to be interpreted. An important condition following this definition is that it needs to be determined that 'activities are carried out'. At what exact point this is the case can, of course, be questioned, and, in practice, legal conflicts might occur. For example, is a continuation of only 10% of the activities enough to state that the activities are still being carried out?

Second, the description of a greenhouse gas installation as given by the Executing Order needs to be interpreted as well.[31] In particular, the several thresholds being included in the Executive Order can lead to different interpretations and thus to uncertainty regarding the question of whether it can be concluded that the installation cannot qualify any longer as a greenhouse gas installation. Moreover, it is not clear how long the non-fulfillment of a certain threshold (like the thermal input of 20 MW) must last before it can be concluded that the installation is no longer to be viewed as a greenhouse gas installation and thus that the permit should be withdrawn. What would be an acceptable duration of non-fulfillment to the threshold criteria: one week, one month, a quarter of a year, or another period of time?

Besides the determination that the installation is no longer to be seen as a greenhouse gas installation in view of the legislative framework, there is also another circumstance in which the competent authority may decide to withdraw the permit: that is when the integrated permit as being regulated in chapter 8 of the Dutch Environmental Management Act has been withdrawn, which can also be a part-withdrawal.[32] The Act gives much discretion to the authority to decide whether or not the permit will be withdrawn from the moment that an installation no longer qualifies anymore as a greenhouse gas installation. The Act says that the authority may decide to withdraw, but does not require it to withdraw in cases of closure. It is obvious that it is the intention of the legislator that the competent authority should withdraw in cases of closure, but this has not been firmly expressed within the Act. Furthermore, the authority should conduct its margin of discretion included in the competence to withdraw the permit in such a way that administrative law principles will be met and that thus any arbitrary and unlawful decision will be avoided. The principle of equality seems in this respect a very relevant principle. The competent authority could issue a specific guideline explaining in which

---

30 Art. 16.1 par. 2 and art. 16.2 Dutch Environmental Management Act, and Besluit handel emissierechten, art. 2.

31 Besluit handel in emissierechten zoals gewijzigd, art. 2.

32 Art. 16.20b par. 1a. The exact moment of the decision-making is not relevant here, but the moment at which this decision enters into force.

specific circumstances a permit will be withdrawn. Thus far, no specific policy has been announced regarding this competence.[33]

As long as the permit has not been withdrawn, the allowances should ultimately be issued to the account of the operator by 28 February annually. This issuance should be in conformity with the national allocation decision. The Dutch legislator assumes that the issuance of allowances is not to be seen as an administrative decision that can be appealed before the administrative court (a claim should be addressed to the civil court). However, whether this view is right can be questioned, and case law is needed to clarify this point.

In sum, it is fair to conclude that in practice the competent authority can face legal problems when it aims to withdraw a permit. Moreover, things get even more complicated as soon as the procedures for, on the one hand, the greenhouse gas permit and, on the other hand, the allocation of allowances are not fine-tuned. These procedures are totally different, with possibly different authorities (like in The Netherlands). Problems might especially arise when one of the procedures is not well-operated. Quite a few Member States, among which The Netherlands, have indeed failed to comply with the time-limit for the allocation procedure for the second trading period, meaning that the deadline for transferring allowances in 2008 has been exceeded. It is not unthinkable that some operators will feel damaged by the late decision-making, and that they will try to hold the authority responsible for not delivering the national allocation decision and the issuance of the allowances in the year 2008 according to the dates set by legislation. This might for instance be at hand in the hypothetical case where the decision to withdraw a permit has been issued on a date after 29 February 2008, while the national allocation decision still had to be adopted and thus the allowances could not have been transferred yet. In such a case the former operator has a disadvantageous position, because, if the government had been issuing allowances before 1 March and before the withdrawal of the permit, he would have been able to get (and to sell) those allowances. This specific case shows that procedures to withdraw the greenhouse gas permit are not to be seen isolated from the allocation procedure, and that co-ordination between both procedures is indicated.

3.3 Germany

3.3.1 Introduction

Contrary to Dutch law, German law for the first trading period contained a rich variety of ex post adjustments. Generally speaking, only lowering of issued

33 The draft national allocation plan only gives a very short explanation about the closure provision on p. 20 and p. 36. www.co2allocatie.nl.

allowances is possible. When considering German law, one has to know that the final rules for the period 2005–2007, which are laid down in the Zuteilungsgesetz 2007 (hereafter ZuG 2007) and the Zuteilungsverordnung 2007, significantly differ from those the Court of First Instance has passed his judgment on. Without going into much detail, withdrawal of an allocation decision during the first trading period was possible in six situations,[34] each following different rules. These are:

1. closure or substantial reduction of production capacity (§ 9 ZuG 2007);
2. lower production after a transfer of allowances to a different installation of the same operator (§ 10 ZuG 2007);
3. general rule on lower production of installations (§ 7 IX ZuG);
4. lower production of installations which began to work in 2003 or 2004 (§ 8 ZuG 2007);
5. lower production than foreseen of newcomers (§ 11 ZuG 2007);
6. lower production of cogeneration installations, producing combined heat and power (CHP) (§ 14 ZuG 2007).

Upward adjustments are forbidden in all cases. Such upward adjustments were intensively discussed within Germany, but more broadly on the EC-level, too. However, in the end, and different from the draft allocation plan the CFI ruled on, the German law did not contain any possibility for ex post increase of allowances. We do not know whether Germany chose to go ahead with the allocation process in a way the Commission insisted on, and did not want to delay the allocation process in order to incorporate the outcome of the appeal against the decision of the Commission, or whether there were other reasons to change the draft allocation plan in that sense.

3.3.2 Some details

The rules for those six cases of possible adjustments are, roughly speaking, as follows:

(a) If operation of an installation terminates, the operator is obliged to return the allowances allocated, but not yet handed out.[35]

---

[34] To a certain extent the division into six reasons for ex post adjustments is ambivalent. For example, the Umweltbundesamt subdivides in a different way into five reasons. We think the subdivision chosen here offers the most clarity for the reader not common with the German Zuteilungsgesetz.

[35] The rule that the operation of an installation is deemed to have terminated when its emissions during the year in question are less than 10% of the average annual emissions recorded during the base period, mentioned in case T-374/04, has not become law in the end, but was eliminated in the last phase of the parliamentary discussion.

(b) Upon application, allowances allocated to a closed installation are not withdrawn where the operator commences operation of a new installation within a period of three months (and in some cases within two years) from closure of the old installation. In such a case, the allowances are allocated for four years on the basis of the historic emissions of the closed installation and then, for a period of 14 years, their allocation is calculated on the basis of a compliance factor of 1, this rule having the objective of encouraging operators to close their obsolete and inefficient installations. If the actual production happens to be lower than the production of the installation which was closed, restitution of allowances for the corresponding difference has to be made.

(c) A general rule on major decreases of production capacity, which was not discussed in the judgment of the CFI, is to be found in § 7 IX ZuG. According to this rule restitution has to be made for allowances already handed out if the actual production is more than 60% lower than the average yearly production the allocation was based on. In earlier drafts of the German Act, this rule counted only for new installations and recently started installations. In the definitive law, this rule applies generally to all installations, including old ones.

(d) The amount of emission allowances allocated to installations that began to operate in 2003 or 2004 had to be adjusted if the actual production volume fell below the production volumes which were declared for the purposes of calculating the amount of allowances that was initially allocated. When the tranche of allowances for the following year is issued, the quantity of allowances will be proportionately reduced.[36]

(e) The amount of emission allowances allocated to new entrants who started to operate after 1 January 2005, or increase in the production capacity of existing installations had to be adjusted according to whether, in the course of operation of the installation in question, the actual level of activity is below the level of activity which was declared for the purposes of calculating the amount of allowances that was initially allocated. When the tranche of allowances for the following year is issued, the quantity of allowances will be proportionately reduced.

---

36 In the judgment of the CFI of 7 November 2007 (T-374/04) about the German allocation plan the possibility of an increase of allowances is mentioned, if the actual production capacity in the cases (c) and (d) is higher than predicted. However, only the draft German act mentioned such an ex post increase of allowances. In the definite act the possibility of an upward-adjustment was deleted.

(f) Cogeneration installations, producing combined heat and power (CHP) (Kraft-Wärme-Kopplung) receive emission allowances by way of a special allocation (Sonderzuteilung) in the first allocation year according to the actual volume of electricity production. The amount of allowances may, however, be subsequently corrected on the basis of the volume of electricity production that is established in the following year.

Common to all six cases of ex post-reduction of allowances is that allowances that are withdrawn will be transferred to the new entrant reserve. Furthermore, it is important that all existing installations may opt to be treated as newcomers (§ 7 XII ZuG). Thus, every operator may choose between grandfathering on the basis of historical data (§ 7 ZuG) and grandfathering on the basis of his/her own predictions and benchmarks, combined with more far reaching possibilities of ex post adjustments (§ 11 ZuG), as described here under (e). According to the Federal Environmental Agency (Umweltbundesamt, hereafter UBA), many operators have chosen the last option mentioned.[37]

Motivation for and discussion about introducing ex post adjustments

The reasons given for the introduction of the ex post adjustments were diverse. On the one hand, the German government posited that the functioning of the emissions trading system itself urges some ex post adjustments. That would, for instance, be the case for new (d and partly b) or nearly new (c) installations, where grandfathering on the basis of historical data is not possible. Grandfathering is usually based on historical data, with several possibilities for correction factors. Providing allowances without payment only on the basis of predictions of the owner of the installations, is, from a market-oriented point of view, an imperfect instrument, which requires the possibility for an ex post correction. In such cases, ex post adjustments prevent these installations from getting far more allowances than they need. That could otherwise interfere with the functioning of the market and could prevent the drivers of the installation concerned from taking emission reduction measures. Other arguments were justice and acceptance. It would not be fair, and could cause competitive distortion, to provide generous allowances to the drivers of new installations generously, on the basis of their own predictions only, whilst deducting the allowances of the drivers of existing installations on the basis of hard production facts. It could distort competition, too, if high premiums (in the form of allowances which are no longer needed) were to be paid for closures of installations.

---

[37] Ziesing (2007, p. 201), http://www.umweltdaten.de/publikationen/fpdf-l/3254.pdf.

The rules for ex post adjustments were intensively discussed during the drafting of the statute. Major changes have been made to the legislative draft. For example, the draft contained the possibility of an ex post increase of allowances in cases where the production of new installations is significantly higher finally than the driver of the installation predicted.

Many changes have been made to the detailed conditions for applying the ex post rules. However, the whole system of allocation of allowances is highly coherent. Thus, changes in the details of one rule can easily disrupt this coherence. This has been the case, at least to some extent. An analysis of the allocation rules, written on the request of UBA, concludes that, in the end, at least some of the reasons for ex post adjustments are not consistent which each other and 'poorly rational'.[38] Generally speaking, this report is not principally against any ex post adjustments, but is in the end quite critical of the need and the effects of this instrument. First of all, according to this report, ex post adjustments principally do not fit into the existing EU-emissions trading system, with its strong emphasis on ex ante assessments. That was the main argument for the Commission's decision to reject the German allocation plan, too. Secondly, some of the adjustments hinder the price-incentives for lowering emissions. Thus these ex post adjustments are not only principally contradicting the system, but are even actually diminishing its results. The report differentiates here between different ex post adjustments. Whilst the rules for new entrants and nearly new installations (d and e) seem to have almost no negative effect on price-incentives, the general 60%-rule on lower production volume (c) seems to have. The authors of the report do see good alternatives for most (but not all) of the situations where ex post adjustments are used and therefore recommend the revision of the ex post rules for the following allocation periods.[39] In that respect the report refers, for example, to Denmark and the UK, where special methods of benchmarking were used as an alternative for ex post adjustments.[40] However, on the other hand, the same report is critical about too many differentiated and process-oriented instead of product-oriented benchmarks as a basis for allocation of allowances for new installations. The author argues that differentiated (process-oriented) benchmarks also contravene the price-incentives of the trading system.[41]

---

38 Ziesing (2007, p. 202 and p. 184), http://www.umweltdaten.de/publikationen/fpdf-l/3254.pdf.

39 Ziesing (2007, p. 202 and p. 196), http://www.umweltdaten.de/publikationen/fpdf-l/3254.pdf.

40 Ziesing (2007, p. 196), http://www.umweltdaten.de/publikationen/fpdf-l/3254.pdf>.

41 Ziesing (2007, p. 182),<http://www.umweltdaten.de/publikationen/fpdf-l/3254.pdf.

3.3.3 Withdrawal of the greenhouse gas permit?

German law does not include rules for the withdrawal of the permit. This is not that strange, because German law does not even use the instrument of a special permit. To fulfill the requirements of the directive, § 4 I Treibhausgas-Emissionshandelsgesetz asks for a permit to emit greenhouse gases and § 4 II-IV sum up the conditions for granting and for the content of the permit. However, Article 4 VI Treibhausgas-Emissionshandelsgesetz states that such a permit is an integral part of the environmental permit for each installation covered by the directive, respectively, the German Treihausgashandelsgesetz, has to have. §§ 17–21 Bundesimmissionsschutzgesetz contain a detailed regulation about adjustment and withdrawal of permits, which applies here, too. However, withdrawal or adjustment of permits as a means of regulating the amount of allowances is not needed and was not a point of discussion in Germany until now, because there are enough sufficient less interventionist possibilities for adjusting only the amount of allowances, and not the permit itself.

The second allocation plan for the period 2008–2012 and its regulation

In the meantime the allocation of allowances for the second trade-period had to take place. Therefore, a new Zuteilungsgesetz 2008–2012 came into force in August 2007, thus some months before the CFI ruled about the rejection of the first allocation plan by the European Commission. That may explain why nearly all possibilities for ex post adjustments have been eliminated. The possibility of a single ex post adjustment remains. If an installation closes down, restitution must be made for all allocated or even issued allowances corresponding with the period after closure (§ 10 I Zuteilungsgesetz 2012). This obligatory ex post adjustment is even stricter than the corresponding Article 9 I ZuG 2007, which left allowances already issued. Thus, in an extreme case, the operator of an installation which closed down almost immediately after allowances for a calendar year were issued, could retain allowances corresponding to unrealized emissions of nearly an entire year. This 'far reaching generosity'[42] has now been corrected. As has been said, all other ex post adjustment rules have been diminished. It seems that the rejection of the first German allocation plan by the Commission, although unlawful, has had its proposed effect.

3.4 Belgium

In Belgium's federal system, the implementation of the EU ETS is compli-

42 Vierhaus (2005).

cated because of the different levels of government. In accordance with the division of power, responsibility is spread between the federal state and the Flemish, Brussels and Walloon regions. Each of these authorities defines its own priorities for environmental and climate change policy and implements its own relevant actions. With regard to the transposition of EU ETS and given the division of powers in Belgium, the regional authorities have the competence to draft an allocation plan for GHG installations on their territories and to allocate allowances and to grant a GHG permit to them. Below we will discuss relevant rules on the federal level and within the Flemish region, with an emphasis on practical difficulties with the adjustment of the registered allowances within the formal registry.

3.4.1 The federal state

The federal state is competent to regulate the *national registry* on emission trading. The European Regulation No. 2216/2004 has been transposed by the Royal Decree of 14 October 2005. The registry operates on the basis of the national allocation plan table, which is based on the NAP. In cases where ex post adjustments of the already taken allocation decision are deemed necessary, the national registry entry has to be modified as well. However, the current Royal Decree (which has not yet been adapted to the amended European Regulation) is not very flexible towards facilitating ex post adjustments. We give two illustrative examples in the following paragraphs.

A transfer of the operator account is only allowed in cases where *all activities* of the GHG installation are to be transferred from one company to another (e.g. in case of merger of full acquisition by another GHG company). When activities are only partially transferred (e.g. transfer of a branch which includes the GHG installation), the allocated allowances cannot be transferred to the account of the other company. This means that the new company that took over the GHG branch will not automatically have the right to claim these allowances, which belong to the first GHG company that sold its GHG branch to the company that took it over. In practice the first GHG company will take the initiative to sell its allowances to the new company. But issues might rise if the GHG company is not willing to do so.

In case of an erroneous surrender of allowances by a company, it is not always possible to correct this mistake. This can be illustrated by the following case: two sister companies, both operating a GHG installation, opened separate accounts. One of the GHG installations requested a prolongation of its GHG permit. The regional competent authority assessed that both GHG installations had to be considered as one technical environmental entity, as being regulated by the Decree on general environmental policy of 5 April 1995 ('milieutechnische eenheid').[43] Consequently, the regional competent authority suggested merging both accounts. However, if the environmental permit (which embodies the GHG permit) does not cease, the holding operational account cannot close down. Nor was it possible to let both accounts merge, because a transfer of accounts is only allowed in case of transfer of activities, which was not the issue. Following the suggestion of the regional competent authority, one of the GHG installations surrendered too many allowances, thinking it had to surrender them as a technical environmental entity. However, the federal competent authority was not able to return the erroneously surrendered allowances. The European Commission even confirmed to Belgium that an ex post adjustment in such cases was not possible.[44] The regional competent authority sought another solution to correct the mistake, and allowed a rectification through the emission monitoring report.

3.5 The Flemish Region

The Flemish legislation to implement the EU ETS is the Decree of 2 April 2004 on the promotion of rational energy use and the Executive Order of 4 February 2005, which lays down the basis of the Flemish emission trading rules. The legislation establishes a GHG permit scheme, which has been incorporated into the environmental (IPPC) permit. The allowances are allocated to GHG installations, which are listed in the Flemish NAP. This allocation is based on benchmarking in combination with historical emission of production data.

Initially the Flemish legislation foresaw during the first trading period in (previous) Article 14, §4 of the Executive Order of 4 February 2005 the possibility by the competent authority of making ex post adjustments downwards to the allocated emission rights in the following cases:

1. in cases where a GHG installation which holds a GHG permit no longer requires a GHG permit through modification of the installation;
2. in cases of annulment or expiration of a GHG permit (incorporated in the environmental permit). According to Flemish law (Vlarem I) an environmental permit can become lapsed in cases where the operation of an IPPC installation (thus a GHG installation) ceases for more than two consecu-

---

[43] Implementing the IPPC directive (Directive 96/61/EC concerning Integrated Pollution Prevention and Control (IPPC), OJ L 257, 10 October 1996, pp. 26–40. The IPPC Directive has recently been codified by Directive 2008/1/EC of the European Parliament and of the Council of 15 January 2008 concerning integrated pollution prevention and control.

[44] Not published, on file with author Schurmans.

tive years, also in case of fire, and in cases when the operation has not begun within three years of the date that a GHG permit was granted;
3. in cases where a GHG permit has been suspended. A suspension is ordered by the civil court or the Highest Administrative Court in case the permit is unlawful.

In these cases, the issued allowances will be annulled.

According to Article 21 of the Executive Order of 4 February 2005 a GHG installation could also request the annulment of its emission allowances. Emissions allowances that have been annulled (at the initiative of the GHG installation, or at the initiative of the competent authority) are no longer valid.

These cases are listed exhaustively. This means that in no other cases would ex post adjustments be allowed. In practice, this leads to some practical issues. For example, if a competent authority erroneously allocated the allowances (e.g. based on erroneous information about the GHG installation), the authority has no competence to correct the amendment. The European Commission communicated in several cases that such corrections were not allowed. The only solution to solve such situations is by amending the NAP, which had to be communicated and approved by the European Commission.

On 7 December 2007 the Executive Order was modified[45] in order to establish a regulation for mergers and acquisitions together with the possibility of ex post adjustments. The GHG installation has a duty to notify the situation (e.g. merger) to the competent authority which will make the agreed adjustments based on information about the GHG installation. However, it must be emphasized that the adjustments cannot lead to an increase of allowances, except in a case where the merged GHG installation can prove that it falls beneath the scope of new entrants. A new entrant in the second trading period 2008–2012 will be considered as

> a modification of the nature or function of a GHG installation by physical enlargement of the GHG installation or an increase of the permitted capacity of a GHG installation, which in any of the forementioned cases has the consequence that the CO_2 emissions of the GHG installation increases with at least 10% or increases with at least 50.000 ton per year compared to the average of the reported CO_2 emissions of the last three years of the GHG installation.

The amended Executive Order also iterates that the competent authority can stop issuing the allowances in cases of annulment, expiring, decease or decline of the environmental permit (which incorporates the GHG permit). The

---

45 Belgisch Staatsblad, 27 December 2007, which entered into force on 27 December 2007.

allowances will then not be issued in the following calendar years of a trading period. The GHG installation at issue still has to fulfil its GHG duties (reporting, monitoring, etc.) in the year where the decision has been taken to stop delivery of emission allowances. The competent authority will not stop issuing the allowances in cases of closure, no matter what the reason for closure is (for example bankruptcy, or closure on voluntary basis). The reason can be found in the fact that in case of closure, the GHG permit does not cease immediately. It will cease in cases where the operation has ceased for more than two consecutive years. During that period, the curator can still activate the operation of the bankrupted GHG installations. Thus, to avoid being in breach with freedom of trade, the legislator decided not to cease the allowances in cases of closure.

4. CASE LAW REGARDING EX POST ADJUSTMENTS

4.1 Introduction

This section gives an overview of the most important case law regarding ex post adjustments. First, we discuss some national case law as delivered in Belgium, The Netherlands and Germany, after which we elaborate on case law from the Court of First Instance in which the rejection by the Commission of ex post adjustments as proposed by Germany was found not acceptable.

4.2 National Case Law

4.2.1 The Netherlands

Within the Netherlands, the issue of ex post interventions was considered in the ruling of the Administrative Court of the Council of State when it reviewed more than forty appeals from industries against the Dutch National Allocation decision for the period 2005–2007.[46] In this ruling the Court also considered the legality of the National Allocation Plan, being the basis for the later National Allocation Decision. Some industries argued that the allocation method should not be based on historical production data, because that would be in conflict with Article 9(1) of the ETS Directive, the fifth criterion of Annex III to this directive (meaning that the plan shall not discriminate between companies or sectors in such a way as to unduly favour certain undertakings or activities in accordance with the requirements of the Treaty, in

---

[46] Tijdschrift voor Milieu en Recht 2006, jr.n,. 53, Jurisprudentie Bestuursrecht 2005, nr. 291.

particular Articles 87 and 88 thereof), and with the preamble. They argued that the NAP should have incorporated the possibility of (upwards and downwards) ex post adjustments. The court, however, rejected the need to ask the European Court of Justice for a preliminary ruling on this matter, and followed the text of Article 11(1), which states that each Member State shall decide upon the total quantity of allowances it will allocate for that period and the allocation of those allowances to the operator of each installation. Furthermore, criterion 10 of annex III also indicates that the NAP shall contain a list of the installations covered by this Directive with the quantities of allowances intended to be allocated to each. As we shall see below, the Court of First Instance in its ruling on *Germany v Commission* annulled the decision of the Commission to reject the German NAP containing ex post adjustments. In view of this ruling, the interpretation of the Dutch Court and therefore its conclusion that there was no reason to ask for a preliminary ruling can be questioned. In contrary to the brief reasoning of the Dutch Court, we can find an extensive consideration in the ruling of the Court of First Instance, thereby conducting a literal, historical, contextual and teleological interpretation of the Directive when questioning the possibility of ex post downwards adjustments.

4.2.2 Germany

As far as we know, there is no German jurisprudence on ex post adjustments, which is as such no surprise. In their reaction to the rejection of the first allo cation plan by the Commission, the German authorities did not execute possible or necessary ex post adjustments until the issuing of the judgment by the Court of First Instance in November 2007. We suggest that this is the reason that none of the enterprises confronted with (possible) ex post adjustments went to court, at least not until the end of 2007. After the judgment was issued, the UBA announced that the planned ex post adjustments would be enforced in due time. However, at that time the actual price of the allowances was and still is very low. Thus, although the German government was justified, in the end, ex post adjustments seem not to have had any real substantial impact in Germany. With regard to the ex post adjustments because of the closure of an installation, they won't have much impact in future, too.

We announce here some jurisprudence which concerns the regulations discussed above and might be relevant in a broader sense. As mentioned above, installations already in existence before 2003 were (for the period 2005–2007) allowed to choose whether they are treated as existing installations (and get their allowances on the basis of former production data) or are treated as new installations (and get their allowances on the basis of predictions and benchmarking) (§ 7 XII ZuG). In the last-mentioned case, the regulations about ex post adjustments for new installations would apply. German authorities argued that the (possibility of) a shortage of allowances, necessary to remain within the total cap of allowances allocatable (the so-called Erfüllungsfaktor), would apply to both kinds of existing installations, independent of which allocation method they have chosen. However, the Bundesverwaltungsgericht decided that the Erfüllungsfaktor only applies to the installations which did not opt for being treated as new installations.[47] Another interesting judgment of the same day concerns the legality of the shortage of allowances (Erfüllungsfaktor). The BVerwG argues that this possibility is not in breach of the Directive, the German constitution or other general law principles.[48]

4.2.3 Belgium

The Belgian courts have not yet delivered much case law regarding emissions trading. However, the Highest Constitutional Court ruled an interesting case on 7 June 2007 with regard to the competence of the government to amend its allocation decision, for instance in cases of closure of an industry or in cases of decease of the GH permit.[49] The company group Arcelor, producer of pig iron and steel in France, Spain, Germany and Belgium brought on 2 June 2005 an action against the Walloon region before the Highest Constitutional Court. Arcelor argued, besides other arguments, that the Walloon decree of 10 November 2004 providing for adjustments of the allocation of allowances inter alia in cases of closure infringed its fundamental rights to property. The Highest Constitutional Court ruled that a GHG installation could claim the ownership of an allowance that has been issued by the competent authority to the GHG installation, but it could not claim ownership of the allowances that had been addressed in a NAP to a GHG installation, but which had not yet been issued to that GHG installation. This judgment is interesting because for the first time a court ruled upon the legal status of a greenhouse gas allowance in terms of property rights. In fact, the court allows for ex post adjustments of the decisions of a national allocation plan, in circumstances as designated by law, for example in cases of closure or decease of the GHG permit.

4.3 EU: Court of First Instance

4.3.1 *Germany v Commission*

4.3.1.1 *Essence of the case* On 7 November 2007 the Court of First Instance ruled a very interesting case with regard to the original German draft NAP.[50] The ruling concerned the acceptability in the German plan of the

---

47 BVerwG 16 October 2006, BVerwG 7 C 29.07.
48 BVerwG 16 October 2006, BVerwG 7 C 33.07.
49 Arrest van het Arbitragehof van 7 juni 2007, no. 92/2006.
50 T-374/04, 7 November 2004.

proposed ex post adjustments to the amount of allowances allocated to installations. As indicated above, Germany wanted to be able to withdraw allowances in cases where a company's production – and therefore emissions – turned out to be much lower than projected. The Commission said operators should be free to reduce their emissions by cutting production levels while keeping emission allowances. But in its ruling the Court of First Instance declared that the commission had 'misconstrued' the ETS Directive and that ex post adjustments did not contradict the essence of the trading scheme: 'The mere fact that the ex-post adjustments at issue are liable to deter operators from reducing their production volume and, therefore, their emission rates is not sufficient to call into question the adjustments' legality in light of the directive's objectives as a whole'.

4.3.1.2 *Elaboration of the case*

First of all, the Court stressed that the Commission is empowered to verify whether the measures adopted by Member States are consistent with the criteria set out in Annex III to the Directive and with Article 10 thereof and, in carrying out that review it itself has a discretion in so far as the review entails complex economic and ecological assessments carried out in the light of the general objective of reducing greenhouse gas emissions by means of a cost-effective and economically efficient allowance trading scheme (Article 1 of ETS Directive and recital 5 in its preamble). The motivation of the Court's findings can be found in Article 249, para. 3 EC that states as follows: 'a directive shall be binding, as to the result to be achieved, upon each Member State to which it is addressed, but shall leave to the national authorities the choice of form and methods'. It follows that, when the directive in question does not prescribe the form and methods for achieving a particular result, the freedom of action of the Member States as to the choice of the appropriate forms and methods for obtaining that result remains, in principle, complete. Nevertheless, the Member States are required, within the bounds of the freedom left to them by the third paragraph of Article 249 EC, to choose the most appropriate forms and methods to ensure the effectiveness of directives.[51] It also follows that, where there is no Community rule prescribing clearly and precisely the form and methods that must be employed by the Member State, the Commission has the task, when exercising its supervisory power, pursuant in particular to Articles 211 EC and 226 EC, of proving to the required legal standard that the instruments used by the Member State in that respect are contrary to Community law. The Court iterates that *the European Commission has the burden of proving* the

[51] See Case C-40/04 Yonemoto [2005] ECR I-7755, paragraph 58, and the case cited there.

extent to which the powers of the Member State and, therefore, its freedom of action, are limited in light of the above-mentioned conditions.[52]

Based on different interpretation methods of criterion 10 of Annex III of ETS Directive and taking the evaluation of the Commission's guidance into account, the Court concludes that, in its guidance, the Commission does not provide any additional explanation beyond the wording of the relevant provisions of the ETS Directive, with regard to the effect of criterion 10 of Annex III to the ETS Directive. Such explanation would be capable of supporting the validity of its interpretation that the ex post adjustments at issue are contrary to that criterion. Nor does the Commission guidance contain anything that helps to determine whether the Member State is able to change the individual allocation of the allowances *after* the adoption of its NAP or of the allocation decision under Article 11(1) of ETS Directive. The Court also stresses that after the national allocation decision has been taken, subsequent modification of the individual allocations of allowances is still possible, and therefore refers to the case T-178/05, *United Kingdom v Commission* 2005, ECR II-4807.

Furthermore, according to the Commission the ex post adjustments relating to the amount of allowances allocated to new entrants are contrary to criterion 5 of Annex III of ETS Directive which requires non-discrimination, since new entrants are unjustifiably favoured compared with operators of installations who are already covered by the German NAP, and who do not benefit from such adjustments. Criterion 5 of Annex III to Directive 2003/87 states that 'the [NAP] shall not discriminate between companies or sectors in such a way as to unduly favour certain undertakings or activities in accordance with the requirements of the Treaty, in particular Articles 87 [EC] and 88 [EC]'. As regards the prohibition of discrimination, paragraph 51 of the Commission guidance, concerning criterion 6 which relates specifically to new entrants, adds that the principle of equal treatment is the guiding principle relating to new entrants' access to allowances. Finally, paragraph 61 of the Commission guidance specifies that, 'in order to respect the principle of equal treatment, the methodology that a Member State uses in order to allocate allowances to new entrants should as far as possible be the same as the one used for comparable incumbents', while acknowledging, however, that 'adaptations may be made for justified reasons'.

However, according to the Court the Commission did not apply the principle of equal treatment correctly to the case at issue. Contrary to what the Commission believes, the German NAP provides for the application of ex post adjustments in respect not only of new entrants but also of certain operators of installations who are already present in the market and covered by the German

52 See para. 79.

NAP (e.g. closure). Even more, the fact that a subsequent downward correction of the allowances is granted to an operator, deprives, according to the Court, the operator of its 'property' having commercial value.[53] This cannot be seen as an advantage compared to other operators.

4.3.1.3 *The future relevance of the case*

The ruling of the Court surprisingly enough stresses the fact that the downwards ex post adjustments as proposed by Germany could not have been rejected by the Commission. Within the original NAP, however, upwards adjustments were also foreseen. Finally, the court only reviewed downwards adjustments, as Germany explained that according to its implementing legislation only such adjustments would be possible.[54] It is however interesting to note that the initial thought was to have a system of both upwards and downwards adjustments, based on actual production data.

One important reason for the outcome of the case is, strikingly enough, that the Commission failed to incorporate its condition to accept only ex ante allocation decisions in its first guidelines, except in cases of closures and new entrants. If the Commission were to have incorporated its strong view on ex ante allocations in its guidelines, the outcome of the case could have been different, although this would depend on the strength of the motivation by the Commission. As stated in the ruling, the Court would make the assessment.

For the moment, the Commission upholds its resistance against ex post adjustments. The Commission stated in its second guidelines, adopted before the ruling of the Court of First Instance:

> The non-acceptance of ex-post adjustments is essential for the allowance market development. The Commission did not approve the so-called ex-post adjustments envisaged by a number of Member States for the first trading period. This plays a vital role in the development of an efficient and liquid allowance market. The good functioning of the allowance market depends crucially on a stable and predictable allocation for the entire trading period in order to create stable incentives for installations to reduce emissions. For compliance purposes, companies can use the full flexibility of the scheme, be it via the allowance market or via company-internal transfers across borders.[55]

The Commission thinks that it is premature to draw conclusions and identify best practice regarding new entrants and closures.[56] In fact, the Commission has no convincing reasoning, let alone hard evidence, that would

53 See no. 161.
54 Section 34.
55 Com 2005(703) final, p. 16.
56 Com 2005(703) final , p. 19.

justify its opinion of not allowing ex post adjustments, even though this would be ex post downwards and upwards adjustments as originally proposed by Germany. A more detailed analysis of the effects of ex post adjustments would be needed before adopting such a strong point of view and thus restricting the discretion of Member States. This current point of view of the Commission lacks sustainable motivation.

After all, the court ruling came too late to have any effect in practice yet. Nevertheless, with a view on the necessary revisions of the greenhouse gas trading scheme, the consideration of ex post adjustments is still relevant. However, the Commission did not incorporate ex post adjustments within its proposal on revisions of the ETS, as it firmly states that no ex post adjustments will be allowed. As we have seen before, the need for ex post adjustments is determined by the allocation method chosen: particularly in a cap-and-trade regime with free allocation on historical data, Germany, at least, felt the need for (downwards) ex post corrections, which view is supported by the Court of First Instance. What is relevant, thus, is how the allocation in the new regime will be done. The Commission proposes the use of benchmarks, e.g. a number of allowances per quantity of historical output. We first would like to see an elaborated explanation before being able to judge whether the exclusion of ex post adjustments is reasonable or even justifiable. If we understand this correctly, the Commission still wants to use historical data, meaning that allowances will be done possibly via a number of allowances per quantity of historical output, and that the emissions would no longer depend on historical emissions.[57]

In its teleological interpretation, the Court however expressed some interesting views regarding the question whether a (downwards) ex post adjustment mechanism linked to changes in production volume would run counter to the objectives of the ETS Directive. First, it says that the Commission has not put forward evidence or arguments capable of establishing that the ex post adjustments would harm the principal objective of the ETS Directive, namely the reduction of greenhouse gas emissions as a whole (sec. 134). The Court then incorporates the objectives of maintaining cost-effective and economically efficient conditions, and that the trading market must cause the least possible diminution of economic development and employment. The court recalls that a fall in production volume does not necessarily lead to a reduction in the overall emission rate. Consequently, the deterrent effect of ex post adjustments linked to falls in production volume is not to be seen as contrary

57 European Commission, Questions and Answers on the Commisson's proposal to revise the European Emissions Trading Scheme, sub 14. Memo-08-35, Brussels, 23 January 2008.

to the objective of maintaining cost-effective and economically efficient conditions. The Commission was even wrong in asserting that the encouragement of the use of more energy-efficient technologies . . . producing fewer emissions per unit of output was only a subordinate objective (sec. 139). The courts considers that

> . . . investment in more energy-efficient technologies constitutes an instrument at least equivalent, if not superior, to that of reducing production volume, for the purpose of successfully reconciling the objective of substantially reducing emissions and that of safe-guarding cost-effective and economically efficient conditions both on the trading market and on the market for goods in question (sec. 139).

The Court rejects the view of the Commission that ex post adjustments discourage operators from investing in more energy-efficient technologies (para. 140). The Commission's argument that ex post adjustments are environmentally neutral or even harmful is thus not well founded. Moreover, the Court cannot see why the Commission does not follow the argument that ex post adjustments are needed to tackle the risk of over-allocation, because there is a natural tendency for operators to seek to obtain the greatest possible quantity of allowances. With ex post adjustments, linked to actual production data, overestimations made by operators can be corrected, and the mere fact that a competence to correct overestimations exists could have a preventive function.

The considerations of the Court upon the possible effects of a downwards ex post adjustment mechanism linked to production volumes are convincing, and should be taken into account for further consideration when reviewing the EU ETS scheme and even other emissions trading schemes. One possible point of view is that the reasoning of the Court does not preclude that even a system of ex post downwards and upwards adjustments, as formerly intended by Germany, could be in line with the Directive. The crucial point would be that the upwards adjustments may not impede the environmental goal of the directive, and thus may not endanger the overall greenhouse gas reduction goal. However, by ensuring in the national allocation plan that a maximum amount of allowances will be allocated, meaning that ex post upwards adjustments can no longer be granted when the total quantity of allowable emissions is reached, this concern seems to be addressed.

On the other hand, it is hard to predict how the Court would have judged or would judge the acceptability of ex post upwards adjustments. Article 29 of the directive establishes the possibility of granting extra allowances in case of a force majeure, and specifies that this can only be done after approval by the Commission. One interpretation could be that this article precludes, as such, any other upwards ex post adjustments (thus, ex post upwards adjustments without force majeure).

However, one could also argue that ex post adjustments, explicitly foreseen by the national allocation plan, and adequately designed in order to ensure that the total quantity of allowances in the specified trading period will not be breached, are not meant to be forbidden by this article. In this respect, we see that the Court of First Instance in its systematic interpretation of the Directive and, more specifically, in its reasoning regarding Article 29 of the Directive, links the article to *the total quantity of the allowances* as being mentioned in the national allocation plan.[58] The Court then says that the possibility of an increase in the amount of individually allocated allowances bears out the proposition that a Member State is not permitted, in principle, to allocate additional allowances. The precise meaning of the wording 'in principle' is hard to interpret; one could say that this leaves the possibility that a specific and justifiable allocation method which is intended by a Member State could be found acceptable. In particular Article 249 EC Treaty, meaning that Member States would have the choice of form and methods, would support this point of view. So, in this view, an explicit downwards and upwards ex post system proposed in a national allocation plan would not necessarily be in conflict with Article 29 of the Directive. As a final observation, taking into account that the court could take another point of view, we consider that Article 29, as well, of the Directive should not be interpreted as a prohibition to include within a national allocation plan the possibility of upwards and downwards adjustments, as long as the total quantity is not endangered. Of course, we can imagine that such a system in itself raises new legal questions, notably regarding the method through which the total cap will be ensured. Obviously, such legal questions deserve further consideration.

4.3.2 *Buzzi Unichem v Commission*

Buzzi Unichem, an Italian private association, brought an interesting case before the European Court of First Instance.[59] The appeal requests the annulment of Commission Decision of 15 May 2007 concerning the NAP notified by Italy for infringement of the EC Treaty and the principles and rules of law adopted in its application. The NAP must be altered so as to render no longer permissible rationalization measures which envisage that the operator may maintain part of the allocated allowances in the event of 'closure due to processes of production rationalization' (Article 1(4) and Article 2(4) of the Decision). According to Buzzi Unichem, the European Commission failed to

---

[58] T-374/04, 2007, §107; here the reasoning refers clearly to 'derogation from the total quantity of allowances that is laid down'.

[59] T-241/07, OJ C 211, 8 September 2007, p. 38.

explain the reasoning which led it to hold that the scheme was incompatible as 'ex-post adjustment'. The Court has not yet ruled upon this case. We can assume, however, that the Court will not review the substance of the claim, as it is expected to hold that the firm lacks standing in view of not having a direct and personal interest.[60] The case again illustrates that ex post adjustments are deemed important enough to start legal procedures.

5. CONCLUSION: LESSONS FOR THEORY AND PRACTICE

Within the initial period of the EU ETS, some interesting initial experiences have been gained with the possibility of adjusting the allocation of allowances. We have seen that the legislative framework for dealing with adjustments has a highly technical-administrative character. The Belgian practical experiences show that the applicable legislative framework, notably the registering of the allowances, was in fact not well suited to cases occurring in practice. In particular, there was a need for more flexibility in order to adjust the accounts of the operators, inter alia in the case of erroneous transfers and in the case of mergers. The discussion of the implementing legislation in The Netherlands regarding the withdrawal of allowances in the case of closures shows that the legislative framing of the administrative competence to withdraw ex officio a greenhouse gas permit raises some questions regarding its applicability, which might complicate the effectiveness of this provision. The first lesson to be drawn is thus that it is a challenge to design a well-suited legislative framework for ex post adjustments, which are administrative-technical by nature.

The second lesson to be drawn is that the question of whether ex post adjustments can be used to correct any perverse consequences of allocation within a cap-and-trade scheme should not be overlooked. This remark even concerns courts: the Dutch administrative court was quite superficial in its reasoning that ex post adjustments as proposed by the claimants would not be possible in view of the EU ETS. Also the government, in particular the European Commission, should not overlook the possible usefulness of ex post adjustments. Germany presented an interesting legislative approach for dealing with ex post adjustments, and it dropped the possibility of upwards adjustments mainly because of opposition by the Commission. Germany however insisted on the possibility of using downwards ex post adjustments in cases of

60 Case T-28/o7 Fels-Werke a.o. v Commission, judgment of 11 September 2007: undertakings have no locus standi in an action for annulment of the decision of the Commission.

a production decrease. The strong view of the Commission against these downwards ex post adjustments has, however, not been followed by the Court of First Instance. This practical example shows that, especially in a cap-and-trade scheme, the possibility of using ex post interventions in order to correct any perverse consequences of ex ante allocation, notably the incentive to forecast a high amount of production in order to get ample allowances, should not be overlooked.

In view of the future EU ETS, the question whether ex post adjustments of allocation decisions or moreover other possible ex post interventions should be part of the legislative framework needs to be considered within the specific context of, on the one hand, auctioning and on the other hand the free allocation of allowances on the EU level. First, in the case of auctioning, we refer to the fact that Dales, who proposed an original design of the emissions trading instrument in 1968, stressed the importance of an active role for the government, in fact envisioning a government acting like a broker.[61] Dales proposed that the government should consider intervention within the emissions market by buying and selling rights inter alia in cases of unexpected market developments. The possibility of governmental ex post interventions thus seems relevant also in cases of auctioning, and should at least be considered when designing the auction scheme. At the same time, the possible need for ex post adjustments should be discussed as well when designing the free allocation, which in the view of the Commission will still be part of the EU ETS from 2013 onwards, with a gradual decline. The Commission proposes to adopt, ultimately by 30 June 2011, 'wide and fully-harmonised implementing measures for allocating allowances', and the need for ex post adjustments should thus be part of the considerations when designing those implementing measures.[62] One specific rule has already been put forward within the text of the amending directive, which is that an installation which ceases to operate shall receive no further allowances.[63] First, it should be questioned whether such a closure rule should be supported, as this could be an incentive for keeping open old installations. This question should already be taken up during the co-decision procedure, as the rule is already included in the text of the proposal to amend the Directive. Furthermore, if this closure rule is adopted, the specific administrative-technical design of the competences needed to implement the closure rule needs close attention. For instance, as we have seen, the situation should be avoided where the competence to withdraw a

---

[61] Dales (1968), republishd in 2002, Edward Elgar Publishing, p. 95.

[62] European Commission, Proposal for a Directive of the European Parliament and of the Council amending Directive 2003/87, Brussels 23.1.2008, COM(2008)16, in particular the new proposed article 10a.

[63] Article 11, last sentence, as being proposed by the Commission.

greenhouse gas permit ex officio is surrounded by uncertainty as to when exactly the administration can execute this provision (when does the installation really close, and what will be done with installations that temporarily close, or only keep on 'burning' with only 10% of the maximum capacity?). As the tradable allowances will increase in value as soon as the cap becomes more severe as compared to the first trading periods, the question of closure is important to address in such a way that legal conflicts may be avoided as much as possible. Apart from the closure rule, the matter of ex post adjustments should specifically be addressed when designing the free allocation criteria. Particularly in cases of the possible use of benchmarks as a criterion of free allocation, and in cases of new entrants, the Commission should at least seriously take into account the possibility of ex post adjustments as formerly intended by Germany.

Apart from these recommendations for the practical design of the EU ETS, we also want to draw also a general lesson in theory. In fact, there is no systematic analysis of different forms of governmental intervention within the available models of the emissions trading instrument. This chapter aimed to give some starting points for such a further examination. The need for and design of such competences requires further exploration from an economic and legal perspective, not only within greenhouse gas emissions trading schemes, but regarding any emissions trading scheme under consideration.

REFERENCES

Dalcs, J.H. (1968), *Pollution, Property and Prices. An Essay in Policy-making and Economics*, republished in 2002, Edward Elgar Publishing, p. 95.

Peeters, M., J. de Cendra de Larragán and S. Weishaar (2007), 'A Governance Perspective on the Choice between "Cap and Trade" and "Credit and Trade" for an Emissions Trading Regime', *European Environmental Law Review*, July 2007, 7, pp. 191–202.

Vierhaus (2005), Anmerking § 9 ZuG, Anm. 3, in: Körner/Vierhaus/Schweinitz, Treibhausgas-Emissionshandelsgesetz/Zuteilungsgesetz 2007, Kommentar, München 2005.

Ziesing, H.J. (2007), (Deutsches Institut für Wirtschaftsforschung), Entwicklung eines nationalen Allokationsplans im Rahmen des EU-Emissionshandels, Umweltbundesamt, Forschungsbericht 202 41 186/03, UBA-FB 000994, 2007, http://www.umweltdaten.de/publikationen/fpdf-l/3254.pdf.

8. Economic impacts of the EU ETS: preliminary evidence

Onno Kuik and Frans Oosterhuis

1. INTRODUCTION

Emissions trading has always been advocated by many economists as a more cost-effective instrument to reduce emissions in comparison with direct regulation. The basic idea, developed by Dales,[1] is that a cap on total emissions, combined with free trading in emission allowances between polluters, ensures that pollution abatement will take place where it can be done at the lowest costs. In principle, therefore, emissions trading will always lead to net efficiency gains and have a positive impact on overall welfare, unless transaction costs are very high or serious market failures exist.

Nevertheless, there has been much discussion about possible *negative* economic impacts of the EU ETS. Clearly, it was not the instrument of emissions trading itself that was expected to adversely affect industry's production costs and competitiveness. What raised concern was the mere fact that restrictions were imposed on emitting CO_2, whereas in the past this could be done freely, and whereas the global competitors as well as most sectors outside the large energy intensive industry were not confronted with such restrictions. This competitive advantage for industries outside the EU might lead to a shift of 'carbon-intensive' production to those countries, implying 'carbon leakage' with no net reduction of global CO_2 emissions as a result.

The (almost completely) free allocation of allowances in the first stages of the EU ETS has done much to make emissions trading an acceptable instrument to industry.[2] Moreover, the private sector has discovered the inherent 'business opportunity' and has learnt how to turn the efficiency gains of the system into financial profit.

1 Dales (1968).

2 The discussions on competitiveness are currently starting anew as the Commission has indicated it intends to gradually introduce a system of auctioning.

The present chapter addresses the economic impact of the EU ETS by comparing the expectations and predictions that accompanied the introduction of the system with the actual results after the initial years of emissions trading. In addition to impacts on production costs and competitiveness, we also devote some attention to trading patterns (between sectors and countries) and the extent to which the EU ETS leads to actual investments in emission reduction and innovations. Obviously, given the fact that the ETS has started just three years ago, the evidence and findings are still of a preliminary nature.

2. *EX ANTE* ESTIMATES OF ECONOMIC IMPACTS

2.1 Estimates Underlying the Directive

In the Economic Analysis of the Green Paper[3] and the Impact Assessment of the proposed ETS Directive[4] the European Commission used the results of simulations for the EU15 with two different models: the PRIMES energy systems model[5] and the POLES world energy model.[6] Both models were used to calculate the cost *reductions* that an EU-wide trading system could bring about. The reference situation in these calculations was a situation with fixed national emission caps without the possibility of trading between Member States (but with optimal allocation at Member State level, e.g. through a national trading system).

Results of the PRIMES model simulations for different scenarios are summarized in Table 8.1. Compared to a 'business as usual' scenario, the EU-15's CO_2 emissions from energy should be reduced by some 373 Mt in 2010 so as to comply with the Kyoto target. EU-wide emissions trading (with the energy sector and energy intensive industries included in the trading system, as in the eventual Directive) could reduce total compliance cost by 24% when compared to a scenario with least cost allocation at Member State level, and by 66.5% when compared to a 'cheese slicer' scenario. The price of emission allowances would be 33 per tonne (equal to the marginal abatement costs in the trading sector). Inclusion of all sectors (including other industries, services, transport and households) in the EU ETS would have led to limited further reductions in compliance costs and marginal abatement costs.

The PRIMES analysis suggested that under EU15-wide emissions trading The Netherlands, Belgium and Finland would be the main net buyers, and Germany and France the main net sellers of allowances. Compared to the

3 COM(2000) 87 final, Annex I.
4 COM(2001) 581 final.
5 Developed by the E3M Lab, National Technical University of Athens; Capros and Mantzos (2000).
6 Developed by CNRS-IEPE; IPTS (2000).

Table 8.1 Summary of simulations with the PRIMES model

Scenario	Emissions in 2010 (Mt CO_2)				Total compliance cost (mln)	Marginal abatement cost (per tonne CO_2)	
	Total	ESS*	EII*	OTH*		TS*	OTH*
Baseline ('Business as usual')	3193	1182	268	1743	–	n.a.	n.a.
'Cheese slicer' (uniform allocation of emission reductions among sectors within Member States)	2820	1063	257	1501	20508	n.a.	125.8
Reference case (least cost allocation within Member States)	2821	967	240	1614	9026	n.a.	54.3
EU wide trading (energy suppliers and energy intensive sectors)	2821	960	247	1615	6863	33.3	43.3
EU wide trading (all sectors)	2821	956	244	1621	5957	32.6	32.6

* ESS: Energy Supply Sector; EII: Energy Intensive Industries; OTH: Other demand sectors; TS: Trading Sectors.

Source: Capros and Mantzos (2000)

reference scenario, the energy supply sectors (ESS) would reduce their emissions further, whereas energy intensive industries (EII) would reduce less. Assuming an 'efficient' initial allocation of allowances (i.e. equal marginal abatement costs at Member State level), this would mean that the ESS would be a net seller and the EII a net buyer of allowances.

The calculations with the POLES model arrived at an overall cost reduction from emissions trading of 25% (6.1 billion). Germany and the southern EU Member States (except Italy) would be the main sellers of allowances and also enjoy the highest cost savings. The UK would also be a main seller, whereas the other Member States (including Italy and France) would be net buyers, with cost savings below the average of 25%. The equilibrium price of allowances was estimated at 49 per tonne of CO_2.

The differences in outcomes between the two models can largely be explained by differences in their structure and specifications (e.g. the number of countries and sectors distinguished).[7]

Table 8.2 Production cost increases (in %) in energy-intensive industries due to EU ETS according to various sources

	A	B	C	D
Electricity		2.2	12–49	0–93
Steel (basic oxygen furnace)	0.7–1.3	1.1	3–16	17.3
Steel (electric arc furnace)	0.8–0.9	4.8		2.9
Cement	1.9–3.4	2.2	27–136	36.5
Pulp & paper/Newsprint	1.1–1.6	0.8–2.2	1–3	1.0–7.5
Aluminium	3.7	8.6	3–13	0.5–11.4

Notes:
A: Reinaud (2005). Allowance price 10 per tonne of CO_2. Low estimate is for a scenario in which industry receives 98% of its allowance needs; high estimate for a scenario in which it receives 90%.

B: Lund (2007). Allowance price 25 per tonne of CO_2. Low estimate for paper industry relates to case with 66% self-production of electricity; high estimate to case with 0% self-production.

C: Oxera (2004) and Carbon Trust (2004). Figures relate to marginal costs for UK industry. Low estimates are for allowance price of 5 per tonne of CO_2; high estimates for a price of 25.

D: McKinsey and Ecofys (2006). Figures relate to marginal costs. Allowance price 20 per tonne of CO_2; associated electricity price increase 10 per MWh. Estimates for electricity and pulp & paper sectors vary with technique used. Low estimate for aluminium is for secondary production; high estimate for primary production.

[7] See the Commission's Green Paper, COM(2000)87 final, footnote 48.

2.2 Other Estimates

Several studies have estimated the impact of the EU ETS on the production cost of energy intensive industry in Europe. Generally, they do not compare it with CO_2 caps without EU-wide trading, but rather with a 'no policy' scenario. Their results are summarized in Table 8.2 for the most common industries that have been considered. The estimates differ widely, which may be explained by differences in definitions, assumptions and methodology. Generally, however, the highest cost increases are projected to occur in the electricity, cement, and pulp & paper industries.

The impact on competitiveness and profitability depends to a large extent on the ability of a sector to pass on the cost increases to its customers, given its exposure to international competition.

Oxera and Smale et al.,[8] using a Cournot oligopoly model, found positive impacts of the EU ETS on the profitability of most energy intensive industries in the UK. For these industries (which mainly operate on the European market), increases in production costs are more than compensated by higher output prices and they also benefit from the free allocation of grandfathered allowances. The only exception is the aluminium industry, which faces global competition and does not receive any free allowances (because most of its CO_2 is not emitted directly, but indirectly through its electricity use). This industry would be unable to bear the cost increases[9] and was expected to relocate its production to outside the EU.

The consultants McKinsey and Ecofys[10] also expected relocation of (primary) aluminium production. In addition, they considered shifts in production capacity to non-EU countries (and thus carbon leakage) to be a real possibility in the case of cement and steel.

International Energy Agency's researcher Reinaud[11] pointed out that any carbon leakage would be considerably lower than previously expected (at least in the near term) due to the free allocation of the vast majority of allowances under the EU ETS and the lower production cost increase that this allocation mode entails (when compared with a carbon tax).

The German economists Klepper and Peterson[12] have emphasized the important role of the new Member States in mitigating the potential negative competitiveness impact of the EU ETS. Using the DART model (a dynamic CGE model) they found that allowance prices by 2012 would be 11 per

---

8 Oxera (2004) and Smale et al. (2006).

9 Unless protected via long-term contracts or associations with electricity producers.

10 McKinsey and Ecofys (2006).

11 Reinaud (2005).

12 Klepper and Peterson (2004).

tonne of CO_2 if the new Member States are included, and 21 per tonne if they were not to participate.

Demailly et al.[13] stressed the importance of lower levels of aggregation (i.e., a more detailed sectoral split) when analysing the competitiveness effects of the EU ETS. An analysis at 4-digit SIC[14] level showed some electricity-intensive and carbon-intensive sub-sectors to be exceptionally exposed, e.g. the production of precious metals and the manufacturing of lime.

2.3 Expected Impact on Innovation

The European Commission has flagged the ETS as 'an open scheme promoting global innovation to combat climate change'.[15] The claim that the ETS would (or could) induce technological innovation was not substantiated thoroughly, but it is true that economic theory tends to support it.[16] Emissions trading puts a price tag on emitting,[17] and thus makes it attractive to invest in the development and application of new technologies that reduce emissions.

Cames and Weidlich, in a paper on the possible impact of the ETS on innovation in the German electricity industry, argued that different *design options* for an emissions trading scheme would create different innovation incentives and thus influence the level and structure of innovation and technological change.[18] In particular, new entrants should be treated in a similar way to existing installations and therefore receive their allowances free of charge, so as to avoid a competitive disadvantage for new, innovative firms vis-à-vis incumbents. Existing plants should be entitled to retain their allowances, even in case of plant closure, until the end of the commitment period, so as to avoid 'perverse incentives' to postpone the closure of old, inefficient plants.

Gagelmann and Frondel have made an *ex ante* assessment of the potential innovation impact of the EU ETS, based upon theoretical considerations as well as empirical evidence from previous emissions trading schemes in the USA.[19] They concluded that no substantial innovation effects could be expected during the pilot phase (2005–2007), given the generous allocation of allowances, the use of grandfathering rather than auctioning as an allocation mechanism, and the opportunity to use JI and CDM credits. However, for the

13 Demailly et al. (2007).

14 Standard Industrial Classification.

15 Subtitle of the EU's brochure on Emissions trading (http://ec.europa.eu/environment/climat/pdf/emission_trading2_en.pdf).

16 See for example Milliman and Prince (1989).

17 This is true irrespective of the method of allocation. Even if the allowances are distributed freely ('grandfathering'), they represent a monetary value as long as there is a positive market price on the emissions trading market.

18 Cames and Weidlich (2004).

19 Gagelmann and Frondel (2005).

period 2008–2012 and beyond they expected this picture to change, as tighter targets would be likely to apply.

3. *EX POST* ANALYSIS OF THE FIRST PHASE OF EU ETS

Has the EU ETS met its expectations by delivering emissions abatement at least cost? And has the EU ETS negatively affected the competitiveness of European industry or industrial sectors? The first phase of the EU ETS (2005–2007) has come to an end, and as data of this phase are being collected, processed and published, researchers can start addressing these questions in a quantitative way.

3.1 A Closer Look at the Market for EU Allowances (EUAs): Prices and Volumes

Three major types of markets for EUAs have emerged: the over-the-counter (OTC), spot and futures markets. Transactions on the OTC market are between firms, with or without the help of brokers. Price data on this market are confidential. A spot market is a commodities or securities market in which goods are sold for cash and delivered immediately. The most important spot market for EUAs is Powernext Carbon, which quotes daily spot prices. Trade in the futures market concerns options to buy or sell EUAs at a certain date in the future, at a specified price. The most important futures market is the European Climate Exchange. In our overview of price developments, we will primarily focus on developments in the spot market as this market directly reflects day-to-day buying and selling decisions of the market participants.

Figure 8.1 presents daily EUA prices from June 2005 until December 2007. Examining the price developments, three distinct sub-periods can be distinguished.[20] The first period stretched from the start of the EU ETS, January 2005, to the end of April 2006. In this period, the EUA price (per tonne of CO_2), starting from 8 on 1 January, quickly increased to around 30 in July, fluctuating around 20–25 in the following six months, rising to 30 again in April 2006, before collapsing to 13 (a decrease of 54%) in four days between 24 April and 28 April. This sudden collapse coincided with the gradual release of verified emissions data over 2005 that showed that verified emissions from the installations covered by the EU ETS in 2005 had, on balance, fallen short of their allocations. This signalled an oversupply of allowances to the market participants.[21] The European Commission confirmed on 15 May that in 2005

20 Alberola et al. (2008).
21 Ibid.

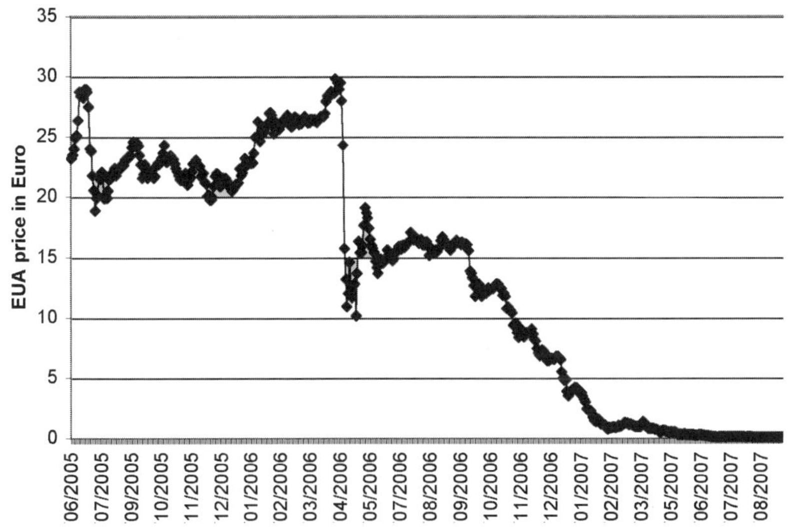

Source: http://www.powernext.fr/ (accessed 14/02/08).

Figure 8.1 Daily EUA prices on Powernext Carbon (June 2005–December 2007).

there had been 93 million or 4.6% more allowances available than actually needed.[22]

The second sub-period stretched from May 2006 until October 2006. In this period, the EUA price moved in a range between 15 and 20. In October 2006, the European Commission announced a stricter validation of National Allocation Plans for the second phase of the EU ETS (2008–2012).

In the third sub-period, after October 2006, the EU ETS price of allowances of the first phase started declining towards zero and did not recover until the end of the first phase trading period in December 2007. In this third sub-period, while EUA spot prices fell to zero, EUA futures prices for delivery in Phase II (2008–2012) remained between 15–20 and were thus, in this third period, totally disconnected from the Phase I spot and futures prices.

On average, daily prices of EUA decreased from 22.51 in 2005, to 17.36 in 2006, and to a low of 0.66 in 2007. The daily volume of trades at Powernext increased from an exchange of 33,891 allowances in 2005 to 125,538 in 2006, and decreased again to 94,060 in 2007.

22 CITL (2006); Kettner et al. (2007).

The most important causes of the price movements of EUAs on the EU ETS market have been (relative) fuel prices, weather conditions, and market information. Of prime importance were the relative prices of coal and gas. With higher market prices of natural gas, as in late 2005 and early 2006, power plant operators may find it profitable to switch to coal (at the margin). They thereby increase their demand for EUAs because coal emits more CO_2 per unit of energy than natural gas. Higher demand for EUAs increases their market price, and makes CO2-reducing (e.g., energy-saving) measures more attractive at the margin, thereby offsetting the rise in emissions because of the switch to coal. The price of EUAs has also been shown to be affected by unanticipated changes in weather conditions during extremely cold weather. Information disclosure on overall supply and demand in the market has been shown to have had the largest effect on prices. Finally, the expectation by the market participants of an oversupply of EUAs at the end of the first compliance period, in combination with the prohibition on banking the allowances for use in the second compliance period (2008–2012), forcefully drove the price of EUAs down to almost zero in the final year of the first compliance period.[23]

3.2 The Direction of Trade: Buyers and Sellers

In the *ex ante* assessments of the EU ETS, some predictions were made about the direction of trade: who traded with whom? While it is not possible to identify buyers and sellers of each individual transaction, some inferences can be made from examining the difference between allocated allowances and verified emissions across Member States and sectors. Kettner et al.[24] have made a detailed analysis of these differences for the first operational year of the EU ETS (2005). From this analysis, they draw the following three important conclusions.

For 19 Member States their initial allocation of allowances exceeded their emissions. In financial market jargon, their position in the EUA market was 'long'. For six Member States their emissions exceeded their allocations, i.e., their position was 'short'. The latter group of Member States must have been net buyers on the EUA market. They are: UK, Ireland, Spain, Italy, Austria, and Greece. In the aggregate, the EUA market was long (allowances exceeded emissions by 93 million or 4.6%).

On a sectoral level, there is a marked difference between the power and heat sector on the one hand, and the other sectors on the other hand. On balance, the power and heat sector was short, while the other sectors were long.

---

23 Alberola and Chevallier (2007).

24 Kettner et al. (2007).

There are also marked differences between large and small installations. On balance, large installations were short, while small installations were long. In addition, the variance of the difference between allocation and emissions is much smaller for large installations than for small installations.

3.3 Over-allocation or Abatement?

There can be several reasons for an apparent surplus of allowances in the EUA market. A reason that is often put forward is that national authorities were too lenient in allocating allowances to their national industries, because of industry pressure and fears for loss of competitiveness and jobs. It can be added that the first National Allocation Plans (NAPs) had to be prepared under severe time pressure and often based on (emissions) data of dubious quality. A second set of reasons concerns projection errors in the allocation process, related to future fuel prices, weather conditions, and other variables that are likely to affect CO_2 emissions. A third reason is that authorities underestimated the level of abatement that would be forthcoming because of the ETS. Ellerman and Buchner examined the level of abatement that can be attributed to the EU ETS in its first year of operation (2005).[25] Based on historical trends and known fuel prices and weather conditions in 2005, Ellerman and Buchner tentatively conclude that without the EU ETS (business-as-usual), CO_2 emissions in the EU25 would have been between 50 and 200 million tonnes higher than what they actually were. Hence, EU ETS would have led to an additional abatement of between 50 and 200 million tonnes. By means of a more qualitative analysis, examining the range of options for abatement that would have been realistically available within such a short time span, Kettner et al. are more sceptical, and argue that the reduction potential of the available options (basically fuel switching and good housekeeping) would have been 'rather limited'. This leads us to tentatively hypothesize that abatement would probably be on the lower end of the Ellerman-Buchner estimate (say, 50 million tonnes), explaining possibly half the aggregate surplus of allowances in 2005 (93 million tonnes).

3.4 Effects on Costs, Prices, and Competitiveness

The power and heat sector has a special position in the EU ETS. It was allocated more than half of all EUAs, but, as we saw in Section 3.2, its allocation was less generous than the allocation for the other sectors. An explanation for this less generous allocation is that the power and heat sector is less exposed to international (extra-EU) competition than the other sectors in the EU ETS, such as refineries, iron and steel, cement, glass, lime, ceramics, and pulp and

25 Ellerman and Buchner (2006).

paper. All else being equal, being less exposed to international competition makes it easier to pass on additional costs to consumers. By now there is a fair amount of evidence that many power producers did indeed pass on the opportunity costs of their EUA to consumers.[26] As almost all EUAs had been given free to the power and heat sector, this inevitably increased gross margins, leading to what is called 'windfall profits'. Estimates of the effect of the EU ETS on wholesale electricity prices range from 1–5 /MWh in France to 13–19 /MWh in Germany.[27]

There are many studies and reports on the expected effects of the EU ETS on costs and competitiveness (see Section 2.2), but the empirical evidence *ex post* did not reveal major effects as yet. Partly this is undoubtedly due to a deliberate or accidental over-allocation of allowances to energy-intensive manufacturing sectors in the first trading period. A study of the European Trade Union Confederation and others[28] reported significant over-allocations in the iron and steel sector and in the cement sector. The reasons for an estimated 24% over-allocation in the iron and steel sector in 2005 were: an (unexpected) reduction in European steel production between 2004 and 2005; the attribution of quota to installations undergoing expansion of capacity which had not yet gone into operation; some technical difficulties in the attribution of emissions; and the fear [of the authorities] for negative effects on competitiveness of the European iron and steel industry.[29] The cement sector also was 'a winner in the European system for the allocation and trading of CO_2 emissions', with an over-allocation of more than 7%.[30] The value of the excess allowances in both sectors in 2005 represented 0.7% and 0.9% of their respective turnovers.[31]

The International Energy Agency examined the competitiveness of the European aluminium sector.[32] Against the background of the long-time trend of a diminishing market share of European producers of primary aluminium, the study found no statistical relationship between EUA prices and net trade flows of primary aluminium in Europe. As possible explanations for this result, the author points to the facts that EU electricity costs in the past ten years have not increased more than the global average; and that most

26 Sijm et al. (2006), estimated pass-through rates of 60% to 100% of allowance costs for Germany and The Netherlands.
27 Ibid., Table 2, page 60.
28 ETUC et al. (2007).
29 Ibid., page 117.
30 Ibid., page 132.
31 Ibid. The excess allowances were valued at an average market price of 12.5.
32 Reinaud (2008).

European smelters operate under long-term, fixed-price electricity contracts, many of which will not expire before 2010.[33]

4. A PRELIMINARY COMPARISON BETWEEN *EX ANTE* AND *EX POST* ASSESSMENTS OF THE EU ETS

Ex ante assessments, such as those presented in Section 2, assumed that EUAs would be treated as 'normal' factors of production by the firms covered by the EU ETS, even if the allowances were given free of charge. This means that decision-makers in these firms would consider the opportunity costs of the allowances in their production decisions. Evidence (at least from the power sector) supports this assumption.

The *ex ante* assessments commonly modeled the EU ETS as being equivalent to the least-cost solution to an optimization problem, solved by one, fully informed, European planner. What the assessments therefore did not foresee was the significant impact of asymmetric information by market participants on trade and prices, and hence on the volatility of the EUA price. In addition, the assessment assumed a binding cap on emissions, while the cap in the 'learning phase' of the EU ETS was, in effect, and on balance, not binding.[34]

The *ex ante* assessments also did not pay specific attention to the banking restrictions in the EU ETS. There is increasing evidence that banking restrictions between trading periods increase price volatility and seriously threaten the intertemporal (and therefore the overall) efficiency of the system.

Table 8.1 above showed *ex ante* assessments of CO_2 emissions in EU15 under different climate change regimes. A comparison between the 'Baseline' (business-as-usual) and 'EU-wide trading' suggested a difference of emissions of 372 million tonnes of CO_2 between the scenarios, of which 243 million tonnes are from the EU ETS trading sectors.[35] The EUA equilibrium price was estimated to be 33.3. We assume that this difference would be caused by abatement activities within the sectors. For the year 2005, our best estimate of abatement is 50 million tonnes of CO_2 in the EU ETS trading sectors of EU27. We may assume, however, that if abatement has occurred, it has occurred in the 'old' EU15, and not in the new Member States. Therefore, our best empirical, *ex post* evidence is that 50 million tonnes of CO_2 have been abated in

---

33 Today, only 18% of EU smelter capacity has no long-term, fixed-price electricity contracts. All long-term contracts will expire by 2016, though. Reinaud (2008).

34 Although not yet required by the Kyoto Protocol, the emission quotas in the EU ETS 'learning phase' were binding in a legal sense. Because of the oversupply of allowances they were not binding in an economic sense, however.

35 The sectors ESS (Energy Supply Sector) and EII (Energy Intensive Industries).

EU15 at an 'average' EUA price of 22.5. There are, of course, obvious difficulties in comparing the *ex ante* and *ex post* estimates, because of, for example, the different time periods (2010 versus 2005). Nevertheless, for 2005 at least, the *ex ante* estimate is of the same order of magnitude as the *ex post* estimate.

There are also some differences. One *ex ante* assessment (POLES) showed the UK as a main seller and France as buyer. In 2005, the UK was biggest buyer and France possibly a seller (at least France had a long position). What this shows is that the *ex ante* assessments made the wrong assumptions on the (over-) allocations of allowances in Member States. This nicely illustrates the broader issue that it is very difficult to make predictions on the distributional consequences of policy initiatives before the details of the initiatives are known.

Ex ante assessments on costs and competitiveness were somewhat mixed. The *ex ante* estimates of costs are probably correct,[36] but cost increases do not automatically translate into effects on competitiveness, as, for example, argued by Oxera and Smale et al.[37] *Ex post* evidence so far does not reveal any negative impacts on competitiveness of European industries, but it does suggest that the EU ETS has favoured at least some industries with 'windfall profits'.

Judging by industry's own opinions, the EU ETS has already had a strong influence on investment and innovation behaviour right from its start. A survey by McKinsey and Ecofys in 2005 revealed that for 50% of the companies concerned the EU ETS played a key role in their long-term decisions. Likewise, about half of the companies claimed that the EU ETS had a strong or medium impact on decisions to develop innovative technologies, with the strongest impact in the steel industry.[38] A study for the American NGO Environmental Defense describes various cases illustrating the kind of innovations that have been stimulated by the introduction of a price tag for CO_2 emissions.[39] These impacts on innovation have occurred even though the first phase of the EU ETS did not fulfill all conditions for 'innovation-friendliness': e.g., most Member States have chosen to cancel EUAs upon plant closure, which can be seen as discouraging innovation (see Section 2.3).

---

[36] Basic cost estimates (share of the (opportunity) cost of allowances in total production costs) are basically engineering type assessments.

[37] See Section 2.2, Oxera (2004); Smale et al. (2006).

[38] McKinsey & Company and Ecofys (2005).

[39] Petsonk and Cozijnsen (2007).

5. CONCLUSION

In many respects, it is still too early to draw conclusions on the economic impacts of the EU ETS. The preliminary evidence, however, clearly suggests that the scheme is actually influencing operational and strategic decisions of energy intensive industry in the EU. Even though some details (e.g. regarding prices and trading patterns) of *ex post* outcomes may differ from *ex ante* estimates, one can say that the system is, by and large, delivering the expected economic results. There are also strong indications that the EU ETS actually contributes to emission reductions and to investment in the development and application of new, low-carbon technology. Nevertheless, the 'real' test for the scheme will be in the current trading period 2008–2012, and beyond. Growing experience, a broader scope (in terms of sectors and countries), a permanent 'scarcity signal', the introduction of auctioning, more harmonization of allocation and trading rules, and fewer restrictions on trading (e.g. no *ex post* adjustments): all this will probably contribute to a better economic performance, enabling the instrument to reveal its full potential in terms of cost-effective emission reduction.

REFERENCES

Alberola, E. and J. Chevallier (2007), 'European Carbon Prices and Banking Restrictions: Evidence from Phase 1 (2005–2007)', Economix Working Paper 2007–32.

Alberola, E., J. Chevallier and B. Chèze (2008), 'Price Drivers and Structural Breaks in European Carbon Prices 2005–2007', *Energy Policy* 36, 787–97.

Betz, R. and M. Sato (2006), 'Emissions Trading: Lessons Learnt From the 1st Phase of the EU ETS and Prospects for the Second Phase', *Climate Policy* 6, 351–59.

Cames, M. and A. Weidlich (2004), *Emissions trading and innovation in the German electricity industry – Impact of possible design options for an emissions trading scheme on innovation strategies in the German electricity industry.* http://www.iw.uni-karlsruhe.de/Publications/CamesWeidlich_2004.pdf.

Capros, P. and L. Mantzos (2000), *The Economic Effects of EU-Wide Industry-Level Emission Trading to Reduce Greenhouse Gases. Results from PRIMES Energy Systems Model.* E3M Lab, Institute of Communication and Computer Systems of National Technical University of Athens, May 2000.

CITL (Community Independent Transaction Log) (2006), Data on Allocation and Compliance. European Commission, Brussels [available at http://ec.europa.eu/environment/ets/].

Dales, J.H. (1968), *Pollution Property and Prices: an Essay in Policy-making and Economics.* University of Toronto Press, Toronto.

Demailly, D., M. Grubb, J.-C. Hourcade, K. Neuhoff and M. Sato (2007), *Differentiation and Dynamics of EU ETS Competitiveness Impacts.* Climate Strategies, Research Theme 1.3, Interim Report, CIRED and University of Cambridge, 30 March 2007.

Ellerman, D. and B. Buchner (2006), 'Over-Allocation or Abatement? A Preliminary Analysis of the EU ETS Based on the 2005 Emissions Data', FEEM Working Paper 139.2006.

ETUC (European Trade Union Confederation), Instituto Sindical de Trabajo, Ambiente y Salud (ISTAS), Social Development Agency (SDA), Syndex, Wuppertal Institute (2007), *Climate Change and Employment – Impact on Employment in the European Union-25 of Climate Change and CO_2 Emission Reduction Measures by 2030.*

Gagelmann, F. and M. Frondel (2005), 'The Impact of Emission Trading on Innovation – Science Fiction or Reality?' *European Environment* 15, 203–11.

IPTS (2000), *Preliminary Analysis of the Implementation of an EU-Wide Permit Trading Scheme on CO_2 Emissions Abatement Costs. Results from the POLES model.* Institute for Prospective Technical Studies, Joint Research Centre, Sevilla, April 2000.

Kettner, C., A. Köppl, S.P. Schleicher and G. Thenius (2007), 'Stringency and Distribution in the EU Emissions Trading Scheme – The 2005 Evidence', FEEM Working Paper 22.2007.

Klepper, G. and S. Peterson (2004), 'The EU Emissions Trading Scheme: Allowance Prices, Trade Flows and Competitiveness Effects', *European Environment* 14, 201–18.

Lund, P. (2007), 'Impacts of EU Carbon Emission Trade Directive on Energy-intensive Industries – Indicative Micro-economic Analyses', *Ecological Economics* 63, 799–806.

McKinsey & Company and Ecofys (2005), *Review of EU Emissions Trading Scheme. Survey Highlights*. European Commission, DG Environment, November 2005.

McKinsey & Company and Ecofys (2006), *EU ETS Review. Report on International Competitiveness*. European Commission, DG Environment, December 2006.

Milliman S.R. and R. Prince (1989), 'Firm Incentives to Promote Technological Change in Pollution Control', *Journal of Environmental Economics and Management* 17, 247–65.

Oxera (2004), *CO_2 Emissions Trading: How Will it Affect UK Industry?* Report prepared for The Carbon Trust. Oxford, July 2004.

Petsonk, A. and J. Cozijnsen (2007), *Harvesting the Low-Carbon Cornucopia: How the European Union Emissions Trading System (EU-ETS) is Spurring Innovation and Scoring Results*. Environmental Defense, March 14, 2007. http://www.edf.org/documents/5852_HarvestingtheLowCarbonCornucopiaMarch2007.pdf.

Reinaud, J. (2005), *Industrial Competitiveness under the European Union Emission Trading Scheme*. International Energy Agency, Paris, February 2005.

Reinaud, J. (2008), Ex-Post Evaluation of the EU ETS: Impacts on the Primary Aluminium Sector, Paper presented at the workshop 'Tackling Leakage in a World of Une-qual Carbon Prices', Climate Strategies, Paris, and February 4, 2008.

Sijm, J., K. Neuhoff and Y. Chen (2006), 'CO2 Cost Pass-Through and Windfall Profits in the Power Sector', *Climate Policy* 6, 49–72.

Smale, R., M. Hartley, C. Hepburn, J. Ward and M. Grubb (2006), 'The Impact of CO_2 Emissions Trading on Firm Profits and Market Prices', *Climate Policy* 6, 31–48.

PART III

Alternatives and new developments

9. Regional regulatory initiatives addressing GHG leakage in the USA

Erik B. Bluemel*

1. INTRODUCTION

While many may view the United States of America's refusal to participate in the Kyoto Protocol as a strategic effort to undermine mandatory greenhouse gas (GHG) reduction targets – a view not without some merit, regional efforts to combat climate change within the United States can provide valuable information about the design of effective regional GHG emissions trading schemes outside the United States. Indeed, despite the United States' reluctance to join the Kyoto Protocol and accept mandatory national GHG emissions limits, a variety of states and localities have taken it upon themselves to impose mandatory GHG caps in their jurisdictions. These regional regimes are sprouting up with the explicit goal of inducing action by the national government,[1] and they appear to have some effectiveness in promoting industry support for a national program. Congress is now seriously considering legislation to develop a nationwide GHG cap-and-trade system similar in nature to the Kyoto Protocol and the regional regimes developed by the various states and localities.[2]

* Assistant Professor of Law, University of Denver Sturm College of Law; Member, 2005–present, Commission on Environmental Law, IUCN-World Conservation Union. This Chapter reflects the views of the author only and does not necessarily reflect the views of any of the author's institutional affiliates, their composite organs, or their staffs. The author can be reached at ebluemel@law.do.edu.

1 See, e.g., RGGI (20/12/2005), Memorandum of Understanding, art. 6(C) (hereinafter, 'RGGI MOU'); California Health & Safety Code § 38501(d).

2 Parker and Yacobucci (24/04/2007). At the time of this writing, the leading presidential candidates all endorse a national cap-and-trade system, and the United States Deputy National Security Adviser, Daniel M. Price, told government officials in Paris on 25 February 2008, that the United States is ready to accept binding international obligations to reduce GHGs if other major economies, including China and India, do the same. How the regional and state programs will interact with any national program that may develop is unclear. See Committee on Energy & Commerce (February 2008). For a discussion of some of the problems inherent in a climate change

Although the United States is something of a latecomer in the battle against climate change, it nevertheless has a history of innovation when it comes to using market-based mechanisms in environmental regulation. It developed the first major emissions trading program: the Acid Rain Program created by the Clean Air Act Amendments of 1990.[3] The Acid Rain Program established a cap on the total allowable amount of sulfur dioxide emissions and enabled sources of pollution such as power plants to purchase pollution credits, or allowances, from one another as a means of achieving overall pollution reductions at lower cost. The design of that program provided a number of useful insights for the design of some of the current GHG cap-and-trade programs around the world.[4] But the Acid Rain Program was designed to address downwind pollution and acid deposition in the eastern portion of the United States. It included all of the major sources that contributed to the problem of acid rain. In that sense, it was not a 'regional' regime or 'open' system: it was a complete and holistic regime.

GHG pollution, on the other hand, is inherently global in nature. Mixing of local GHG pollution in the global atmosphere occurs within weeks, and GHGs can last in the atmosphere for years and even centuries.[5] Accordingly, a program designed to address the problems resulting from GHG pollution must address all the major global sources of such pollution.

Despite the global nature of climate change, however, a number of sub-global, or regional regimes have sprouted up to tackle the problem of GHG pollution. The European Union Emissions Trading Scheme, the United Kingdom Emissions Trading Scheme, and the forthcoming trading regimes of Australia and Canada are just a few examples of some leading sub-global and regional GHG cap-and-trade programs that exist to combat global climate change.

A regional regime that does not address all sources of GHG pollution risks emissions 'leakage' by promoting emigration of industry to countries and regions not regulated by the GHG regime. The relocated industry may then export energy or GHG-intensive products back into the regulated regime. A regional regime that fails to address all sources contributing to GHG pollution – including pollution from imports, therefore, may prove ineffective at reducing actual GHG emissions associated with consumption in the region.

---

regime with overlapping jurisdictions and differing requirements, see Bluemel, E.B. (2007a).

[3] Clean Air Act Amendments of 1990, Pub. L. No. 101-549, 104 Stat. 2399 (codified as amended at 42 U.S.C. §§ 7401-7671q (2000)).

[4] See generally Pring (2006); Ellerman et al. (2003); Tietenberg et al. (2003).

[5] See United Nations Env't Program (2007), p. 43 fig. 2.1.

Addressing this regulatory challenge given the current framework of international trade law is no easy task,[6] but the regional GHG initiatives in the United States – the authority of which is constrained by a system that bears many of the hallmarks of international trade law – provides many useful and innovative approaches to tackle the problem of leakage in a regional regime. This chapter describes some of the major regional GHG cap-and-trade initiatives in the United States, defines the problem of emissions leakage in the context of regional GHG regulation, identifies the various approaches employed by the regional GHG cap-and-trade regimes in the United States to address leakage, and draws some basic conclusions about the design of regional GHG regimes outside the United States, including in the European Union.

2. REGIONAL GHG REGIMES IN THE UNITED STATES

2.1 Regional Greenhouse Gas Initiative

The Regional Greenhouse Gas Initiative (RGGI, pronounced 'REGGIE') was the first regional cap-and-trade program established in the United States. On 20 December 2005, the eastern states of Connecticut, Delaware, Maine, New Hampshire, New Jersey, New York, and Vermont agreed to establish statewide carbon dioxide (CO_2) emission caps. Massachusetts, Rhode Island, and Maryland later joined RGGI, adopting similar CO_2 emission caps.

RGGI operates within a few different regional transmission organizations (RTOs) and independent system operators (ISO) – entities that administer the electricity transmission grids in the RGGI region. The major RTOs and ISOs include PJM (Pennsylvania-New Jersey-Maryland, which provides for the movement of wholesale electricity to eleven states), ISO New England (covering six New England states), and New York ISO. Trading regularly occurs between these various power pools, as well as with Canada. Not all of the state members of these RTOs and ISOs are members of RGGI, so it is possible for the RGGI states to import electricity from non-RGGI states. Table 9.1 identifies some of the primary importers of electricity within the RGGI region and exporters outside the RGGI region within the PJM RTO, along with their primary fuel source and contribution to GHG pollution from the electricity generated by their combined sources of electricity. Currently, the RGGI states

[6] See Rabe et al. (2005, pp. 34–36), (noting that attempts to address leakage may be viewed as disguised economic protectionism).

collectively import approximately 90 million megawatt hours of electricity, or about 21% of their total consumed electricity.[7]

Although electricity used within the RGGI region may be imported from outside the RGGI region, the RGGI states agreed to adopt mandatory CO_2 limits that stabilize emissions from sources located within RGGI. The RGGI states agreed to stabilize CO_2 emissions at 2009 levels between 2009 and 2014 (effectively reducing emissions from projected economic growth in the region), and to reduce their CO_2 emissions by 2.5% annually thereafter, through 2018. This amounts to a fairly modest emission reduction (10% below 2009 levels) by 2018. Given projected growth, this translates into GHG emissions approximately 35% lower than under a business-as-usual scenario.

In addition to requiring only modest emission reductions, the impact of RGGI is further constrained by other limitations of the program's scope. RGGI only covers CO_2 emissions; it does not regulate the other GHGs. It also is limited to electricity-generating units over 25 MW that use more than 50%

Table 9.1 PJM Generation and Sales in 2006[8]

	State (Primary Fuel Source)	Net Generation (MWh)	Retail Sales (MWh)	Net Generation (% of Sales)	CO_2 Emissions (1000 tons)	Average Retail Price (¢/kWh)
RGGI	New Jersey (Nuclear)	60,700,139	79,680,947	76%	19,861	11.88
RGGI	Delaware (Coal)	7,182,179	11,554,672	62%	5,885	10.13
RGGI	Maryland (Coal)	48,956,880	63,173,143	77%	30,497	9.95
Non-RGGI	West Virginia (Coal)	93,815,804	32,312,126	290%	85,075	5.04
Non-RGGI	Pennsylvania (Coal)	218,811,595	146,150,358	150%	125,864	8.68
Non-RGGI	Indiana (Coal)	130,489,788	105,664,484	123%	121,950	6.46

[7] See United States Energy Information Administration (2007), *State Electricity Profiles 2006*, DOE/EIA-0348(01)/2, pp. 261–2, tbl. A1 (Selected Electric Industry Summary Statistics by State, 2006), http://www.eia.doe.gov/cneaf/electricity/st_profiles/sep2006.pdf.

[8] Ibid. The CO_2 emissions identified in the table relate to emissions from all electricity sources within each state, not merely the emissions from the primary fuel source.

fossil fuels for combustion. This is expected to cover approximately 750 sources generating CO_2 pollution.[9] It does not, however, cover electricity consumption by industrial processes.

Like the EU ETS cap-and-trade program discussed throughout this book, each electricity-generating unit (source) has a CO_2 allocation, distributed as allowances (also known as credits). RGGI permits sources to obtain additional pollution allowances by paying other sources to reduce their emissions or funding emissions-reducing projects known as 'offsets.'[10] To prevent sources from polluting in excess of their allowances and remaining in compliance with RGGI by funding offsets – an outcome that would limit the potential of the program to force the development and deployment of newer clean technology – RGGI limits the amount of offsets that a source may use to ensure compliance with its net CO_2 allocation.

Under a business-as-usual scenario, CO_2 reductions are recognized from offset projects located within the RGGI states, non-RGGI jurisdictions of the United States (if those jurisdictions have a cap-and-trade program with specific tonnage limitations for GHGs imposed on significant economic sectors), and non-RGGI jurisdictions within the United States that have entered into a memorandum of agreement with the RGGI state to ensure the credibility of the offsets.[11] Offsets are credited at a ratio of one ton of CO_2 reduced for a one ton allowance of CO_2 that can be emitted by the source.[12] Under a business-as-usual scenario, a source can purchase offsets totaling up to 3.3% of its total CO_2 emissions allocation.

If the price of CO_2 on the RGGI market exceeds certain price thresholds after the 'market settling period' – the first 14 months of the compliance period – RGGI attempts to minimize the economic burden imposed by the higher price of CO_2 on the regulated sources by using what is essentially a safety

---

9 RGGI adopted a source-based approach to regulating CO_2 pollution because it concluded that such an approach is most consistent with RGGI's goal of reducing the carbon intensity of generators. Peress and Booher (2007, p. 8).

10 Currently, regulated entities can only earn emission offsets from certain categories of projects, including: natural gas, heating oil and propane energy efficiency, landfill gas and combustion, methane capture from animal operations, forestation of non-forested lands, and reductions in sulfur hexafluoride (SF6) emissions from transmission and distribution equipment. RGGI MOU; RGGI (8 August 2006), Memorandum of Understanding Amendment, art. 2 (hereinafter, 'RGGI, MOU Amendment').

11 RGGI, MOU Amendment, n. 10, art. 3(a).

12 Initially, the RGGI states concluded that the potential uncertainties associated with the lack of regulatory control over offset projects outside the RGGI region warranted discounting the value of offset credits earned in non-RGGI jurisdictions by half. RGGI MOU, n. 10, art. 2(F)(2)(a)(2). After further discussions, however, the RGGI states ultimately determined that such a discounting mechanism was inappropriate, and the MOU was amended accordingly.

valve that expands the use of offsets. If the price of CO_2 exceeds the modest price of \$7 (2005\$) per ton, a source may purchase offsets totaling up to 5% of the source's total CO_2 allocation, and offsets throughout North America become eligible. If the price of CO_2 reaches \$10 (2005\$)/ton (as adjusted for inflation), international offsets become eligible, and sources can purchase offsets up to 10% of the source's total CO_2 allowances.[13]

Because RGGI imposes only modest CO_2 reductions over a decade and its offset safety valve is triggered at a relatively modest CO_2 price, the projected cost of compliance for industry and consumer cost increases are expected to be minimal. Through the stabilization period, RGGI is expected to increase the cost of compliance for industries by less than 5% of the average wholesale electricity price.[14] These costs will not be passed on to consumers to any significant degree. Average household electricity bills are expected to experience an annual increase of \$3–16, which is approximately a 0.3–0.6% increase.[15] After factoring in increased energy efficiency that should result from RGGI, RGGI should actually result in net energy cost savings to consumers.[16] RGGI's overall impact on the economy is similarly expected to be positive, as RGGI should promote investment in new technologies, including nuclear energy, which exist throughout the RGGI region.[17]

Since RGGI only imposes modest requirements on electricity generators in the region, the cost of compliance (cost adder) for those generators is not expected to be substantial. As a result, modeling performed for RGGI suggests that RGGI should not promote a large increase in the amount of lower-cost electricity imported from non-RGGI states. This is in part the result of long-term contracts, which account for approximately 14% of electricity generated in the RGGI region, reducing the near-term leakage potential of the program.[18] Indeed, the RGGI modeling concluded that its design should result in acceptable levels of emissions leakage resulting from increased electricity imports from unregulated sources.

---

13 RGGI, MOU Amendment, n. 10, art. 5(a)(2). Initially, the RGGI states wanted to increase the use of offsets under this 'Safety Valve Trigger' to 20% in the extended year of compliance. See RGGI MOU, n. 10, art. 2(F)(4)(a)(3). After further discussions, however, the RGGI states decided to limit the use of offset credits to 10%, and the MOU was amended accordingly.

14 RGGI (14/03/2007), Initial Report of the RGGI Emissions Leakage Multi-State Staff Working Group to the RGGI Agency Heads, p. 6 (hereinafter, 'RGGI, Initial Leakage Report').

15 RGGI (2005), 'RGGI Region Projected Household Bill Impacts,' http://www.rggi.org/docs/ rggi_house_hold_bill_impacts12_12_05.ppt, December 12.

16 Sussman (2006, p. 47).

17 Ibid.

18 RGGI, Initial Leakage Report, n. 14, p. 7.

Nevertheless, as described below, there are a variety of factors that contribute to emissions leakage. While the modeling suggests acceptable levels of leakage, the RGGI states have always recognized that the competitive electricity markets within which the program will operate create a potential leakage problem.[19] If the RGGI states cannot resolve the potential leakage problem, the states will try to mitigate those leakage-induced emissions. A report released by RGGI in March 2008 suggests that demand-reduction strategies should help reduce such leakage.[20]

Currently, the RGGI states appear marginally concerned about the impact of potential leakage on the electricity market, the region's competitiveness, or uncontrolled emissions. RGGI staff has concluded that the leakage problem is a near- to medium-term concern because the staff believes the political momentum in the United States is toward a national program, which staff expects will equalize regulatory inequities among the regions.[21] This view, however, is somewhat myopic because so long as regional regimes are permitted to exist and differentiate themselves with stricter requirements than the national regime, leakage concerns will remain, even if reduced. Potentially more troubling, however, is that RGGI is designed to be a model for a national program. If RGGI cannot control leakage, it may very well hinder the development of a national program, which must also concern itself with leakage abroad.

Indeed, concerns about leakage are evident in the national program context, as evidenced by discussions over the Lieberman-Warner climate change bill. That bill seeks to address leakage to protect industry from cheaper, unregulated competition abroad.[22] National legislation may curtail leakage from abroad while remaining consistent with the rules of international trade, but it is no easy task to design legislation consistent with those rules. Given the difficult road ahead for national legislation, a regional regime that fails to address leakage in the domestic context, despite efforts to minimize such leakage and a more hospitable legal environment within which to do so, could cause national regulators and legislators to look for mechanisms other than a cap-and-trade system to curb GHG pollution.

---

[19] RGGI, MOU, n. 10, art. 6(A) ('The Signatory States recognize the potential that the Program may lead to increased electricity imports and associated emissions leakage.').

[20] RGGI (March 2008), Potential Emissions Leakage and the Regional Greenhouse Gas Initiative (RGGI), at 41-42. But see Tanton (2008).

[21] RGGI, Initial Leakage Report, n. 14, p. ES-2.

[22] See Brevetti (2008).

2.2 California and the Western Climate Initiative

Following the model established by RGGI, California, after a number of intermediary steps, is working to establish a mandatory cap-and-trade regime for GHGs. On 1 June 2005, Governor Arnold Schwarzenegger signed Executive Order S-3-05, calling for statewide GHG emission reductions by 80% below 1990 levels by 2050. California started the process of meeting this goal with the passage of the California Global Warming Solutions Act of 2006, Assembly Bill 32 (AB32), in fall 2006.

AB32 establishes a GHG emission limit of 1990 levels by 2020, and permits the use of market-based mechanisms to achieve those levels.[23] It covers all six GHGs regulated by the Kyoto Protocol under Annex A: CO_2, methane, nitrous oxide, hydrofluorocarbons, perfluorocarbons, and sulfur hexafluoride. AB32 imposes GHG reduction requirements for sources 'whose emissions are at a level of significance, as determined by the state board.'[24] The program scope is left somewhat undefined by the legislation, which explicitly mentions only 'electrical generation, petroleum refining, and statewide fuel supplies' as having GHG reduction requirements.[25] The program is nevertheless expected to cover other major industrial sources as well.[26]

Notably, AB32 gives official approval to load-based standards. Load-based standards regulate electricity delivered into a region from any source, whether domestic or imported. In contrast to generator-based standards, which apply only to generators located within the region's boundaries, load-based standards target consumption of electricity within the region's boundaries. Accordingly, AB32 defines 'statewide greenhouse gas emissions' as:

> the total annual emissions of greenhouse gases in the state, *including all emissions of greenhouse gases from the generation of electricity delivered to and consumed in California*, accounting for transmission and distribution line losses, *whether the electricity is generated in state or imported*. Statewide emissions shall be expressed in tons of carbon dioxide equivalents.[27]

AB32, then, gave official sanction to the California Public Utility Commission's (CPUC) decision to use a load-based standard as the unifying framework for utility procurement incentives.[28]

---

23 California Health & Safety Code §§ 38550, 38570.
24 Ibid. § 38505(i).
25 Ibid. § 38561(a).
26 Patrick (2006, p. 3).
27 California Health & Safety Code § 38505(m) (emphasis added).
28 California Public Utilities Commission (16/02/2006), Decision 06-02-032; see Fitch (2007).

While AB32 does not require a cap-and-trade scheme,[29] on 16 October 2006, just a few weeks after signing AB32, Governor Schwarzenegger issued Executive Order S-17-06, calling for the creation of a 'market-based compliance program with the goal of creating a program that permits trading with the European Union, the Regional Greenhouse Gas Initiative and other jurisdictions.' Like the RGGI states, the California legislature believes passage of AB32 may induce federal and international action to address global climate change.[30] In the interim, AB32 calls on California to 'minimize leakage.'

California is concerned about leakage because although its net electricity imports approximate the net import ratio of the RGGI states from non-RGGI states, the imported electricity in California is substantially more carbon-intensive, and half as costly, as electricity generated within California. Indeed, while California imports between 22–32% of its electricity, those imports account for 39–57% of electricity-related CO_2 emissions.[31] This differential is attributed primarily to stricter environmental regulations in California than surrounding jurisdictions and California's substantial reliance on natural gas (and absence of coal-fired power plants) for electricity generation. Accordingly, emissions leakage is a potential problem for California, especially for its energy-intensive industries such as refining.

Given California's concern regarding leakage, it is attempting to incorporate all the states in the Western Interconnection – an alternating current power grid stretching from Western Canada south to Baja California in Mexico – in a regional GHG regime. California's efforts yielded the creation of the Western Regional Climate Action Initiative on 26 February 2007, by a group of Western states (Arizona, California, New Mexico, Oregon, and Washington), with the purpose to 'collaborate in identifying, evaluating, and implementing ways to reduce GHG emissions.'[32] Utah, Montana, and the Canadian provinces of British Columbia, Ontario, Quebec, and Manitoba have since joined the Initiative, now known as the Western Climate Initiative (WCI). The WCI was established to develop 'a regional market-based multi-sector

---

[29] Although the California law does not require a cap-and-trade scheme, and there is some opposition to a cap-and-trade scheme from the environmental justice lobby, it nevertheless appears, at the time of this writing, that California will ultimately implement a cap-and-trade program. See Whetzel (2008, p. A–5).

[30] California Health & Safety Code § 38501(d) ('[A]ction taken by California to reduce emissions of greenhouse gases will have far-reaching effects by encouraging other states, the federal government, and other countries to act.').

[31] California Energy Commission (2006), 'Inventory of California Greenhouse Gas Emissions and Sinks: 1990 to 2004,' http://www.energy.ca.gov/2006publications/CEC-600-2006-013/CEC-600-2006-013-SF.PDF, p. 12.

[32] Western Regional Climate Action Initiative, Memorandum of Understanding, p. 1.

mechanism, such as a load-based cap and trade program, to achieve the regional GHG reduction goal.'[33]

The WCI partners agreed to establish a regional GHG emission reduction goal of 15% below 2005 levels by 2020.[34] Currently, the WCI is very early in the design phase. The scope and regulatory structure of, allocation and reporting of allowances within, and use of offsets within the program have not been finalized at the time of publication. The WCI partners have released draft documents establishing a draft scope of the regulatory program, the intention to auction a minimum amount of allowances, concluding that a generator-based approach designed to minimize leakage is the preferable approach, and determining that offsets should be incorporated into the system.[35] Of the WCI participating jurisdictions, however, Oregon, Washington, Arizona, and New Mexico have signaled that they will follow California's lead by developing a regional load-based cap-and-trade program.

2.3 Midwest States

On 15 November 2007, a group of Midwestern states (Wisconsin, Minnesota, Illinois, Iowa, Michigan, Kansas, and the Canadian province of Manitoba), through the Midwestern Governors Association, combined to form the Midwestern Regional Greenhouse Gas Reduction Accord (MR 66RA). The group agreed to 'develop a market-based and multi-sector cap-and-trade mechanism to help achieve GHG reduction targets.'[36] The program will be designed to 'enable linkage to other jurisdictions' systems to create economies of scale,' and 'address potential interaction or integration with a future federal program.'[37] While the Midwestern Greenhouse Gas Accord states may be able to design their cap-and-trade program more quickly than was possible under RGGI and WCI given their ability to learn from the design of those regimes, the program is still under design and is not expected to be operational before 2013.

2.4 Voluntary Programs

In addition to the mandatory regional GHG emission reduction regimes

---

33 Ibid., p. 2.

34 WCI (22/08/2007), Statement of Regional Goal, p. 1.

35 WCI documents, including drafts, are available at: http://www.westernclimateinitiative.org/WCI_Documents.cfm.

36 Midwestern Governors Association (2007), Midwestern Greenhouse Gas Accord, p. 3.

37 Ibid., p. 4.

discussed above, two other major voluntary regional GHG initiatives exist in the United States.[38]

On 8 May 2007, 31 states, a Native American tribe, and two Canadian provinces agreed to establish 'The Climate Registry,' a multi-state GHG emissions tracking system.[39] The Climate Registry is designed to '[d]evelop and manage a common greenhouse gas emissions reporting system,' and '[p]rovide an accurate, complete, consistent, transparent, and verified set of greenhouse gas emissions data from reporting entities, supported by a robust accounting and verification infrastructure.'[40] The Climate Registry is designed to enable the participating states 'to incorporate these minimum data quantification standards into any mandated greenhouse gas reporting and emissions reduction program.'[41] On 1 February 2008, The Climate Registry issued its Fourth Draft of the General Verification Protocol for public comment, and in May 2008, the Registry is expected to issue its final General Reporting Protocol.

The other primary voluntary regional regime in the United States is the Chicago Climate Exchange (CCX). The CCX was established in 2003 by Richard Sandor, who has been dubbed the 'father of carbon trading.' Participating entities under the CCX voluntarily agree to undertake legally binding commitments to meet GHG emission reduction targets.[42] The CCX, while imposing relatively modest emission reduction targets on participating entities, is important to regional GHG reduction regimes because it helps build capacity for future carbon trading regimes, whether voluntary or mandatory.

---

[38] Some of the mandatory regional regimes may develop a voluntary element similar to the Acid Rain Program, whereby participants may voluntarily join the mandatory regime through an opt-in program that provides incentives for those entities to join. Environmental Protection Agency (2 Feb. 2007), 'Opt-in Program Fact Sheet,' http://www.epa.gov/airmarkets/progsregs/arp/opt-in.html. Such a regime would operate similarly to the United Kingdom Emissions Trading Scheme, which provides reductions in payments due to the Climate Change Levy for participating entities. For a discussion of the United Kingdom's trading scheme, see Bluemel (2007a, pp. 2021–2025). Generally, opt-in programs are a good way to increase participation in a regulatory regime, but they are difficult to manage to ensure beneficial environmental outcomes. See Aulisi, et al. (2005, pp. 22--23). Despite the problems associated with opt-in programs, some have nevertheless called for their use in GHG emission reduction regimes. Ellerman (2003, p. 35 and pp. 41–43).

[39] Jones and McIntyre (2007, p. 1640).

[40] The Climate Registry (2008), 'Principles and Goals,' http://www.theclimateregistry.org/principlesgoals.html (accessed 22/02/2008).

[41] Ibid.

[42] For a good discussion of the Chicago Climate Exchange, see Yang (2006, pp. 274–82).

3. THE PROBLEM OF EMISSIONS LEAKAGE

Emissions leakage occurs when there is 'a reduction in emissions of greenhouse gases within the state that is offset by an increase in emissions of greenhouse gases outside the state.'[43] The United States Environmental Protection Agency (EPA) notes that emissions leakage 'occurs when economic activity is shifted as a result of the emission control regulation and, as a result, emission abatement achieved in one location that is subject to emission control regulation is offset by increased emissions in unregulated locations.'[44] Emissions leakage is a serious concern for most sub-global and sub-national GHG emissions regimes, including the RGGI states and California, since leakage can undermine the effectiveness of such regimes.[45] Figure 1 depicts the reach, or lack thereof, of the major regional cap-and-trade regimes in the United States.

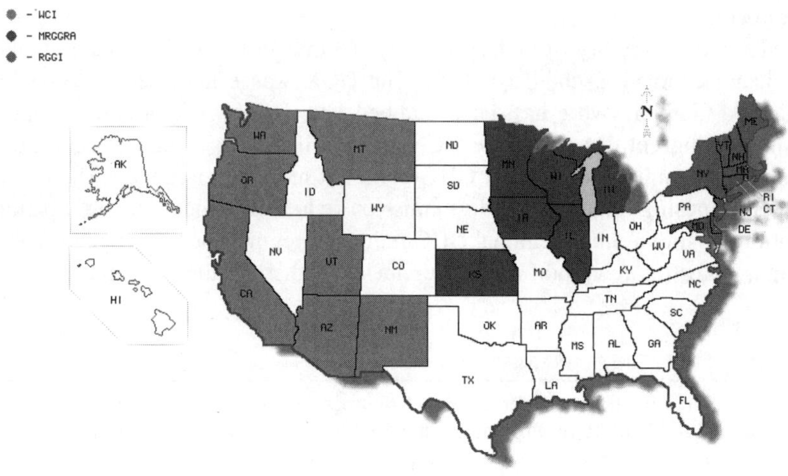

Figure 9.1 Regional Initiatives in the United States

[43] California Health & Safety Code § 38505(j).

[44] Environmental Protection Agency (2003), *Tools of the Trade: A Guide to Designing and Operating a Cap and Trade Program For Pollution Control*, EPA 430-B-03-002, p. Glossary-3. The Kyoto Protocol uses a similar definition of leakage: 'the portion of cuts in greenhouse-gas emissions by developed countries-countries trying to meet mandatory limits under the Kyoto Protocol- that may reappear in other countries not bound by such limits. For example, multinational corporations may shift factories from developed countries to developing countries to escape restrictions on emissions.' See United Nations Framework Convention on Climate Change (21 March 2007), Glossary, http://unfccc.int/essential_ background/glossary/items/3666.php.

[45] See Wiener (2007, pp. 1967–73); Burtraw et al. (2006, pp. 5–12 to 5–13); Wiener (1999, pp. 693–7).

Emission leakage results from the higher cost of compliance associated with the GHG reduction requirements in the regulated region than in surrounding jurisdictions.[46] As entities within regional organizations such as RGGI and WCI face higher production costs than neighboring, unregulated states, they may decide it advantageous to shift their production to plants outside the regulated region, or decide to build a greater number of new plants outside the region, than they would have absent the GHG regional regime.

Leakage, however, only occurs when the GHG regime cost adder is sufficiently large to undermine other advantages associated with the facility's current location. There are a number of factors that contribute to, or lessen the impact of, leakage. Primarily, emission leakage depends on how strict the emission cap is, or how 'short' the allowances are compared to existing emissions. This is highly sensitive to a variety of factors, including geographic and facility-specific factors.[47] The sensitivity of the market to various exogenous factors makes it difficult to quantify the extent of leakage resulting from a particular regime.[48]

The electricity sector is generally a fairly stable market that does not depend on a significant amount of imports in part because of domestic price supports,[49] line losses through transmission inefficiencies,[50] and congestion charges in some areas (e.g., the PJM).[51] Emission leakage is, therefore, generally not a substantial concern in the electricity sector. In a competitive electricity market, however, where a number of states or countries are linked by efficient transmission lines that extend outside the regulated region, leakage can be problematic.

The difficulty associated with estimating leakage is compounded by the wide variety of mitigating factors that lessen the impact of a GHG reduction regime on the decision-making of regulated entities. The availability of new transmission capacity within and into the regulated region can lessen the interregional price differentials, thereby minimizing leakage. Similarly, other regulatory restrictions and private decisions, including those related to siting power

---

46 Richard Cowart has convincingly argued that while leaked emissions are only those caused by the cost adder of an emission reduction regime, the regime should nevertheless seek to avoid any increase in emissions outside the regulated region without worrying about whether such an increase is the result of leakage. Cowart argues this approach is best because increased emissions are inconsistent with the fundamentals of a cap-and-trade system and focusing on causation may permit double-counting. Cowart (2006, p. 5).

47 RGGI, Initial Leakage Report, n. 14, pp. ES-2, 6.

48 For a discussion of leakage modeling, see Wiener (1999, n. 45, p. 695, n. 70).

49 See Bluemel (2007b, pp. 695–6).

50 RGGI, Initial Leakage Report, n. 14, p. 6.

51 Ibid.

plants (including nuclear power plants) and emissions portfolio standards, the desire of load-serving entities (LSEs) to add coal-fired power plants to their mix of electricity generators to stabilize their generation output, improved capacity at existing generators, as well as a variety of other tax and revenue policies, may influence emissions leakage.[52] Given the wide variety of mitigating factors, some have suggested that the leakage problem may, in fact, be overstated, given the United States' experience in its nitrous oxide trading program, where 'the economic incentive to avoid environmental regulation is small compared with other financial incentives.'[53]

Despite the variety of mitigating factors, most analysts agree that leakage is a substantial concern for GHG emission reduction regimes. This is in large part because GHG pollution is a necessary byproduct of electricity production using fossil fuels. Given this, emission and carbon capture control technologies necessary to reduce GHG emissions may impose substantial additional costs on electricity generators in the United States. Accordingly, a GHG regime that is 'short' on allowances is not akin to the nitrous oxide trading program, where the additional cost of reducing nitrous oxide was fairly small.

Unfortunately, even a minimal amount of leakage can have substantial impacts on the effectiveness of these budding GHG emission reduction regimes. RGGI set GHG reduction targets of 10% over 2009 levels by 2018. If RGGI promotes 1.5–2% more new coal-fired electricity generated in non-RGGI states over each of the next ten years, the GHG emissions from that electricity will completely offset RGGI's expected GHG reductions.[54] Although the initial RGGI modeling projected acceptable levels of leakage, it nevertheless projected leakage through a shift in the location of new natural gas-fired power plants, resulting in 27% leakage of net CO_2 emissions under the 'middle-of-the-road' scenario.[55] Since RGGI only regulates CO_2, this leakage does not account for unregulated GHG emissions, such as methane, that result from natural gas production. Because electricity generated from natural gas emits methane – a GHG 20 times more potent than CO_2 – increases in natural gas production, and leakage more generally, can substantially offset the benefits of RGGI and could promote GHG emission increases.[56]

California relies on imports for over 20% of its electricity. With its prohibition on coal-fired plants and reliance on natural gas for electricity, approximately half of the CO_2 emissions attributable to in-state electricity demand come from imported coal-fired power. Since California's GHG emission

52 Cowart (2006, p. 2).
53 Aulisi et al. (2005, pp. 13–14).
54 Cowart (2006, p. 3).
55 RGGI, Initial Leakage Report, p. 9.
56 Wiener (2001, p. 1322).

regime covers methane, there is little concern about leakage to natural gas production within California, though there remains the concern that leakage may occur by promoting electricity imports from cheaper, dirty coal-fired power plants located outside the region. Given existing regulatory controls in California, the electricity imported by California is nearly twice as GHG-intensive as electricity generated outside California.[57] Accordingly, a shift in electricity production to areas outside California can have a significant impact on the effectiveness of California's GHG emission reduction strategies.

Finally, by increasing operating costs for industry, the cost of electricity may increase for consumers as generators and distributors of electricity pass those costs on to them. The cost of compliance may, indeed, be significant for GHG-intensive industries such as steel production, refining, and other industries. The increased cost these industries may face can result in their relocation to areas where such GHG controls do not exist, creating a second pathway to emission leakage. This is a particular concern in highly competitive industries. As these products are then imported back into the regulated region, there will presumably be greater transportation-related emissions of those products, creating a third pathway to emission leakage.

The effectiveness of any sub-global GHG emission reduction regime, therefore, must be measured by the GHG reductions associated with consumption within the region. Emissions reduction strategies in a regulated region should not depend on whether consumption is of imported or domestic goods. Emissions are emissions, wherever generated. Calculating and controlling emission leakage is important to the success of any sub-global GHG emission reduction regime. Yet designing a regime to control leakage is no easy task and can have substantial implications for both short-term and long-term economic competitiveness. As demonstrated in the next section, the RGGI states and California have attempted to address leakage in very different ways.

4. ADDRESSING LEAKAGE IN REGIONAL REGIMES

Addressing leakage is important not only to protect the economic competitiveness of industries located within the regulated regime, but also to protect states outside the regulated region and the future trajectory of national and global GHG emissions reduction regimes. Regional GHG emission reduction regimes that promote leakage reduce their internally generated emissions but increase the emissions of neighboring, unregulated states. Increasing the emissions of neighboring states may make it difficult for those states to join a

57 Davis (2005, p. 6).

regional regime later, since they increased their electricity production to increase exports to the neighboring regulated region. Entering the regional regime may put the late entrants at a disadvantage if their emissions reduction targets do not account for their increased production and export of electricity to serve the regulated region.

The failure to address leakage also makes it unlikely that non-member states will join the regional regime for other reasons. Non-member states that decide to join a leakage-prone regional regime will lose their ability to export electricity into the regional regime at a price advantage and will face the problem of cheaper electricity imports competing with domestic electricity generators. Finally, a leakage-prone regional regime, if designed as a model for a broader, coordinated emission reduction regime, can hinder the development of such a program by infusing doubt about the effectiveness of such a regime into the minds of legislators. Given the high stakes presented by emissions leakage, the RGGI states and the Western states, including California, have attempted a number of different approaches to minimize or eliminate leakage.

4.1 The Eastern Approach: Cost-Containment

Recognizing that leakage is a product of compliance cost differentials between entities within the RGGI region and surrounding jurisdictions, the RGGI states seek to address leakage by reducing the RGGI cost adder. RGGI's modeling analysis concluded that given the unique nature of CO_2 pollution in the region and the various incentives that promote or mitigate leakage, RGGI will increase the cost of compliance by less than 5% of the average wholesale electricity price.[58] Given the minimal existing price differences between RGGI-generated electricity and non-RGGI-generated electricity, this cost adder is not expected to promote significant leakage. Based on these data, the RGGI states concluded that they need not eliminate leakage completely, so long as they can keep it within acceptable levels. Their primary strategy to mitigate leakage is to contain the additional compliance cost faced by regulated entities.

4.1.1 Safety valve

The primary cost containment measure employed by the RGGI states is termed a 'safety valve.' A safety valve is a relief mechanism designed to minimize some of the cost implications of a regulation if the compliance costs become unacceptable or exceed a pre-determined threshold. RGGI builds on the Acid Rain Program's idea of a safety valve, which releases reserve

---

58 RGGI, Initial Leakage Report, n. 14, p. 6.

allowances into the market if the price of allowances exceeds a particular price threshold, with an offsets safety valve.[59]

A safety valve, while effective at keeping the cost of compliance within modeled expectations, may have unintended negative consequences. A safety valve, if not properly designed, can establish a 'price cap' on allowances. Indeed, the more effective the safety valve, the more likely it is to establish such a price cap and ensure that costs are contained below the trigger price level or cost threshold. While effective at containing costs, a price cap can keep the cost of compliance sufficiently low to discourage investment in technologies that might achieve sustained emissions reductions. Ultimately, RGGI rejected applying a safety valve in a way that would create a robust price cap.

The RGGI states took a slightly different approach than the Acid Rain Program when establishing a safety valve for the regime. Rather than increasing the number of available allowances when the price of allowances reaches a particular trigger level, the RGGI states created a safety valve that incrementally increases the availability of offsets if the price of allowances exceeds various trigger levels. Offset credits and allowance tracking are managed by a non-profit regional organization (RGGI, Inc.), which has no regulatory or enforcement authority, established by the RGGI states in 2007. This difference in design has substantial implications for the incentive structure of the regime.

The Acid Rain Program of the Clean Air Act essentially creates a price cap by increasing the supply of sulfur dioxide allowances available to regulated entities – and thereby reducing the cost of those allowances – if the price of allowances exceeds a particular threshold.[60] This provides great clarity for regulated entities to determine whether the cost of investing in new emission-reducing technologies is economically advantageous, but it is not necessarily technology-forcing. It also does not ensure overall emissions reductions by a particular regulated entity, since it promotes a game of chicken, whereby all facilities seek to exceed their allowances, thereby driving up the allowance price, in the hope that the regime will release additional allowances. The Acid Rain Program's safety valve, therefore, undermines the Program's goal to ensure sulfur dioxide emission reductions.

RGGI, on the other hand, does not have reserve allowances that are released into the market if allowances prices (e.g., the compliance cost) are deemed too high. Instead, if meeting the RGGI emission reduction targets becomes too expensive for regulated entities (i.e., exceeds the pre-determined allowance price triggers known as the 'Offsets Trigger Event' and the 'Safety Valve Trigger Event'), regulated entities can decide either to purchase

59 See Pring (2006, n. 4, p. 193).

60 42 U.S.C. §§ 7651–7651o.

allowances on the market or offset their emissions in areas where cheaper emissions reductions are possible. This increased ability to use offset emissions helps keep the demand for, and therefore price of, allowances down.

Because RGGI limits the use of offsets to five or ten percent of a source's total allowances depending on the price level trigger, RGGI does not set an allowance price cap in the same way that the Acid Rain Program does. The allowance price under RGGI is variable and could increase fairly substantially depending on a variety of uncontrollable circumstances. Given this situation, RGGI encourages regulated entities to develop and install new technologies that will achieve lasting emissions reductions, which may result in total regional emissions below the overall target.

Given the potential volatility in the allowance market, RGGI's safety valve provision is only triggered when the price level exceeds the trigger level. When the allowance price exceeds the highest price triggering level ($10 (2005$)/ton CO_2, as adjusted for inflation) for twelve consecutive months, then the compliance period may be extended for a year, for a maximum compliance period of four years.[61] After a trigger of the various safety valve provisions, RGGI resets the compliance requirements and offset limitations at the beginning of the next three-year compliance period.[62] Overall, RGGI's offsets safety valve serves to minimize potential leakage because it sets a rather low trigger price, though leakage can still be fairly significant if allowance price volatility is not controlled.

While the safety valve provisions related to the use of offsets are designed primarily to reduce the compliance cost to sources, RGGI also seeks to lessen the potential economic burden of the program suffered by consumers. RGGI requires that at least 25% of each state's CO_2 allowances be awarded for a 'consumer benefit or strategic energy purpose.'[63] These allowances can be used as a means of 'fostering renewable energy, offering consumer rebates, stimulating innovative carbon-reduction technologies, and funding the administration of the program.'[64] The RGGI states have determined that the easiest way to allocate the allowances for consumer benefit or strategic purposes is through an auction. While the RGGI states have discretion how they will allocate their remaining allowances, most of the states (New York, Massachusetts, Vermont, Rhode Island, Connecticut, and Maine) have declared their inten-

---

61 RGGI, MOU Amendment, n. 10, art. 1(a). This is similar to the safety valve provision in California's AB32, which allows the Governor to extend compliance deadlines up to one year in 'extraordinary circumstances.' California Health & Safety Code § 38599.

62 RGGI, MOU Amendment, n. 10, art. 5(b).

63 RGGI, MOU, n. 10, art. 2(G)(1).

64 Sussman (2006, p. 46).

tions to auction 100% of their allowances, and use the proceeds of the auctions in programs designed to reduce consumer costs, such as demand-reduction strategies.

4.1.2 Early reduction credits

RGGI also seeks to reduce the compliance cost imposed on regulated entities by giving them an opportunity to earn additional allowances before the compliance period begins. Since RGGI uses 2009 as the baseline year against which future reductions are compared, the RGGI states recognized that the program may create perverse incentives for a state to over-pollute in 2009 to inflate its emissions baseline artificially. Inflating the baseline in this way would prevent the RGGI market from being 'short' on allowances – essentially creating a surplus of cheap allowances post-2009 – that would enable regulated entities to meet their GHG emission reduction targets cheaply.

To lessen this problem, RGGI enables sources to earn 'early reduction credits' for actions taken prior to 2009, but after the state hosting the sources entered into the RGGI Memorandum of Understanding, to reduce GHG emissions.[65] By creating additional allowances, rather than carving them out of the existing budget, RGGI effectively increased the supply of allowances and helped to alleviate speculation during the 'market settling period' and start-up phases of the program.[66] This approach will help to reduce price volatility and lower the allowance price, thereby reducing potential leakage.

4.1.3 Other cost containment measures

The RGGI states also employ a variety other mechanisms to help contain compliance costs and increase smooth price discovery of allowance prices during the 'market settling period.'[67] First, RGGI uses a three-year compliance period unless a safety valve triggering event occurs to lengthen the

65 RGGI, MOU, n. 10, art. 2(H). Similarly, California's AB32 provides an opportunity for regulated entities to earn early action credits. Ca. Health & Safety Code § 38562(b)(3).

66 Aulisi et al. (2005, p. 27).

67 Some claim that RGGI uses a circuit-breaker to help contain costs. This is technically inaccurate. A circuit-breaker cost containment provision operates in a manner similar to a safety valve provision in that if an allowance price trigger is exceeded, the schedule providing for a declining overall emissions cap is frozen until the spike in the allowance price abates. RGGI has no such direct cost containment provision, despite the request of some commenters for such a provision. Nevertheless, one might view the possible extension of the compliance period for a year on the occurrence of a Safety Valve Trigger Event as a type of circuit-breaker cost containment provision, since the compliance deadline extension may prevent a lower cap from becoming effective during that year.

compliance period. This multi-year compliance period smooths out CO_2 emissions spikes relating to unforeseeable events, such as heat waves that result in more air conditioning use than expected. The multi-year compliance period, then, reduces the cost of compliance for regulated entities. The first three-year compliance period begins on 1 January 2009.

Second, RGGI enables participating states and regulated entities to 'bank' their surplus credits into future years of the program, thereby using surplus credits from one year to offset increased emissions in subsequent years. This temporal flexibility appears scientifically permissible in a GHG emission reduction regime given the long-time horizon of GHG pollution and its expected impacts. Banking credits helps smooth out weather-related compliance crunches, thereby reducing the cost of compliance for regulated entities.

Finally, most of the RGGI states plan to auction all of their allowances. Auctioning allowances can help ensure smooth discovery of allowance prices, and thereby minimize price volatility. Reducing volatility of the market price for allowances can help reduce costs to regulated entities, and thereby reduce a potential cause of emissions leakage.

4.2 The Western Approach: Load-based Emissions Caps Plus

In the western United States, there is greater diversity of primary energy fuels and environmental controls among states, which results in states having highly divergent per kilowatt hour GHG emissions. California, relying heavily on natural gas, generally emits low levels of GHGs per kilowatt hour – at a relatively high price – while its neighboring states either have an abundant supply of cheap hydroelectric power (Pacific Northwest) or rely on cheaper, but GHG-emitting, coal-fired power plants. This diversity results in a substantial price difference between electricity generated in California and its surrounding states – a price differential that spans from a minimum of over three cents per kilowatt hour to almost eight cents per kilowatt hour (nearly two-thirds of California's retail electricity price of 12.82¢/kWh in 2006).

Given the large electricity price differentials between California and the surrounding WCI and non-WCI states identified in Table 9.2, and California's need to import electricity to satisfy its insatiable energy demands, emissions leakage is a significant concern for California. Accordingly, California has taken a variety of steps to try to minimize and eliminate emissions leakage.

4.2.1 Load-based emissions caps

Beginning in 2005, California began discussing use of a load-based emission cap to control emission leakage. California's Public Utility Commission ultimately determined that it prefers to use a load-based emission cap to implement the cap-and-trade regime authorized by AB32. This load-based emission

Table 9.2 Western Interconnection Generation and Sales in 2006[68]

	State (Primary Fuel Source)	Net Generation (MWh)	Retail Sales (MWh)	Net Generation (% of Sales)	CO_2 Emissions (1000 tons)	Average Retail Price (¢/kWh)
WCI	New Jersey (Nuclear)	60,700,139	79,680,947	76%	19,861	11.88
WCI	Arizona (Coal)	104,392,528	73,252,776	143%	28,494	8.24
WCI	California (Gas)	216,798,688	262,958,528	82%	59,389	12.82
WCI	New Mexico (Coal)	37,265,625	21,434,957	174%	33,051	7.37
WCI	Oregon (Hydro)	53,340,695	48,069,265	111%	7,088	6.53
WCI	Utah (Coal)	41,263,324	26,365,716	157%	36,445	5.99
WCI	Washington (Hydro)	108,203,155	85,033,335	127%	10,360	6.14
Non-WCI	Colorado (Coal)	50,698,353	49,733,698	102%	41,847	7.61
Non-WCI	Idaho (Hydro)	13,386,085	22,761,749	59%	875	4.92
Non-WCI	Montana (Coal)	28,243,536	13,814,980	204%	19,087	6.91
Non-WCI	Nevada (Gas)	31,860,022	34,586,260	92%	16,620	9.63
Non-WCI	Wyoming (Coal)	45,400,370	14,946,612	304%	45,216	5.27

cap is the first of its kind and is designed to capture the GHG emissions from electricity imports in California's regulatory scheme.

California's load-based emission cap imposes GHG emissions caps on load serving entities, which are private companies that sell electricity to end users after purchasing that electricity from sources. This load-based regime targets GHG emissions related to delivery of electricity into California, in contrast to a generator-based regime, which, if adopted, would target the GHG emissions resulting from the production of electricity within California.

By imposing emissions caps on LSEs, California's load-based regime creates an incentive for LSEs to purchase electricity provided to California from low GHG-emitting generators. LSEs are not bound by geography in their provision or purchase of electricity, so a load-based cap provides emissions

[68] United States Energy Information Administration, n. 7, pp. 261–2, tbl. A1 (Selected Electric Industry Summary Statistics by State, 2006).

caps for all LSEs seeking to deliver electricity to the California grid, irrespective of whether they are located in California or purchasing electricity generated within California.[69] This load-based cap regulates the GHG emissions of almost all electricity consumed within California.[70] While California is more prone to problems with leakage than RGGI and other states, most of the other members of the WCI have declared their intention to establish a similar load-based cap, which is specifically sanctioned by WCI.

A load-based emission cap has a number of advantages over a source- or generator-based emission cap. First, a load-based regime is most effective at reducing or eliminating leakage.[71] By focusing on consumer demand, it incorporates industrial process-related emissions as well as generation-related emissions. As with a source-based emission cap, however, it does not address leakage that results from the export of energy-intensive industries (and resultant consumption of electricity) outside the regulated region as a result of increasing electricity costs. Second, by focusing on consumer demand, rather than generator supply, a load-based cap inherently values demand reduction strategies and efficiency improvements.

A load-based emission cap is also preferable to a source-based cap because it places the decision-making power in the hands of LSEs, which have the ability to prioritize low GHG-emitting technologies. A load-based cap creates incentives for LSEs to invest in cleaner technologies and creates a price signal that low-emitting generation is a valuable commodity. This price signal means that the design of a load-based emission cap is simpler than a source-based cap when it comes to incorporating renewable energy strategies, since renewables have inherent value in a load-based system. In a source-based system, on the other hand, creating value and incentives for production of renewable energy requires treating different categories of sources differently or creating a second layer of regulation, in terms of emission portfolio standards.

Finally, a load-based cap ensures greater flexibility than a source-based cap. Many generators are coal-fired power plants that cannot suddenly shift to producing renewable electricity without substantial economic loss.

---

69 California Public Utilities Commission (08/02/2008), Proposed Decision on Rulemaking 06-04-009, Interim Opinion on Greenhouse Gas Regulatory Strategies (adopting the first-seller rule as recommended by the Market Advisory Committee, with a slight modification). For a discussion of the first-seller rule, see generally Market Advisory Committee (2007).

70 California Public Utilities Commission (16/02/2006), Decision 06-02-032 in Regulation 04-04-003, p. 17 ('LSEs would be subject to an emissions cap for all resources procured to serve their load, no matter what the source, including imports.').

71 See Bird et al. (2007, p. 39); RGGI, Initial Leakage Report, n. 10, pp. ES-11, 41; Climate Action Team (March 2006, p. 69), *Report to Governor Schwarzenegger and the Legislature*.

Accordingly, a generator-based emission cap, while effective at reducing per kilowatt hour GHG emissions for each type of generating source, does not create incentives to shift electricity generation to different *types* of generating sources with lower GHG emissions. A load-based cap does. Under a load-based cap, LSEs can meet their load-based allocation by shifting their portfolio to reduce their purchase of high GHG-emitting electricity in favor of low-GHG electricity. The flexibility of a load-based emission cap makes it less likely to increase costs to end-use consumers as significantly as a generator-based cap might.

While a load-based emission cap is desirable for a variety of reasons, it also suffers some potential design problems. A load-based cap may suffer from a problem known as 'contract shuffling.'[72] LSEs purchase electricity from different types of electricity generators in a number of states to provide electricity to a variety of states. Given this dynamic, LSEs can decide to allocate all the low GHG-emitting electricity in its generator portfolio to California and the WCI states and allocate all the highest GHG-emitting electricity to the other states served by the Western Interconnection electricity grid that are not within the load-based regional regime. This contract shuffling may mean that while the electricity consumed in California meets California's GHG targets, no actual emissions reductions were achieved by the regulation.[73]

Governor Schwarzenegger's Climate Action Team concluded that contract shuffling is likely to be a one-time problem that is unlikely to persist given California's declining emission cap.[74] It concluded that 'once all the existing low-emitting units are spoken for, additional emission reduction would need to be achieved through other means.'[75] This conclusion has merit since the states with excess electricity capacity derived from low GHG-emitting sources – Oregon and Washington – have determined that they will join WCI and implement a load-based cap. Accordingly, there is unlikely to be substantial contract shuffling of low GHG-emitting electricity out of those states to California, since each of the states joining a regional load-based regime will need low GHG-emitting electricity to satisfy their emission reduction targets. This means that only the excess low GHG-electricity capacity will likely be sold to California. That excess capacity would only cover about half of

72 Market Advisory Committee (2007, p. 44; RGGI, Initial Leakage Report, n. 10, p. 41); Cap and Trade Subgroup (2006, pp. 23–4).

73 Similarly, a load-based emission cap may create an incentive for high GHG-emitting generators within the regulated region to sell their electricity outside the regulated region, resulting in 'reverse leakage.' Cowart (2004). This problem is of little concern in California, given other environmental protections and performance standards that discourage the development of high GHG-emitting sources.

74 Cap and Trade Subgroup (2006, pp. 23–4).

75 Ibid.

California's import needs. Low-GHG emitting sources in other states simply do not exist in sufficient quantity to enable significant contract shuffling in a declining cap scenario.[76]

Nevertheless, given the potential problem of contract shuffling, a load-based cap must be able to identify the GHG emissions associated with imported power, which may be difficult given existing GHG monitoring and reporting (despite advances made through The Climate Registry). Some have suggested that one way to avoid contract shuffling would be to assign an average GHG emission rate to imported power.[77] This ostensibly would avoid the contract shuffling incentives, but could present problems under the United States Constitution by treating imported power differently from domestic power.

Irrespective of whether such an approach is ultimately adopted, addressing the problem of contract shuffling in a load-based cap-and-trade scheme presents a number of implementation difficulties.[78] One such difficulty arises because a load-based cap requires an emissions tracking system. A tracking system is problematic for long-term electricity supply contracts that do not specify the generating source, as well as spot market purchases, which generally do not identify the generating source. In 2007, the Oregon legislature proposed House Bill 3545 to address this problem by specifying a 'residual emission rate' to unspecified power purchases based on an average emission rate of all generation within the power pool not accounted for in existing contracts, but the bill has not yet gone to vote. The California Energy Commission is considering a similar proposal, but California addressed this concern another way: by prohibiting LSEs from entering long-term power purchase agreements unless the contracting generator meets a GHG emission performance standard.[79]

While solving one problem, such proposals create a complex incentive structure that may require additional study prior to adoption. On the one hand, a 'residual emission rate' could inadvertently create an incentive for dirty

---

[76] For a discussion of contract shuffling in this regard, see Davis (2005, p. 14).

[77] Cowart (year. 46, p. 6).

[78] For a discussion of some of these difficulties, see RGGI, Initial Leakage Report, n. 10, p. ES-11.

[79] California Public Utilities Code § 8341. For more information about emission performance standards, see Davis (2005, pp. 8–11). It is well-recognized that efficiency improvements can reduce leakage by reducing the cost of compliance to generators and LSEs. E.g., Prindle (2006). Accordingly, California employs both emission performance standards and 'decoupling,' which rewards generators for selling less electricity than expected. Bryk & Snyder (2007, p. 99). The effectiveness of providing 'decoupling' rewards to high GHG-emitting generators within a load-based emission cap is yet to be determined, but appears to risk rewarding dirty generators.

generators to provide their electricity through the unspecified spot market or long-term power purchase agreements to receive a more favorable GHG emission rate.[80] This unintended consequence, however, may be offset by a similar incentive for low GHG-emitting generators to avoid the spot market and unspecified generator contracts so they might receive the full value of their low rate of GHG emissions. Since LSEs are price takers in the spot market, a 'residual emission rate' may ultimately provide a clear decision rule for LSEs deciding whether to purchase electricity on the spot market, which may discourage carbon-intensive spot market purchases, but more information is needed about the incentives created by a 'residual emission rate' to understand its full implications.

Another concern with a load-based emission cap is that LSEs may currently lack sufficient information to make informed decisions about the GHG emissions from their existing contracts. As a result, it may be difficult for LSEs to discover the price of allowances on the market, which could result in significant allowance price volatility in the early years of operation. To overcome this problem, some have suggested that allowances be sub-distributed to generators based on LSE emission allocations, since it is presumed that generators will have greater information about their GHG emissions than LSEs. This solution is unsatisfactory, however, because this would enable LSEs to earn windfall profits from 'flipping' their allowances to generators, and it would undermine a core benefit of the load-based cap: providing flexibility in meeting the cap through demand-reduction strategies and portfolio adjustments.

To address this concern, California adopted an innovative approach to ensuring smooth price discovery in a load-based emission cap without an existing GHG emission tracking system. California imposed a carbon procurement adder as a near-term bridge to a load-based cap. A carbon procurement adder requires LSEs to consider the 'shadow price' of carbon (assumed to be \$8/CO_2) in their planning decisions.[81] While a carbon procurement adder may

[80] Gillenwater and Breidenich (2007, p. 5). In the PJM, the spot market accounted for about 40% of the total electricity load. RGGI, Initial Leakage Report, n. 10, p. 7 n.13.

[81] California Public Utilities Commission (07/04/2005), Decision 05-04-025, 'Order Instituting Rulemaking to Promote Consistency in Methodology and Input Assumptions in Commission Applications of Short-Run and Long-Run Avoided Costs, Including Pricing for Qualifying Facilities,' p. 44; California Public Utilities Commission (16/12/ 2004), Decision 04-12-048. In addition to the carbon procurement adder, the California Public Utilities Commission is considering charging a fee for GHG pollution. California Public Utilities Commission (31/01/ 2008), BAQMD Regulation 3-334 & Schedule T (draft). It is unclear, however, how such a fee will interact with other GHG pollution abatement requirements.

not in and of itself minimize leakage, as a near-term bridge to a load-based cap, it does help LSEs discover the price of allowances in the cap-and-trade system, which will help minimize price volatility.

4.2.2 Emission portfolio standards

California and other western states also seek to reduce leakage by imposing emission portfolio standards (EPS) – essentially a performance requirement – requiring that LSEs meet an output-based GHG emissions standard per kilowatt hour.[82] Such EPSs encourage investment in low GHG-emitting technologies to reduce the overall GHG emission rate of the LSE's mix of generated electricity.

An EPS will improve the per kilowatt hour emission of GHGs, but without an emissions cap, absolute GHG emissions can continue to grow as demand grows. Accordingly, an EPS is insufficient, standing alone, to ensure GHG emission reductions.

Even with an emissions cap, however, an EPS will only marginally impact emission leakage by reducing the supply of GHG-intensive electricity. In combination with a load-based emission cap, however, a GHG EPS could promote investment in renewable energy technologies and would capture GHG emissions from sources outside the regulated region. It could also promote energy efficiency initiatives by crediting such initiatives as zero emission generation.

Similar to the contract shuffling problem faced by a load-based emissions cap, an EPS may suffer from 'attribute shuffling.'[83] An EPS would identify environmental attributes of electricity generation and separate those attributes from the underlying commodity. Since WCI does not yet cover all the sources serving the Western Interconnection, it is possible for LSEs to shuffle their low GHG-emitting attributes to ensure compliance with the EPS within the WCI region or California, while sending their high GHG-emitting attributes outside the regulated region.

As with contract shuffling, however, there is unlikely to be a sufficient surplus of low GHG-emission attributes available to meet California's demand, so attribute shuffling is not expected to be a significant problem in California. Nevertheless, California established a carbon procurement emis-

---

[82] Another type of EPS is the renewable portfolio standard (RPS), which requires that a certain percentage of an LSE's electricity comes from renewable electricity. California, Oregon, Washington, Arizona, Nevada, Colorado, Montana, and New Mexico have mandatory RPSs of varying degrees of stringency, while Idaho, Utah, and Wyoming do not. Pew Center on Global Climate Change (August 2007), 'States with Renewable Portfolio Standards,' http://www.pewclimate.org/what_s_being_done/in_the_states/rps.cfm.

[83] RGGI, Initial Leakage Report, n. 10, pp. ES-10, 37.

sion rate – requiring long-term contracting *generators* providing electricity to California to meet a specified CO_2 emission rate, in an effort to avoid attribute shuffling.[84] California's carbon procurement emission rate is based on the CO_2 emission rate of a combined cycle natural gas turbine (approximately 1,000 pounds of CO_2/MWh).[85] By tying the emission rate to the bundled commodity, California has successfully avoided the attribute-shuffling problem inherent with an EPS that is not contract-driven.[86] This attempted solution, however, may create a disincentive for establishing long-term contracts, resulting in an increase in spot market purchases. As discussed above, the potential for dirty generators to hide behind cleaner generators in the spot market creates a potential regulatory problem, though one that is solvable.

5. CONCLUSION

The eastern and western states of the United States have taken different approaches to addressing GHG emissions leakage, resulting in rather divergent systems.[87] Given these differences, it is unclear whether or how emissions trading might occur between RGGI and WCI, if WCI ultimately adopts a load-based emissions cap. Despite potential linking complications (WCI's draft documents suggest that a generator-based system would be preferable to promote linking with a national generator-based regime), the experimentation by the various states and regions provides a number of important lessons that will benefit the development of a national emission reduction regime and other regional GHG emission reduction programs.

---

[84] California Public Utilities Code § 8341; California Public Utilities Commission (25/01/2007), Decision 07-01-039, 'Order Instituting Rulemaking to Implement the Commission's Procurement Incentive Framework Rulemaking 06-04-009 and to Examine the Integration of Greenhouse Gas Emissions Standards into Procurement Policies.'

[85] California Public Utilities Code § 8341. Brian Potts concluded that this may improperly disadvantage out-of-state generators since they are less likely to meet the CO_2 emission rate of a natural gas generating source. Potts (2006, pp. 45–9). This analysis is fairly limited and does not account for a multiplicity of different sources of electricity generation within the surrounding jurisdictions, including Washington and Oregon, which primarily rely on hydroelectric power, and Nevada, which primarily relies on natural gas power. Further, if an EPS or an RPS is permissible, it is unlikely that making an EPS contract-based would run afoul of any constitutional protection designed to protect states from unfair competition. For a good discussion of how California's EPS does not improperly disadvantage out-of-state generators, see Weisselberg (2007, pp. 210–24).

[86] RGGI, Initial Leakage Report, n. 10, pp. ES-8, 35 & n.52.

[87] See Peress and Booher (2007, p. 7).

First, each system attempts to address the specific circumstances faced by the participating jurisdictions. In the eastern states, leakage is of minimal concern given the modest emissions cap and cost of compliance, relatively minor electricity price differentials, relative uniformity of GHG emissions from generators, and other mitigating factors. Operating in a deregulated electricity industry, the RGGI states have fewer regulatory options available to address leakage. As a result, RGGI established a source-based cap that creates incentives to reduce the emission rate for each generator and minimizes, rather than eliminates, leakage.

By contrast, electricity generated in the WCI region and California has a relatively low GHG emission rate compared to neighboring jurisdictions. California, as a significant importer of electricity, is particularly concerned with emission leakage resulting from a GHG emission reduction regime. Having a re-regulated electricity industry, however, has enabled California to pursue unique strategies to address leakage. A load-based cap, for instance, is expected to be most effective in a regulated electricity industry.[88] Given California's concerns about leakage, it should not be surprising that California adopted a load-based cap in an attempt to eliminate most leakage, rather than pursuing cost containment measures to minimize such leakage.

Second, there exist synergistic combinations of policies that can be effective at addressing leakage. California has identified a number of policies that are likely to minimize the costs of an ambitious, multi-pollutant load-based emissions cap worthy of consideration by regions considering adopting a GHG cap-and-trade regime. For instance, California established a carbon procurement adder to help ensure smooth price discovery and reduce market volatility, which may lessen the need to auction allowances for that purpose. Allowance auctioning is generally considered to be an effective way to reduce end-user costs (through re-investment policies or consumer rebates), and such auctioning can help with smooth price discovery of allowance prices, thereby reducing price volatility and costs.[89] Accordingly, most of the RGGI states have decided to auction 100% of their allowances. Yet, it is important to recognize that RGGI only covers the CO_2 emissions of electricity generators. It does not cover carbon-intensive industries that may suffer from the increased cost of electricity. Because RGGI is not expected to increase the cost of compliance substantially, the regulated generators do not need free allowances that can serve as a source of income to help offset the cost of compliance. Auctions in this setting help avoid windfalls.

88 See RGGI, Initial Leakage Report, n. 10, pp. ES-7 to ES-8.
89 Aulisi et al. (2005, p. 2).

In another setting, however, auctioning may not be the most effective way to protect consumer prices and ensure smooth price discovery. Determining whether allowance auctioning is an appropriate method of allocation will depend on the reach of the regime, the competitiveness of the industries affected, and the cost of technologies required to comply with the new emissions targets, among other things. Since the provision of electricity is generally not as competitive as other industries, auctioning of allowances is less problematic from a leakage perspective. As the varying regional approaches discussed above illustrate, there are ways to promote smooth price discovery and reduce market volatility, such as a safety valve and other mechanisms, which do not depend on auctioning. Such mechanisms may be particularly important for highly competitive industries in the European Union, since auctioning allowances for those industries may prove too costly or undermine long-term competitiveness.

The experiences of the regional regimes in the United States also provide an analogy that may be useful for the European Union as it considers major revisions to the regulatory framework of its Emission Trading Scheme. The United States has rules prohibiting inter-state discrimination in trade and commerce. These rules apply equally to the private sale of products and the private provision of electricity services. Like the international trade rules applied to national governments by the World Trade Organization, states must ensure equal treatment of electricity provided by all states. The various policy initiatives described above are likely to overcome challenges to such anti-discrimination rules because they do not facially discriminate against inter-state commerce and treat in-state and out-of-state electricity similarly. While the provision of electricity through LSEs is somewhat unique in the United States,[90] the above regimes nevertheless provide lessons for how the European Union might structure its GHG regime to ensure compliance with the rules of international trade.

Although the United States federal government is, for the first time, seriously contemplating a national GHG cap-and-trade regime, the next phase of the Kyoto Protocol is still unlikely to incorporate all developing countries. Accordingly, the international system will, in all likelihood, remain an 'open' system, with some regulated jurisdictions and some unregulated jurisdictions. In highly competitive industries, emissions leakage presents a very real and difficult problem to solve.[91] Until the California and Midwestern regimes are

[90] For a discussion of some of the policy measures that are available to electricity regulators in the United States, see Palmer and Burtraw (2007).

[91] See Morganstern, et al. (2007). This problem is complicated by the fact that regulating GHG-intensive industries is difficult to implement. See Committee on Energy and Commerce (October 2007).

fully developed, it is unclear whether any of the United States regional regimes will attempt to address leakage in all major GHG-intensive industries.

Nevertheless, the regional regimes in the United States provide some helpful examples of ways to design a cap-and-trade program to address the leakage problem presented by migration of industry to unregulated jurisdictions. Those regimes demonstrate that there exist a number of policy tools to address leakage in a non-discriminatory manner. It may be possible for the European Union and other nations to control leakage by establishing load-based GHG allocations for imported products. This may be done by requiring allowances for imported goods as a border adjustment under the World Trade Organization rules. Similarly, countries and regions may be able to impose carbon-intensity environmental standards – effectively an EPS – under the Agreement on Technical Barriers to Trade.[92] Solutions to leakage are out there, but it is important to design the cap-and-trade institutions in a manner that will ensure effectiveness in minimizing emissions leakage. Mistakes now can cost the environment and local industry for years to come. It is, therefore, important to identify the particular leakage problems faced by a regulated region and design around those problems. As the regional GHG emission reduction regimes in the United States demonstrate, there is no 'one size fits all' solution to the problem of emission leakage.

REFERENCES

Aulisi, A. et al. (2005), *WRI White Paper: Greenhouse Gas Emissions Trading in U.S. States: Observations and Lessons from the OTC* NO_x *Budget Program*, Washington, US: World Resources Institute.

Bird, L. et al. (2007), *Technical Report No. NREL/TP-640-41076: Implications of Carbon Regulation for Green Power Markets*, Golden, US: National Renewable Energy Laboratory.

Bluemel, E.B. (2007a), 'Unraveling the Global Warming Regime Complex: Competitive Entropy in the Regulation of the Global Public Good', *University of Pennsylvania Law Review*, 155 (6), 1981–2049.

Bluemel, E.B. (2007b), 'Biomass Energy: Ensuring Sustainability Through Conditioned Economic Incentives', Georgetown International Environmental Law Review, 19 (4), 673–97.

Brevetti, R. (19 March 2008), U.S. Trade Representative Tells Dingell of Concerns with Greenhouse Emissions Proposals, *BNA International Environment Daily*.

Bryk, D. and J.J. Snyder (22–23 March 2007), 'Regional and State Programs: Measuring, Allocating, Trading, and Complying, Global Warming: Climate Change and the Law', in American Law Institute-American Bar Association, *SM106 ALI-ABA*, pp. 91–124.

92 See Committee on Energy and Commerce (January 2008).

Burtraw, D. et al. (2006), 'Lessons for a Cap-and-Trade Program', in W. Michael Hanemann and Alexander E. Farrell (eds), *Managing Greenhouse Gas Emissions in California*, Berkeley, US: California Climate Change Center, University of California, Berkeley; pp. 5-1 to 5-47.

Cap and Trade Subgroup (2006), *Cap and Trade Program Design Options: Report of the Cap and Trade Subgroup of the Climate Action Team Final Report*, Sacramento, US: Climate Action Team.

Committee on Energy & Commerce, House of Representatives (2008), 'Climate Change Legislation Design White Paper: Appropriate Roles for Different Levels of Government', http://energycommerce.house.gov/Climate_Change/white%20 paper%20st- lcl%20roles%20final%202-22.pdf, February.

Committee on Energy & Commerce, House of Representatives (2008), 'Climate Change Legislation Design White Paper: Competitiveness Concerns/Engaging Developing Countries', http://energycommerce.house.gov/Climate_Change/White_Paper.Competitiveness.013108.pdf, January.

Committee on Energy & Commerce, House of Representatives (2007), 'Climate Change Legislation Design White Paper: Scope of a Cap-and-Trade Program', http://energycommerce.house.gov/Climate_Change/White_Paper.100307.pdf, October.

Cowart, R. (2006), 'Addressing Leakage in a Cap-and-Trade System: Treating Imports as Sources, Report to RGGI Leakage Subgroup', http://www.raponline.org/Pubs/RC-Leakage-4-06.pdf, April.

Cowart, R. (2004), 'Another Option for Power Sector Carbon Cap and Trade Systems – Allocating to Load, Regional Greenhouse Gas Initiative (RGGI) Concept Memo', http://www.rggi.org/docs/allocating_to_load.pdf, 1 May.

Davis, S.E. (2005), *California Energy Commission Draft Consultant Report No. CEC-600-2005-010-D: Policy Options for Reducing Greenhouse Gas Emissions from Power Imports*, Washington, US: Center for Clean Air Policy.

Ellerman, A.D. et al. (2003), *Emissions Trading in the U.S.: Experience, Lessons, and Considerations for Greenhouse Gases*, Arlington, US: Pew Center on Global Climate Change.

Fitch, J., California Public Utilities Commission (2007), 'Why a Load-Based Emissions Cap for California?', ftp://ftp.cpuc.ca.gov/puc/energy/electric/climate+change/JulieFitchPresentation.ppt, 19 April.

Gillenwater, M. and C. Breidenich (2007), 'Internalizing Carbon Costs in Electricity Markets: Using Certificates in a Load-Based Emissions Trading Scheme, Discussion Paper', http://www.princeton.edu/~mgillenw/Load%20based%20cap%20paper-formatted%20_final.pdf, August.

Jones, S.C. and P.R. McIntyre (27 July 2007), 'Filling the Vacuum: State and Regional Climate Change Initiatives', *BNA Environment Reporter: Analysis and Perspective*, 38 (30), 1640–49.

Market Advisory Committee, California Air Resources Board (2007), 'Recommendations for Designing a Greenhouse Gas Cap-and-Trade System for California', http://www.climatechange.ca.gov/events/2007-06-12_mac_meeting/2007-06-01_MAC_DRAFT_REPORT.PDF, 30 June.

Morganstern, R.D. et al. (2007), *Backgrounder: Competitiveness Impacts on Carbon Dioxide Pricing Policies on Manufacturing*, Washington, US: Resources for the Future.

Palmer, K. and D. Burtraw (2007), *Backgrounder: The Electricity Sector and Climate Policy*, Washington, US: Resources for the Future.

Parker, L.P. and B.D. Yacobucci (2007), *CRS Report RL33846: Climate Change: Greenhouse Gas Reduction Bills in the 110th Congress*, Washington, US: Congressional Research Service.

Patrick, J. (11 October 2006), 'Bicoastal Carbon Trading: California and RGGI Markets Mapped Out', *Evolution Markets Executive Brief*, 29, pp. 1–3.

Peress, N.J. and M. Booher (April 2007), 'East and West Coast States Take Divergent Paths on Carbon Cap and Trade for Suppliers of Electricity', *ABA Air Quality Comm. Newsletter*, 10(2), 7–8.

Potts, B.H. (June 2006), 'Regulating Greenhouse Gas "Leakage": How California Can Evade the Impending Constitutional Attacks', *The Electricity Journal*, 19(5), 43–53.

Prindle, B. (2006), American Council for an Energy Efficient Economy, 'RGGI Leakage Working Group, Energy Efficiency's Role in Limiting RGGI Leakage', http://www.rggi.org/docs/prindle.ppt, 15 June.

Pring, G. (2006), 'A Decade of Emissions Trading in the USA: Experiences and Observations for the EU', in Marjan Peeters and Kurt Deketelaere (eds), *EU Climate Change Policy: The Challenges of New Regulatory Initiatives*, Cheltenham, UK and Northampton, US: Edward Elgar; pp. 188–204.

Rabe, B.G. et al. (2005), 'State Competition as a Source Driving Climate Change Mitigation', *New York University Environmental Law Journal*, 14 (1), 1–53.

Sussman, Edna (May 2006), 'New York Addresses Climate Change with the First Mandatory U.S. Greenhouse Gas Program', *May-New York State Bar Journal*, 78, p. 43–50.

Tanton, T. (2008), *California's Energy Policy: A Cautionary Tale for the Nation*, Washington, US: Competitive Enterprise Institute.

Tietenberg, T. et al. (2003), 'The Tradable-Permits Approach to Protecting the Commons: Lessons for Climate Change', *Oxford Review of Economic Policy*, 19(3), 400–19.

United Nations Env't Program (2007), *Global Environment Outlook 4*, Valletta, Malta: Progress Press.

Weisselberg, P. (2007), 'Shaping the Energy Future in the American West: Can California Curb Greenhouse Gas Emissions from Out-of-State, Coal-Fired Power Plants Without Violating the Dormant Commerce Clause?', *University of San Francisco Law Review*, 42(1), 185–225.

Whetzel, C. (28 February 2008), 'California Law Does Not Guarantee Carbon Cap-and-Trade Scheme', *BNA Daily Environment Report*, 39, p. A–5.

Wiener, J.B. (2007), 'Think Globally, Act Globally: The Limits of Local Climate Policies', *University of Pennsylvania Law Review*, 155(6), 1961–79.

Wiener, J.B. (2001), 'Something Borrowed for Something Blue: Legal Transplants and the Evolution of Global Environmental Law', *Ecology Law Quarterly*, 27(4), 1295–1371.

Wiener, J.B. (1999), 'Global Environmental Regulation: Instrument Choice in Legal Context', *Yale Law Journal*, 108(4), 677–800.

Yang, T. (2006), 'The Problem of Maintaining Emissions "Caps" in Carbon Trading Programs Without Federal Government Involvement: A Brief Examination of the Chicago Climate Exchange and the Northeast Regional Greenhouse Gas Initiative', *Fordham Environmental Law Review*, 17(2), 271–86.

10. Domestic initiatives in the UK

Karen E. Makuch and Zen Makuch[1]

1. GENERAL INTRODUCTION

The United Kingdom of Great Britain and Northern Ireland's (hereinafter 'UK') legal and policy initiatives at the domestic level designed to address climate change and meet EU and international commitments are ubiquitous. They span the range of conventional environmental instrument categories including: command and control regulations; market-based instruments and negotiated agreements. The UK also produced a world first with a national economy-wide greenhouse gas emissions trading scheme (ETS). This chapter discusses several of the climate policy initiatives that have been, are or will imminently be employed in the UK. The UK comprises the four constituent parts of England, Scotland, Wales (the grouping to which the political term Great Britain is attributed) and Northern Ireland. The United Kingdom has been a centralized, unitary state for much of its history, and environmental law and policy developments within the UK tend to be similar throughout each of its constituent parts. The focal point of this work is largely on initiatives within England and Wales, the larger and arguably more politically dominant constituency.

Given the ubiquity of domestic initiatives, there is a key critical point worthy of detailed scrutiny. It concerns the following question: Is there policy cohesion (i.e., joined-up thinking in policy design/implementation) in relation to climate change policy instruments? This is an issue that we have examined in this chapter in relation to five key instruments pertaining to emissions trading, the Climate Change Agreements and the Climate Change Levy. Surveys and interviews have taken place with stakeholders in this regard.

1.1 Domestic Goals

It was in 1993 that the UK Government set the initial target for reducing UK

[1] The authors are grateful to N. Davies for the survey and related aspects of this research related to the Climate Change Agreements.

emissions of carbon dioxide (CO_2) by 20% of 1990 levels by 2010.[2] This target is not currently legally binding and the UK is not currently on track to meet it. There have subsequently been several revised targets proposed, none of which are currently legally binding, though the Climate Change Bill (which, at the time of writing, is currently at proposal stage, more on which is below), will contain a statutory goal of at least 26–32% reduction in carbon dioxide emissions by 2020 and a 60% reduction by 2050 against a 1990 baseline. There have been calls[3] for all three of these targets to be included in the Climate Change Bill and not just the 2020 and 2050 goals.

It is worth noting that including all three targets would firm up the regulatory means to continue to battle for significant GHG emissions reductions at all stakeholder levels within the UK, ranging from ordinary householders to heavy industry. Such a multi-binding approach would also strengthen the message that – as evidence of its international commitments – the UK Government is taking action on global warming seriously. The UK also has a legally binding target under the United Nations Framework Convention on Climate Change (UNFCCC) 1997 Kyoto Protocol to reduce its GHG emissions to 12.5% below base year levels between 2008 and 2012 and will also be bound under the European Union GHG emissions[4] reduction target of 20% from 1990 levels by 2020 once this is legislated for[5] along with the proposed Decision for a reduction of 16% for the period 2013 to 2020.[6] According to the Office for National Statistics, in 2004, UK emissions were estimated to be about 14.6% below base year levels, at 179 million tonnes carbon equivalent.[7] CO_2 emissions fell by 5.6% between 1990 and 2004, to 152 million tonnes carbon equivalent.[8]

---

2 When this target was set there was no EU Emissions Trading Scheme nor was the Kyoto Protocol concluded. The UK is not on track to meet the 2010 target and has since refocused on the 2020 target.

3 See, *Taking Forward the UK Climate Change Bill: The Government Response to Pre-Legislative Scrutiny and Public Consultation*, Presented to Parliament By the Secretary of State for Environment, Food and Rural Affairs By Command of Her Majesty, October 2007, Cm 7225, for example at p. 104.

4 From sources not covered under Directive 2003/87/EC.

5 This target was proposed by the European Council. See, *20 20 by 2020: Europe's Climate Change Opportunity*, Communication From the Commission to the European Parliament, the Council, the European Economic and Social Committee and the Committee of the Regions, Brussels, 23.1.2008, COM(2008)30 final.

6 *Proposal for a Decision of the European Parliament and of the Council on the Effort of Member States to Reduce their Greenhouse Gas Emissions to meet the Community's Greenhouse Gas Emission Reduction Commitments up to 2020.*

7 http://www.statistics.gov.uk/cci/nugget.asp?id=366

8 http://www.statistics.gov.uk/cci/nugget.asp?id=366

1.2 Organizations

By way of scene-setting, there are several bodies which have some level of policy-making or regulatory competency *vis-à-vis* initiatives to address the climate change issue within the UK, a number of which are noted herein. The regulatory competent authority with principle responsibility for the implementation of international and EU environmental legislation is the Department for Environment, Food and Rural Affairs (DEFRA).[9] DEFRA is the central 'ministry'[10] with responsibility for UK environmental issues and policy. The Secretary of State has powers to enact secondary environmental legislation for England and Wales. However, there are also other bodies which have competency for some areas of emissions reduction initiatives, and these are noted, in part, below.

At the time of writing, it has been announced that DEFRA has been allocated a budget of over £400 million over the next three years to advance a low carbon Britain as part of the domestic Environmental Transformation Fund (ETF).[11]

Some additional bodies have been created to address research and policy development goals. The Energy Savings Trust (EST)[12] was established by the UK Government in 1993 following the 1992 Rio United Nations Conference on Environment and Development at which the global communities' attention was drawn to the issue of climate change upon the promulgation of the UNFCCC. The EST's role is advisory and research-focused. The EST provides: pragmatic advice to the public on energy savings and GHG emissions reductions techniques; facilitates local community projects; and, undertakes research. The EST is a non-profit organization, funded both by government (including the Department for the Environment, Food and Rural

---

[9] See: http://www.defra.gov.uk/

[10] The Scottish Executive has competency for the Scottish Executive Environment and Rural Affairs Department (SEERAD) though they are subject to Primary UK and European environmental legislation. In Northern Ireland the Department of Environment Northern Ireland (DOE) (http://www.doeni.gov.uk/) and its Environmental Policy Group (EPG) are responsible for the formulation of policy and legislation on all aspects of Northern Ireland's environment.

[11] *£400 million for low-carbon Britain*; DEFRA News Stories, announced 21 February 2008. The domestic Environmental Transformation Fund is jointly funded by DEFRA and the Department for Business, Enterprise & Regulatory Reform and was announced on 9 October 2007 alongside increases to £2 billion of credits for local authorities to attract private investment in sustainable waste and recycling facilities. Funding through the waste Private Finance Initiative will rise from £280 million in 2007–08 to £700 million in 2010–11. See: http://www.defra.gov.uk/news/latest/2008/defra-0221.htm, accessed 27/2/08.

[12] See: http://www.energysavingtrust.org.uk/.

Affairs, the Department for Transport, the Department for Trade and Industry and the Scottish Executive) and the private sector (including: BP plc; BG Group plc; Centrica plc; EDF Energy plc; Firmus Energy; Innogy plc; Johnson Matthey Catalysts; National Grid Transco plc; Northern Ireland Electricity; Phoenix Natural Gas; Powergen plc; Scottish and Southern Energy plc). The EST claims[13] to be one of the UK's leading organizations set up to address the damaging effects of climate change, with offices in England, Scotland, Northern Ireland and Wales. Its aim is to encourage CO_2 emissions reduction by promoting the sustainable and efficient use of energy. The fact that the ETS combines membership from both the private and public sector adds to the accountability of the organization and is indicative of a genuine aim to work with all stakeholders in addressing the problem of climate change, under the umbrella of sustainable development.

The Office of Climate Change (OCC)[14] unites several government departments in working together on climate change issues and the development of relevant policies and strategies. Such departments include, *inter alia*, DEFRA, the Department for International Development, the Department for Business, Enterprise and Regulatory Reform and the Department for Transport. Such institutional integration at central level is certainly necessary given the myriad of national activities which impact upon, or are affected by, global warming: the pooling of resources and research efforts can only serve to further enhance regulatory and policy initiatives in this regard.

The UK Climate Impacts Programme (UKCIP) was set up in 1997 and is currently funded by DEFRA. It helps organizations assess how they might be affected by climate change so that they can mitigate accordingly. UKCIP is part of the wider programme of research into climate change being undertaken by DEFRA.[15]

The Carbon Trust (CT)[16] was set up by Government in 2001 as an independent company. As such, it aims to facilitate and advance the move to a low-carbon economy through collaboration and cooperation with organizations to reduce GHG emissions and to develop commercially viable low-carbon technologies.

The above initiatives indicate a strong policy and research commitment taken by (and largely funded by) the UK Government in mitigating and adapting to climate change. This is certainly in line with the aims and objectives

---

13 See: http://www.energysavingtrust.org.uk/ (accessed 27/2/08).

14 The Office was set up in September 2006. See: http://www.occ.gov.uk/about/index.htm (accessed 27/2/08).

15 See: http://www.ukcip.org.uk/about/ (accessed 27/2/08).

16 See: http://www.carbontrust.co.uk/ (accessed 27/2/08).

promulgated at the international level under the UNFCCC[17] and is indicative of a strong commitment to achieve real change. In particular, there is an emphasis upon energy efficiency and GHG emissions reductions for the benefit of the nation (and the wider global community) through the advancement of best practice in terms of research, technological innovation and regulatory reform. Such best practice could lead the way for other global initiatives where financial and social capacity is similar to that of the UK for addressing climate change concerns at the national level.

2. POLICY FOCUS

2.1 The 1998 Marshall Report – Setting the Trends? The Creation of the UK Emissions Trading Scheme

In 1998, the Government-commissioned Marshall Report on *Economic Instruments and the Business Use of Energy*[18] examined the role that economic instruments – such as business energy taxation or tradable emissions permits – might play in improving energy efficiency in the business sector and in the reduction of GHG emissions.

One key recommendation was for a national emissions trading scheme. The Government, with input from the Emissions Trading Group[19] (created under the auspices of the Confederation of British Industry and the Advisory Committee on the Environment), followed up on this recommendation and created the 'world's first economy-wide greenhouse gas emissions trading scheme',[20] more on which is found below.

---

[17] 1992 United Nations Framework Convention on Climate Change, Article 2 OBJECTIVE 'The ultimate objective of this Convention and any related legal instruments that the Conference of the Parties may adopt is to achieve, in accordance with the relevant provisions of the Convention, stabilization of greenhouse gas concentrations in the atmosphere at a level that would prevent dangerous anthropogenic interference with the climate system. Such a level should be achieved within a time frame sufficient to allow ecosystems to adapt naturally to climate change, to ensure that food production is not threatened and to enable economic development to proceed in a sustainable manner.' http://unfccc.int/resource/docs/convkp/conveng.pdf (accessed 27/2/08).

[18] Lord Marshall (1998): http://www.hm-treasury.gov.uk/media/F/7/EconomicInstruments.pdf (accessed 27/2/08). Lord Marshall was, at the time, the Chairman of British Airways. See Nye and Owens (2008, p. 4).

[19] http://www.uketg.com/aboutus.asp (accessed 27/2/08).

[20] See DEFRA WebPages on UK Emissions Trading Scheme: http://www.defra.gov.uk/environment/climatechange/trading/uk/index.htm (accessed 27/2/08).

What is particularly interesting about the early period during which policy options were being considered is the political background. Lord Marshall initially proposed a domestic ETS as an experiment: the more serious proposal at the time was for a downstream energy tax. It was proposed by Lord Marshall 'that the Government seriously consider a dry run pilot [domestic emissions trading scheme] with interested players, as soon as possible, as a means of learning lessons for the participation of our industry in the international [Kyoto] scheme'.[21] Such an experimental emissions trading scheme was conceived of as being a '"virtual" pilot with interested players [that] would be an opportunity to experiment with the detail of a trading system and allow firms to practise using it, without the dangers of legislating [...] straight into a particular regime'.[22] The energy tax, which was subsequently adopted as the Climate Change Levy (CCL), was proposed on account of the fact that Lord Marshall doubted that it would 'ever be practical for the majority of small and medium sized enterprises (SMEs) and less intensive users in industrial and commercial sectors to participate in the international trading scheme'.[23] Against this backdrop, it is interesting that SMEs may well have the opportunity to participate in a future iteration of the European Union Emissions Trading Scheme.

Nye and Owens note that 'although business originally supported emission trading as an alternative to taxation, more socio-symbolic motives shaped business interest in emission trading after the announcement of the Climate Change Levy'.[24] In short, Nye and Owens observed distinct division amongst the ranks of business interests. They assert that a UK Emissions Trading Group (ETG)[25] was formed by influential UK businesses to lobby for this particular policy instrument.[26] Tellingly, the Government was favourable to the ETG's contribution to the negotiation and design of such a national ETS scheme and it is certainly arguable that the economic-based incentives of an emissions trading scheme were viewed by key business interests as much more rewarding than the punitive CCL. Data collated and analysed by Nye and Owens indicate that 'interview evidence confirms that there was a very strong anti-taxation agenda driving early interest in UK emission trading'.[27] However, while it may well have been the case that maintaining the business

[21] See Lord Marshall (1998, p. 2); see Nye and Owens (2008, p. 4).

[22] Lord Marshall (1998, p. 16) : See also: Pearson (2004, p. 3).

[23] See Lord Marshall (1998, p. 2).

[24] Nye and Owens (2008, p. 1).

[25] According to Nye and Owens, senior personnel from BP headed the UK ETG steering committee and the secretariat, and Gordon Brown was, at the time, a member of the ETG steering committee: Nye and Owens (2008, p. 5 and p. 9).

[26] Nye and Owens (2008, p. 5).

[27] Nye and Owens (2008, p. 7).

advantage or market stability were thought of as priorities for UK businesses, the UK Government held firm on the climate change levy, and the ETS was, according to Nye and Owens, in fact created around this taxation scheme, something which may well surprise readers of this chapter.[28] Nye and Owens suggest that in lobbying for the ETS, 'industry may have felt compelled to flex its muscles in support of emission trading, in order to bolster its defences against future legislation'[29] (i.e., the CCL). They may have rationalized that the prospective costs of future legislation could have had the potential to be much more devastating than a voluntary emissions trading scheme.

2.2 UK Climate Change Programme

Originally launched on 17 November 2000[30] as a response to facilitating national compliance with the UNFCCC, and described by some as a 'suite of policy instruments',[31] the UK Climate Change Programme set out the Government's proposals for meeting the UK's legally-binding target of a 12.5% reduction in greenhouse gas emissions agreed under the 1997 Kyoto Protocol and for moving towards the Government's domestic target of a 20% reduction in carbon dioxide emissions. Following a 2004 review, the current 'phase' of the Climate Change Programme has been set from 2006. It sets out national policies and priorities for climate action in relation to meeting national and international commitments. Under the 2006 Climate Change Programme, DEFRA committed to introducing an annual report to Parliament on progress made in meeting the said targets. This commitment has since been made legally binding through the Climate Change and Sustainable Energy Act 2006 (more on which is below).

2.3 'The UK Energy Challenge'[32]

The onset of the new millennium saw a flurry of 'soft law' activity at UK level designed to foster a reduction in greenhouse gas (GHG) emissions and to encourage sustainable energy consumption.

The report of the Department for Trade and Industry [DTI] (now Department for Business, Enterprise and Regulatory Reform [BERR]) on the Energy Review entitled 'The Energy Challenge' was released on 11 July 2006.

---

28 Nye and Owens (2008, p. 7).

29 Nye and Owens (2008, p. 10).

30 http://www.defra.gov.uk/environment/climatechange/uk/ukccp/2000/index.htm (accessed 27/02/2008)

31 Pearson (2004, p. 2).

32 http://www.berr.gov.uk/files/file31890.pdf, Cm 6887 (accessed 27/02/2008).

This work aimed to put the UK in a position to meet two major long-term challenges in UK energy policy which are as follows: Climate Change Mitigation and the need to tackle climate change by reducing carbon dioxide emissions; and Energy Security and the need to deliver secure, clean energy at affordable prices, as the UK moves to increasing dependence on imported energy.

The Energy Review largely focused on large-scale energy production and consumption, electricity distribution and road transport. It also addressed GHG emissions reduction objectives. To this end it outlined some aspirations of the Government articulated by way of schemes to reduce carbon emissions such as: offering energy performance certificates for both new and existing buildings (section 2.16.1); promoting energy efficient buildings (section 2.16.5); offering grants or tax concessions to householders that install low-carbon equipment (implied in section 2.16.6); and incentivizing combined heat and power (section 7.31). The Energy Review also considered a new mandatory emissions trading scheme for businesses not covered by the EU ETS.[33]

3. LEGISLATIVE AND POLICY FOCUS

There is a plethora of legal and policy instruments which regulate energy in relation to eco-efficiency and climate change within the UK. The following are examples of some such instruments.

3.1 UK Emissions Trading Scheme (ETS)

We have already mentioned the UK ETS, above, within the context of policy-making. The UK ETS began in March 2002 and ended in December 2006, with the final reconciliation completed in March 2007. Thirty-three organizations were involved in the scheme as Direct Participants (DPs) and voluntarily took on emission reduction targets to reduce their emissions against 1998–2000 levels. The DPs committed to reducing their emissions by 3.96m tonnes of carbon dioxide equivalent (CO_2e) by the end of the Scheme and in fact achieved emissions reductions of over 7.2 million tonnes of CO_2e.[34]

A 2006 appraisal for DEFRA of the UK ETS cites[35] a benefit of the UK

33 Nye and Owens (2008, p. 2).

34 See DEFRA website for information on the scheme and key elements, including appraisals of the scheme: http://www.defra.gov.uk/environment/climatechange/trading/uk/documents.htm (accessed 31/3/08).

35 *Appraisal of Years 1–4 of the UK Emissions Trading Scheme*, A Report by ENVIROS Consulting Limited December 2006, pp. 32–33. http://www.defra.gov.uk/environment/climatechange/trading/uk/pdf/ukets1-4yr-appraisal.pdf (accessed 27/02/2008).

ETS as being the establishment of an emissions trading infrastructure, including: 'the development of the software required to make trading feasible; establishment of standard contract forms which can help to reduce transaction costs; provision of information and assistance to stakeholders; and, a greater understanding of the legal implications and most appropriate financial (tax) treatment of allowances'. Other benefits of the ETS will certainly be the role that it has played in raising environmental and climate change-related awareness amongst the public both in the UK and elsewhere. Other comments offered in relation to the UK ES concern its voluntary nature. In terms of weaknesses, the appraisers assert that a larger number of participants in the scheme would have arguably made the market more efficient;[36] that the timescales within which the scheme was designed and the auction run did not give some organizations sufficient time to absorb and understand the scheme rules and implications;[37] and that verifiable baseline data was not available or was deemed as being too costly to collect, particularly for small companies.[38]

3.2 The Climate Change Levy

In the 1999 Budget the British Government announced its intentions to launch a tax, the Climate Change Levy (CCL), on energy use for certain sectors and organizations in a bid to encourage energy efficiency (Pearce, 2006). Legal provisions for the CCL were enshrined in Schedule 6 of the Finance Act 2000, and the CCL entered into force in April 2001 (Department for Environment, Food and Rural Affairs, 2001a). The Levy is only applied once in the supply chain, usually directly to the end user.

Early expectations of the Government were that the 'revenue neutral' CCL would generate approximately £1 billion per year. This £1 billion would then be recycled by means of reducing employer National Insurance Contributions by 0.3% and through the establishment of the Enhanced Capital Allowances

[36] *Appraisal of Years 1–4 of the UK Emissions Trading Scheme*, A Report by ENVIROS Consulting Limited December 2006, p. 2 and p. 41. http://www.defra.gov.uk/environment/climatechange/trading/uk/pdf/ukets1-4yr-appraisal.pdf (accessed 27/02/2008).

[37] *Appraisal of Years 1–4 of the UK Emissions Trading Scheme*, A Report by ENVIROS Consulting Limited December 2006, p. 9. http://www.defra.gov.uk/environment/climatechange/trading/uk/pdf/ukets1-4yr-appraisal.pdf (accessed 27/02/2008).

[38] *Appraisal of Years 1–4 of the UK Emissions Trading Scheme*, A Report by ENVIROS Consulting Limited December 2006, p. 10. http://www.defra.gov.uk/environment/climatechange/trading/uk/pdf/ukets1-4yr-appraisal.pdf (accessed 27/02/2008).

Table 10.1 *CCL income and approximate National Insurance Contribution rebate 2001–2006*

Financial year	Levy income (£ million)	Approx. NIC rebate (£ million)
2001/02	555	1,035
2002/03	829	1,125
2003/04	832	1,185
2004/05	764	1,215
2005/06	744	1,275

Source: National Audit Office, 2007.

(EHA)[39] with a £150 million fund to promote energy efficiency in business (Carbon Trust, 2002). However, by 2006 the CCL had not managed to meet the income expectations. Income peaked in 2003/04 at £832 million (UK Trade Info, 2004) (see Table 10.1). Conversely, the National Insurance Contributions rebate to employers has gradually increased over the life of the Levy.

As part of the UK's Climate Change Programme, the CCL replaced the Fossil Fuel Levy.[40] Key implementing provisions are contained in Part II, Section 30 (Climate Change Levy)[41] and Schedule 6, which makes provision for a new tax 'that is to be known as climate change levy' and Schedule 7 on 'climate change levy: consequential amendment'. The Levy is administered by HM Revenue & Customs'[42] Commissioners.[43] This Levy is also known as the 'Energy Tax'.

The Levy imposes a 'pollution tax' (perhaps the use of the term 'levy' is strategically applied on account of it appearing more innocuous than the word

39 EHA reward businesses that invest in energy saving equipment. These businesses also pay less tax on profits arising from energy saving equipment.

40 Section 33 Electricity Act 1989, superseded by Fossil Fuel Levy Act 1998 until its repeal by the Utilities Act 2000. http://www.opsi.gov.uk/acts/acts1989/ukpga_19890029_en_4#pt1-pb7-l1g32 (accessed 31/03/2008).

41 Finance Act 2000, Chapter 17, Part II, Para. 30 Climate change levy: '(1) Schedule 6 to this Act (which makes provision for a new tax that is to be known as climate change levy) shall have effect. (2) Schedule 7 to this Act (climate change levy: consequential amendments) shall have effect. (3) Part V of Schedule 6 to this Act (registration for the purposes of climate change levy) shall not come into force until such date as the Treasury may appoint by order made by statutory instrument; and different days may be appointed under this subsection for different purposes'.

42 See: http://www.hmrc.gov.uk/ (accessed 27/02/2008).

43 Para. 1(2).

'tax') on use of energy in industry, commerce and the public sector, with the aim of driving down UK annual CO_2 emissions by 2.5 million tonnes by 2010. The Levy applies to most energy users except the transport sector and is chargeable on the industrial and commercial supply of taxable commodities for lighting, heating and power by consumers in the following sectors of business: industry; commerce; agriculture; public administration; other services. Taxable commodities are listed in Para. 3(1) as: (a) electricity; (b) any gas in a gaseous state that is of a kind supplied by a gas utility; (c) any petroleum gas, or other gaseous hydrocarbon, in a liquid state; (d) coal and lignite; (e) coke, and semi-coke, of coal or lignite; (f) petroleum coke. Hydrocarbon oil or road fuel gas[44] and waste[45] are exempt from the Levy.[46]

The Levy is payable by the UK supplier (Para. 40(1)) or in cases of a non-UK supplier, the recipient of the supply (Para. 40(2)).

The rates from 1 April 2008 will be: electricity – £0.00456 per kilowatt hour; gas supplied by a gas utility or any gas supplied in a gaseous state that is of a kind supplied by a gas utility – £0.00159 per kilowatt hour; any petroleum gas, or other gaseous hydrocarbon, supplied in a liquid state – £0.01018 per kilogram; any other taxable commodity – £0.01242 per kilogram.[47]

The Levy does not apply to taxable commodities used by domestic consumers (Para. 8(a)), or by charities for non-business use (Para. 8(b)). Other exemptions include: supply not for burning in the UK (Para. 11); supply not used in transport (Para. 12); supplies to producers of commodities other than electricity (§13); supplies (other than self-supplies) to electricity producers (Para. 14); supplies (other than self-supplies) to combined heat and power stations (Para. 15); supplies (other than self-supplies) of electricity from partly exempt combined heat and power stations (Para. 16); self-supplies by electricity producers (Para. 17); supply not used as fuel (Para. 18); and, electricity from renewable sources (Para. 18). All revenues raised through the Levy are recycled back to business through a 0.3% cut in employers' national insurance contributions, introduced at the same time as the Levy, and are also used to support the development of energy efficiency and low-carbon technologies within the UK.

44 Within the meaning of the 1979 c. 5 Hydrocarbon Oil Duties Act 1979.

45 Within the meaning of Part II of the Environmental Protection S.I. 1997/2778 Act 1990 or the meaning given by Article 2(2) of the Waste and (N.I. 19) Contaminated Land (Northern Ireland) Order 1997.

46 Para. 3(2).

47 See: http://customs.hmrc.gov.uk/channelsPortalWebApp/channelsPortalWebApp.portal?_nfpb=true&_pageLabel=pageExcise_InfoGuides&propertyType=document&id=HMCE_PROD1_027235#downloadopt (accessed 26/03/2008).

Noting the economic and social significance of the horticulture sector, horticultural producers are able to benefit from a half-rate levy under Para. 43.

The major incentive under the Levy comes from the reduced-rate for supplies covered by the climate change agreement (Para. 44). Climate change agreements are regulated under Paras. 46–50 of the Act. Notwithstanding the reduced rates, however, concerns arise in relation to the competitiveness of UK industry relative to other countries that do not impose such a levy. There are arguments that policy instruments such as negotiated agreements and emissions trading can soften the economic blow while still achieving emissions reductions. Perhaps a global energy tax that would represent 'a global burden sharing system, fair, with solidarity, and legally binding to all nations', as proposed by the Swiss Department of the Environment, Transport, Energy and Communications,[48] is what is needed to alleviate this concern.

3.3 Climate Change Agreements for Energy-intensive Sectors

The UK Government recognized the need for special consideration to be given to the position of *energy-intensive* industries noting their energy usage, the requirements of the Integrated Pollution Prevention and Control regime and their exposure to international competition as part of the UK's Climate Change Programme.

Upon the announced introduction of the CCL in 1999, the Government also made clear that there would be opportunities for businesses to reduce their CCL payments in return for efforts to reduce energy consumption (HM Treasury, 1999) (see Table 10.2). In order to qualify for the reduced CCL rate businesses would have to enter into a voluntary agreement with the Secretary of State for the Environment, Transport and the Regions. This is legally prescribed through Paragraph 44 of the Finance Act 2000.

With the legal basis in the Finance Act 2000,[49] there is a series of Climate Change Agreements (CCAs) for Energy-intensive Installations,[50] designed to offer an 80% discount from the Climate Change Levy for those sectors that agree to meet certain targets for improving their energy efficiency or reducing carbon emissions. The reasoning behind the CCAs certainly appears to be in

---

[48] UVEK 2007: *Global Solidarity in Financing Adaptation*, A Swiss Proposal for a Funding Scheme, Paper for further Discussion, Federal Office for the Environment, Berne. 40 pp, Berne, 11. December 2007. www.environment-switzerland.ch.

[49] Paragraph 52 of Schedule 6 to the Finance Act 2000.

[50] The Climate Change Agreements (Energy-intensive Installations) Regulations: Statutory Instrument (S.I). No. 1139 2001; S.I. No. 662 2006; S.I. No. 1931 2006; S.I. No. 1848; 2006; S.I. No. 60; S.I. No. 59 2006.

Table 10.2 *CCL rates per tonne CO_2 at the 2001–2007 rates*

	CCL equivalent price per tonne of CO_2 (£/tCO_2)	CCL discounted price per tonne CO_2 (£/tCO_2)
Levy on electricity (0.43 p/kWh)	£10.30	£2.06
Levy on natural gas (0.15 p/kWh)	£8.05	£1.61
Levy on coal (0.15 p/kWh)	£5	£1

Source: National Audit Office 2007.

line with current thinking on sustainable development.[51] It is noteworthy that the period for obtaining the discount was curtailed on 31 March 2003, indicating that the economic and social dimensions of the sustainable development triangle were not open to 'abuse' for the sake of the environment. Eligibility for discount from 1 April 2003 now depends on whether the first targets set in the agreements were met.

An 'Energy-intensive Sector' is defined by the UK Government as one which carries out activities which are listed under Part A1 or A2 headings in Part 1 of Schedule 1 to the Pollution Prevention and Control (England and Wales) Regulations 2000.[52] This means that sites operating Part A PPC activities (ten major energy-intensive sectors: aluminium, cement, ceramics, chemicals, food & drink, foundries, glass, non-ferrous metals, paper, and steel) are subject to a legal requirement to use energy efficiently. Small sites which fall below PPC-size thresholds (with the exception of thresholds relating to combustion plants), but which would otherwise be covered by the proposed regulations, were also eligible for the relevant sector agreement.[53] The

[51] The concept of 'sustainable development' has customarily focused on striving for *balance* between *economic* development goals and *environmental* protection efforts as a solution to dealing with developmental needs and environmental conservation goals. More recently, within the Johannesburg Declaration on Sustainable Development and the resultant WSSD Plan of Implementation, the idea of *social* development has been added to the equation. Following, MacDonald (2006, pp. 26–27).

[52] Statutory Instrument 2000 No.1973 as amended by the Pollution Prevention and Control (England and Wales) (Amendment) Regulations 2001 Statutory Instrument 2001 No. 503.

[53] See DEFRA website information on Climate Change Agreements: http://www.defra.gov.uk/environment/climatechange/uk/business/ccl/intro.htm (accessed 24/2/08).

Agreements operate on a sector basis represented by trade associations[54] with whom the Agreements are negotiated. Some Climate Change Agreements have now replaced other Negotiated Agreements.[55]

It should be noted that CCAs may be applied to energy-intensive installations/sites, parts of energy-intensive installations/sites *or* ancillary energy-intensive installations/sites that are identified pursuant to Sections 50 and 51 of the Finance Act, 2000. Further to the Table in Section 51, energy-intensive installations are divided into six categories: energy industries; production and processing of metals; mineral industry; chemical industry; waste management; and other industries including paper, food production and intensive farming. Further to Section 2(1) of the Climate Change Agreements (Energy-intensive Installations) Regulations, 2006 (Statutory Instrument 2006 No. 59) (hereinafter CCA Regulations 2006) the following installations have been added to those which already feature in Section 51 of the Finance Act 2000: compression or liquefaction of nitrogen, oxygen or argon; extraction of process clay; processing of calcium carbonate; treatment of metals with heat; horticulture activities; manufacture of textiles; and production of plastic films.

The Secretary of State originally negotiated Agreements with 44 sectors,

---

54 The Secretary of State certifies that the facilities in the various sectors (some examples are provided below) are to be taken as being covered by a climate change agreement, in accordance with Paragraph 44 of Schedule 6 to the Finance Act 2000: Aluminium Federation (Last amended 11 April 2007); Apparel & Textiles 'Energy Intensive' (Last amended 25 February 2008); British Apparel and Textile Confederation (Last amended 21 January 2008.); British Beer and Pub Association (Last amended 2 April 2007); British Calcium Carbonate Association (Last amended 2 April 2007); British Cement Association (Last amended 6 December 2007); British Ceramic Confederation (Last amended 18 February 2008); British Compressed Gases Association (Last amended 2 April 2007); British Egg Industry (Last amended 2 April 2007) UK Agricultural Supply Trade Association (Last amended 11 February 2008) Heat Treatment Sector (Last amended 18 February 2008); Steel (Last amended 30 May 2007).

55 Various other Negotiated Agreements (NAs) were used at the start of the UK 2000 Climate Change Programme until other forms of environmental policy were formalized, such as the Climate Change Levy. NAs are tailor-made contracts between the regulator and individual firms or groups of firms, which include targets and timetables for action and define rewards and penalties (see de Muizon and Glachant (2004, chapter 13). According to Sairinen and Teittinen (1999), by using NAs, the voluntary implementation of environmental goals gets started as soon as possible. In the UK, Beck notes (*Negotiated Emission Abatement Agreements Principles and Practice*, Australasian Emissions Trading Forum Reference Paper, May 2002: www.aetf.net.au) that NAs 'are linked to other environmental policies in order to attain a more ambitious abatement objective [and] they play an important support function in the policy mix. The UK and Dutch schemes are examples of this use, in both cases linked to taxation and trading regimes.'

representing more than 5000 companies and 12 000 target units[56] (ENDS, 2005). The number of Agreements and the target units they represent has changed continuously since their introduction in 2001. Energy usage targets were negotiated using four possible units: relative energy; relative carbon; absolute energy; and absolute carbon. Relative targets terms refer to the energy use or carbon emissions produced per tonne of unit of output. Absolute targets refer to the total quantity of carbon emissions or energy used, irrespective of produced output. The majority of CCAs were negotiated using relative energy use.[57] The choice of relative or absolute targets would appear to be counterintuitive from the perspective of limiting greenhouse gas emissions. The use of relative targets complies with the general aim of firms to expand production in the interest of seeking greater profits and leads to an increase in overall GHG emissions in relation to any unit expansion of output.

Sectors were able to choose their preferred target type and the sector target was based on the sum of the targets within underlying Agreements – the second tier of Agreements signed by operators of target units and the Secretary of State (National Audit Office, 2007). During the original negotiation of CCAs, provisions were made to allow for a review of three final interim targets (2006, 2008 and 2010) in 2004. This was done to ensure that the targets for each sector would still represent the best option in terms of cost-effective energy savings. This review process was also used as a means to assess whether targets set in the original Agreements were too stringent or too lenient (Future Energy Solutions, 2005).

At each interim target point, individual organizations have to present data concerning energy use and production levels to the sector association. DEFRA subsequently verifies the data to determine whether both the sector and the individual organization are eligible for recertification and the resulting 80% CCL discount during the next interim target period (de Muizon and Glachant, 2004).

As indicated above, eligibility to enter into an Agreement was initially based on the European Council Directive concerning Integrated Pollution Prevention and Control (IPPC Directive). However, the thresholds for inclusion in IPPC are disregarded and all activities irrespective of size are eligible to enter into a CCA (Department for Environment, Food and Rural Affairs, 2001b). An extension of the eligibility criteria pursuant to the CCA Regulations 2006 resulted in 12 new umbrella agreements. These sectors are

56 A target unit refers to individual site or parts of a site that are included in a CCA.

57 On a sectoral (rather than an installation) basis four sectors negotiated absolute energy targets and a further two negotiated absolute carbon targets.

Table 10.3 Results of target periods 1–3

	Target period 1 (2002)	Target period 2 (2004)	Target period 3 (2006)
A Total number of umbrella agreements	44	42	49
B Number of sectors meeting targets	24	21	32 (41)[58]
C Total number of target units	5742	4675	4885
D Target units re-certified (either meeting individual target or whole sector target)	5042	4420	4401
E Target units withdrawing from an agreement	164	228	345
F Target units not re-certified (i.e. failing to meet underlying agreement target if sector as whole fails to meet umbrella target)	219	23	23
G Target units not submitting data at end of target period[59]	319	4	116
H % of Target units re-certified	88%	95%	90%
I Absolute annual carbon saving (MtC)	4.5	3.9	4.5
J Absolute annual carbon saving (excluding steel sector) (MtC)	1.9 (steel = 2.6)	1.9 (steel = 2)	2.5 (steel = 2)
K Proportion of annual savings arising directly from CCAs (MtC)	1.9	1.9	1.9

Sources: Future Energy Solutions, 2004; Future Energy Solutions, 2005; Department for Environment, Food and Rural Affairs, 2007; National Audit Office, 2007.

[58] DEFRA uses both of these figures in Results of the Third Target Period Assessment stating 'overall 32 out of the 49 sectors have met their targets after taking the emissions trading by operators into account. However one of these sectors, whilst

all relatively small in terms of number of companies and energy use compared with the original CCA participating sectors. The Treasury estimated that these sectors will make carbon emissions savings of approximately 0.3 MtC by 2010 (Department for Environment, Food and Rural Affairs, 2006a). Participating target units are subsequently eligible to receive the 80% discount on CCL payments provided the terms and conditions of the CCA are met. Companies that meet the IPPC criteria for participation must do so under the IPPC arrangements further to Schedule 1 of the Pollution Prevention and Control Regulations (England and Wales) 2000 (Statutory Instrument 2000/1973). However, in cases where a single site comprises separate facilities that are eligible under both criteria the site can share one target (Department for Environment, Food and Rural Affairs, 2006a), thus reducing the administrative burden on the operator, the sector association and DEFRA.

In relation to the performance of CCAs to date, in *absolute* terms (i.e. total energy consumption or CO_2 emissions reductions) the CCAs have overachieved in each target period by nearly double the absolute target compared to the initial targets within the umbrella agreements. Table 10.3 presents the performance of CCAs in the first three target periods. It is important to note that this overachievement is largely attributed to a significant downturn in steel production (see Table 10.3, rows I–K). The steel sector accounts for approximately a quarter of the overall CCA targeted energy consumption. The number of target units that were not re-certified following the interim target period assessment has decreased from 12% after the first target period to 10% in the third target period (see Table 10.3, row H).

The 2007 National Audit Office assessment of CCAs highlighted several fundamental issues with the calculation of performance and difficulties in modeling the impacts of the agreements during the negotiation period. Prior to the negotiations, the Global and Atmospheric Division (GAD) of DEFRA modelled the potential carbon savings from 14 of the CCA sectors. Extrapolated to the original 44 sectors the modelling suggested that the CCAs could save 4.5 MtC per year beyond business as usual by 2010 (ETSU–AEA Environment, 2001). In the event, DEFRA negotiated agreements that should produce carbon savings of 2.5 MtC in 2010 (National Audit Office, 2007), despite the evidence of the GAD modelling. This is puzzling because there is no explanation for the decreased negotiated target, perhaps there was insufficient faith in the modelling work or perhaps DEFRA anticipated greater

meeting their CCA target, did not meet their European Union Emissions Trading Scheme corrected target, and a single target unit within the sector was decertified. In effect, however, 41 of the 49 sectors met their targets as all target units have been re-certified'.

[59] Failure to submit data at the end of a target period results in termination of the agreement with the individual target unit.

carbon savings to be incentivized by the CCL. Furthermore, even the 2.5 MtC saving negotiated with the original 44 sectors is unlikely to occur. The current prediction for carbon savings beyond the business as usual scenario is 1.9 MtC in 2010 (National Audit Office, 2007).

In 2006, econometric modelling by Ekins and Etheridge led to some critical observations about the approach taken to negotiate and apply targets within some of the participating sectors:

> on the basis of econometric projections, and with some exceptions, the CCA targets were within what companies would have been likely to achieve anyway. This was especially true of the sectors in the broad Mineral Products and Chemicals sectors. . . . This conclusion is hardly a surprise. The superior knowledge of industrial managers about their business compared to even the most diligent external consultant means that they are always likely to be able to present compelling argument (which they may or may not themselves believe) about why proposed measures will cost more than is actually the case.

Contrary to the industry positioning on the CCL, further analysis undertaken by Ekins and Etheridge led to the conclusion that the CCL is, in fact, beneficial to the economy due to the NIC rebates and resulting small increases in GDP and employment levels. Further to this point, in terms of economic efficiency it is posited that a uniform environmental tax applied across sectors can result in cheaper emission reductions compared to the situation we find with the addition of CCA-type instruments where some companies receive a reduced tax rate as companies not in receipt of a discount are often subject to a higher tax rate.

Putting aside questions of economic efficiency for the moment, in the case of the CCL/CCA package, Ekins and Etheridge assert that the addition of CCAs is likely to lead to higher emission reductions in 2010 compared to the scenario in the absence of CCAs. This assertion was made on the basis that industry might not have been fully aware of energy-saving opportunities prior to the CCAs. This was evidenced by the 2002 target period and the vast over-achievement against targets (Future Energy Solutions, 2004). However, this may also suggest that DEFRA has been too willing to accept industry's position on targets and that target setting might have been better positioned against actual emissions/energy data (e.g. IPPC reporting or utility bills).

At the time of writing, 45 trade associations have signed a Climate Change Agreement covering the emissions of their associated industries and there are three closed sectors.[60] The 'Umbrella Agreements', which are signed between DEFRA and the trade bodies, are publicly available documents, in accordance

---

60 Vehicle Builders and Repairers Association, National Microelectronics Institute (Cathode Ray Tube) and Reprotech.

with the Aarhus Convention[61] and relevant EC[62] and UK implementing legislation.

3.4 The UK Non-Energy-intensive Sector

Early policy initiatives for the non-energy-intensive sector (some are discussed below) appear to be much softer than those for the energy sector. This said, in 2007, DEFRA carried out an extensive consultation to evaluate the emissions reduction potential of the non-energy-intensive sector. The resultant 2007 study[63] evaluated the viability of the following options for this sector: expanding the EU Emissions Trading Scheme; expanding Climate Change Agreements; requiring mandatory reporting or benchmarking; strengthening the implementation of building regulations and the Energy Performance of Buildings Directive;[64] extension of the current voluntary UK ETS to cover additional organizations in the target group; expansion of the Carbon Trust's remit; creation of differentiated business rates for buildings to encourage reduced CO_2 intensity; imposition of an obligation on Energy Suppliers (similar to the EEC for household customers) to reduce target group energy consumption; alternative obligations on energy suppliers such as a cap-and-trade approach.

3.5 Horticulture Assistance Package (HAP)

The Government recognized that the horticulture sector was relatively energy intensive, contained a large number of smaller companies and was exposed to significant international competition. Other countries with energy taxes had also tended to afford their horticulture sectors special treatment in relation to emissions reduction initiatives. Consequently, the UK Government implemented a special package of measures for horticulturists (growers of fruit, certain vegetables, flowers, shrubs, trees, and certain seeds) to improve energy efficiency in the sector.

The assistance package included: a special allocation for the sector from the

61 The Aarhus Convention on Access to Information, Public Participation in Decision-Making and Access to Justice in Environmental Matters.

62 Directive 2003/4/EC of the European Parliament and of the Council of 28 January 2003 on public access to environmental information and repealing Council Directive 90/313/EEC, OJ L 041,14.02.2003, pp. 26–32.

63 *Comparison of Policies to Reduce Carbon Emissions in the Large Non-Energy-Intensive Sector*, May 2007 http://www.defra.gov.uk/environment/climatechange/uk/business/crc/pdf/nera-report.pdf (accessed 23/02/2008).

64 Directive 2002/91/EC of the European Parliament and Council on energy efficiency of buildings.

energy efficiency fund to provide site specific advice; inclusion of thermal screens in the list of technologies qualifying for enhanced capital allowances. Thermal screens are panels used to reduce the volume of a building to be heated during cold periods and are used in greenhouses.

A temporary 50% discount on the climate change levy for a period of up to five years was central to the HAP – this was intended to give the sector some relief while anticipated energy efficiency measures took effect. Some horticulturists may also have benefited from the exemption from the levy for energy from good-quality CHP.

After 1 April 2006, the temporary 50% rate for 2006 energy used in horticulture was abolished. However, since this period, many businesses in the horticulture sector have become eligible to sign climate change agreements and instead benefit from the 80% reduction from the levy in exchange for meeting specific energy efficiency targets.

4. PLANNING AND BUILDING

4.1 The Code for Sustainable Homes

The Code for Sustainable Homes is the national standard for the sustainable design and construction of new homes, launched by the Department of Communities and Local Government on 13 December 2006, commencing in April 2007. It sets sustainability benchmarks for new builds in terms of energy use, waste management, water efficiency and other environmental criteria. Under the scheme, a developer of any new home in England could *voluntarily choose* to be assessed against the Code. As of 1 May 2008, however, there are *mandatory* ratings against the Code for all new homes. In terms of CO_2 emissions, the standards are found in Annex B Category 1, Energy and Carbon Dioxide Emissions, with the aim of limiting 'emissions of carbon dioxide (CO_2) into the atmosphere that arise from the operation of a dwelling and its services'. To achieve this aim, minimum assessment criteria are set based on a credit system. Credits are awarded based on the percentage improvement in the Dwelling Emission Rate (DER),[65] over the Target Emission Rate (TER),[66] for the dwelling. A table in the Annex sets out the improvement of DER over TER and the corresponding level of the Code. Credits are also awarded in relation to measures taken to: stem heat loss; encourage provision of energy effi-

---

[65] The estimated carbon dioxide emissions in kg per m^2 per annum arising from energy use for heating, hot water and lighting for the actual dwelling.

[66] The maximum emission rate permitted by Building Regulations. DER and TER are as defined in AD L1A 20065 Edition of the Building Regulations.

cient internal lighting so as to reduce energy consumption;[67] provide a reduced energy means of drying clothes; encourage the provision or purchase of energy efficient white goods; encourage the provision of energy efficient external lighting; encourage local energy generation from renewable sources; encourage the wider use of bicycles as transport by providing adequate and secure cycle storage facilities; and, reduce the need to commute to work.

In addition to the above, the Department of Communities and Local Government *Planning Policy Statement: Planning and Climate Change*[68] sets out how GHG emissions reduction concerns and climate change issues should be taken into account at all levels of spatial planning.

Building a Greener Future: policy statement, 2007 confirms the Government's intention for all new homes to be zero carbon by 2016[69] with a progressive tightening of the energy efficiency building regulations – by 25% in 2010 and by 44% in 2013 – up to the zero carbon target in 2016.[70] In relation to this the Government has announced the intention to create 15 UK 'Eco-towns'; five to be built by 2016 and the remainder to be built by 2020.[71]

4.2 Transport: Renewable Transport Fuel (RTF)

In terms of the transport and climate change sector, the major policy initiative in this area is the Renewable Transport Fuel (RTF) Obligation Programme which commenced April 2008[72] to oversee a 'cost effective transition to a renewably fuelled long-term transport system'.[73] Modelled

67 The Code is more stringent than current regulations on this, including: Building Regulations England and Wales, Part L1A7 which requires fixed dedicated energy efficient light fittings to be installed in the most frequented locations in the dwelling to a number not less than one per 25m² floor area or one per four fixed light fittings.

68 *Planning Policy Statement: Planning and Climate Change Supplement to Planning Policy Statement 1* http://www.communities.gov.uk/documents/planningand building/pdf/ppsclimatechange (accessed 27/02/2008).

69 Proposals set out in the 2006 consultation document *Building a Greener Future* http://www.communities.gov.uk/documents/planningandbuilding/pdf/153125 (accessed 23/03/2008).

70 The Communities and Local Government, *Building a Greener Future: policy statement*, July 2007, http://www.communities.gov.uk/documents/planningandbuilding/pdf/building-greener (accessed 23/03/2008).

71 See the Communities and Local Government *Eco-towns Living a greener future*, consultation document, April 2008, http://www.communities.gov.uk/documents/housing/pdf/livinggreenerfuture, (accessed 02/04/2008).

72 Initiated under the UK Climate Change Programme 2006.

73 Defra, http://www.defra.gov.uk/environment/climatechange/uk/transport/index.htm (accessed 23/03/2008).

on the electricity supply Renewables Obligation and Renewable Obligations Certificates (ROCs), the Renewable Transport Fuel Obligations Order 2007[74] provides the legal basis to fuel suppliers to ensure that by 2010 renewable fuels make up 5% of their UK road fuel sales. Section 4 outlines the precise obligation and Section 16 governs the issuance of RTF certificates. Suppliers are required to report their road fuel volumes to an Administrator, the Office of the Renewable Fuels Agency, which is set up by the Order.[75] Tradable certificates will be issued in return for the supply of renewable road fuel[76] as evidence of meeting the obligation. If they have insufficient certificates, suppliers are allowed to pay an amount calculated as set out in the Order so as to fulfil the obligation. Such payments are pooled in a fund to be distributed amongst suppliers, according to the number of certificates they redeem or surrender. The underlying emissions saving of the RTFO is expected to be approximately 2.6–3.0 million tonnes per annum by 2010.[77]

4.3 Strategies

Several strategies have been advanced by the Department of Transport in order to address the sustainable development concerns of the transport sector. The 2002 *Powering future vehicles strategy* sets targets for 'making the UK a world leader in the move to a low-carbon transport system' by 2012 and beyond for all types of road vehicles.[78] The July 2004 White Paper, *The Future of Transport: a network for 2030*[79] advocates an integrated, affordable and sustainable transport system.

Further to the above, the UK Climate Change Programme 2006 raised the possible inclusion of surface transport in the EU Emissions Trading Scheme itself or as a separate measure. The Department for Transport has since

---

74 S.I. 2007 No. 3072 Transport Energy Sustainable And Renewable Fuels, The Renewable Transport Fuel Obligations Order 2007 http://www.opsi.gov.uk/si/si2007/uksi_20073072_en_1 (accessed 22/03/2008).

75 See provisions contained in Part 3 of the Order.

76 See provisions contained in Part 4 of the Order.

77 See Explanatory Memorandum to The Renewable Transport Fuel Obligations Order 2007 2007 No. 3072 http://www.opsi.gov.uk/si/si2007/em/uksiem_20073072_en.pdf (accessed 23/03/2008).

78 http://www.dft.gov.uk/pgr/roads/environment/poweringfuturevehicles/poweringfuturevehiclesstrategy?page=3#a1002 (accessed 23/03/2008).

79 Department for Transport, 'The Future of Transport: a network for 2030', Cm 6234, http://www.dft.gov.uk/about/strategy/whitepapers/fot/thefutureoftransportwhitepap5710 (accessed 23/3/08).

produced a discussion paper[80] which, *inter alia,* discusses the types of players in the scheme (there are three proposals: individual drivers; car manufacturers; fuel producers) and the interaction of such a scheme with other policy instruments.

4.4 Aviation

Influenced by the 2003 White Paper *The Future of Air Transport,*[81] the Civil Aviation Act 2006[82] includes environment-related provisions on noise, vibration and emissions (Sections 1–4), the latter amending section 38 of the Civil Aviation Act 1982 (c. 16) and allowing an aerodrome authority[83] to fix its charges in respect of an aircraft or a class of aircraft by reference to *inter alia,* emissions. This includes:

> (b) any fact or matter relevant to the amount or nature of emissions produced by the aircraft or the extent or nature of any atmospheric pollution resulting from such emissions; (c) any fact or matter relevant to the effect of the aircraft on the level of noise or atmospheric pollution at any place in or in the vicinity of the aerodrome; (d) any failure by the operator of the aircraft to secure that any noise or emissions requirements applying to the aircraft are complied with.

Hence, there is certainly a legal basis here for regulating the GHG emissions of the aviation sector through some form of secondary legislation, including 'for encouraging the use of aircraft which produce lower emissions of any substance which contributes to atmospheric pollution'. Following this, Department for Transport (DfT) published in May 2005 the findings of a stakeholder consultation on *Aviation and the environment: Using economic instruments.*[84] There is much debate on whether the aviation sector should be

---

80 *Road Transport and the EU Emissions Trading Scheme,* (not dated), http://www.dft.gov.uk/pgr/sustainable/climatechange/euemistrascheme?page=4#a1009 (accessed 23/03/2008).

81 Department for Transport White Paper 'The Future of Air Transport', published on 16 December 2003, sets out a strategic framework for the development of airport capacity in the United Kingdom over the next 30 years, against the wider context of the air transport sector. http://www.dft.gov.uk/about/strategy/whitepapers/air/.

82 http://www.opsi.gov.uk/acts/acts2006/pdf/ukpga_20060034_en.pdf (accessed 23/03/2008).

83 The Secretary of State also has powers to direct the airport authorities to levy necessary charges.

84 http://www.dft.gov.uk/pgr/aviation/environmentalissues/uei/ (accessed 23/03/2008).

included in the EU ETS, and several studies[85] have been promulgated in this regard by DEFRA and DfT.

4.5 Climate Change and Sustainable Energy Act 2006

The Climate Change and Sustainable Energy Act 2006 places an obligation on DEFRA to report to Parliament on greenhouse gas emissions in the UK and action taken by Government to reduce these emissions. It aimed to: enhance the United Kingdom's contribution to combating climate change; alleviate fuel poverty; secure a diverse and viable long-term energy supply. Obligations under the Act include: annual government reporting on greenhouse gas emissions; local authority regard to information on energy measures in exercising functions; increased microgeneration; increased energy efficiency; building regulations relating to emissions and use of fuel and power; carbon emissions reduction targets; dynamic demand technologies; community energy and renewable heat; electricity from renewable sources.

4.6 Combined Heat and Power – Applications under Section 36 Electricity Act 1989

In bringing forward power station proposals (except renewable energy projects) Section 36 and Section 14 of the Electricity Act 1989 require developers to show they have explored opportunities to use combined heat and power.

The Department of Business, Enterprise & Regulatory Reform administers the provisions of the Electricity Act 1989 and the Transport and Works Act 1992 (for offshore wind farms only) for developers seeking development consents from the Secretary of State for the construction of electricity generating stations and for overhead lines. Such development will also bring into play the requirements for environmental impact assessment through the Electricity Works (Environmental Impact Assessment) (England and Wales) Regulations 2000.

4.7 Non-Fossil Fuel Obligation (NFFO)

The aim of the NFFO, governed by Section 32 of the Electricity Act 1989,[86]

---

[85] See for example: *Including the Aviation Sector in the European Union Emissions Trading Scheme*, http://www.defra.gov.uk/environment/climatechange/trading/eu/pdf/including-aviation-icf.pdf (accessed 23/03/2008).

[86] Section 32, Electricity from non-fossil fuel sources: http://www.opsi.gov.uk/acts/acts1989/ukpga_19890029_en_4#pt1-pb7-l1g32 (accessed 31/03/2008).

was to encourage growth within the renewable energy industry, especially prior to the introduction of the Renewables Obligation. Its aim was to assist industry by providing premium payments for renewables-generated electricity over a fixed period, with contracts being awarded to individual generators. The NFFO applied in England and Wales. In Scotland and Northern Ireland, the Renewables Obligation (Scotland) (ROS) or the Northern Ireland NFFO (NI-NFFO) applied. According to the Department for Business Enterprise and Regulatory Reform (BERR), there are more than 400 NFFO projects currently operational. BERR states that with the introduction of the Renewables Obligation, no new NFFO contracts will be awarded but existing contracts will continue until the last of them expires in 2019. Existing NFFO 3, 4 and 5 contracts will continue in their present form.[87] If electricity from generating stations built under the NFFO arrangements in England and Wales (NFFO 1–5) or in Scotland (ROS) meet prescribed requirements it will be eligible under the Renewables Obligation. Where output continues to be sold under an NFFO contract, the Non-Fossil Purchasing Agency will sell the electricity into the market. Renewables Obligation Certificates will be used to offset the cost of these contracts to consumers through the Fossil Fuel Levy. Electricity generated subject to a qualifying arrangement under the NI-NFFO is not currently eligible for the Renewables Obligation. This is because there is no Northern Ireland Obligation, although one was expected from 1 April 2005.[88]

4.8 Renewables Obligation

The Renewables Obligation is another incentive-based scheme designed to encourage the use of renewable energy within the electricity supply sector. It has since replaced the NFFO as the main support driver for this. The Renewables Obligation Order came into effect[89] in April 2002 in England, Wales and Scotland and 2005 in Northern Ireland. The Utilities Act 2000[90] provides the legal basis for the Renewables Obligation and it is administered by the Gas and Electricity Markets Authority (OfGEM). Section 62 of the Utilities Act 2000 confers powers on the Secretary of State to place an obligation on

[87] http://www.berr.gov.uk/energy/sources/renewables/policy/renewables-obligation/non-fossil-fuel-obligation/page18369.html (accessed 28/03/2008).

[88] See: http://www.berr.gov.uk/consultations/page31914.html. A Proposal to Amend Non-Fossil Fuel Obligation Contracts for Municipal and Industrial Waste Projects with Combined Heat and Power consultation (accessed 28/03/2008).

[89] Introduced by what was at the time, Department of Trade and Industry (England & Wales), the Scottish Executive (Scotland) and the Department of Enterprise, Trade and Investment (Northern Ireland).

[90] Amending the Electricity Act 1989.

electricity suppliers to require that they source a prescribed percentage of their total sales of electricity in the United Kingdom from renewable sources. The Orders are amended annually in order for the levels of the obligation to be increased and for necessary pricing adjustments to be made. In 2005–06 it was 5.5% (2.5% in Northern Ireland). In 2006–07 the obligation was set at 6.7% (2.6% in Northern Ireland). When suppliers source from renewable energy sources, they can obtain a Renewable Obligations Certificate (ROC) as evidence for each one megawatt hour (1000 units) of electricity generated from eligible sources. Suppliers meet their obligations by presenting sufficient ROCs. Suppliers must pay an equivalent amount into a fund when they do not have sufficient ROCs to meet their obligations. Moneys from the fund are paid back on a pro-rated basis to those suppliers that have presented ROCs. At the time of writing, suppliers will be subject to a renewables obligation until 31 March 2027.[91]

Renewable sources that are eligible for ROCs include: landfill gas; sewage gas Hydro exceeding 20 megawatts declared net capacity (dnc) – only stations commissioned after 1 April 2002 Hydro 20 megawatts or less dnc; onshore wind; offshore wind; co-firing of biomass (there are no restrictions on the amount of co-firing a generator can undertake. However, suppliers can only meet 10% of their obligation from co-fired ROCs.); other biomass; geothermal power; tidal and tidal stream power; wave power; photovoltaics; energy crops; and energy from certain types of waste management processes.[92]

4.9 Climate Change Bill (2008)

The Climate Change Bill (the Bill) was published on 13 March 2007 for pre-legislative scrutiny and public consultation and was introduced into the House of Lords on 14 November 2007 after the Government published its response to the parliamentary scrutiny and public consultation in the Command Paper *Taking Forward the UK Climate Change* Bill in late October 2007. The aim with this piece of primary legislation is for the Bill to receive Royal Assent by early summer 2008.[93] At the time of writing the Bill is not in force.

---

[91] For more information see: http://www.ofgem.gov.uk/Sustainability/Environmnt/RenewablObl/Pages/RenewablObl.aspx and http://www.carbontrust.co.uk/climatechange/policy/renewables_obligation.htm (accessed 28/02/2008).

[92] See: http://www.berr.gov.uk/energy/sources/renewables/policy/renewables-obligation/what-is-renewables-obligation/page15633.html (accessed 28/03/2008).

[93] See DEFRA: http://www.defra.gov.uk/environment/climatechange/uk/legislation/index.htm (accessed 31/03/2008). See Taking Forward the UK Climate Change Bill: The Government Response to Pre-Legislative Scrutiny and Public Consultation. October 2007, Cm 7225: http://www.official-documents.gov.uk/document/cm72/7225/7225.pdf (accessed 28/03/2008). Follow progress on the Bill at the time of

Part 1 of the Bill contains key provisions in relation to greenhouse gas emissions levels and reductions in the UK. The principal aim of the – at the time of writing, draft – Bill is to ensure that UK emissions of greenhouse gases do not exceed the level necessary to contribute to limiting the global average temperature increase to not more than 2°C above pre-industrial levels (Section 1(1)). A unilateral, long-term legal target of ensuring that the net UK carbon account for the year 2050 is at least 60%[94] lower than the 1990 baseline is set in Section 2(1). Under Section 3(1), the Secretary of State for the Environment has powers to amend the target and baseline on account of developments in scientific knowledge about climate change or European or international law or policy (Section 3(2)) on climate change. The Secretary of State also has powers to set further legislation in relation to the designation of further greenhouse gases as targeted greenhouse gases (Section 3(2) (b) (i)) or emissions from international passenger travel or imports or exports of goods (Section 3(2) (b) (ii)) and to amend the target and baseline year accordingly if necessary. The power to change the parameters is certainly welcome noting the uncertainties and heavy reliance on scientific data in relation to regulating climate change, and it is hoped that in light of convincing scientific evidence or the strengthening of international law standards that this will not prove to be a discretionary provision.

Under Section 5 of the Bill, the Secretary of State also has powers to set five-yearly carbon budgets through secondary legislation for the net UK carbon account, with the level of the carbon budget to be prescribed based on Section 6.[95] Seventy per cent of the UK carbon budget is to be met domesti-

writing: http://services.parliament.uk/bills/2007-08/climatechangehl.html (accessed 31/03/2008).

[94] A statutory review by the Committee on Climate Change will take place under the Bill to determine whether the target should be 80%. See: http://www.defra.gov.uk/environment/climatechange/uk/legislation/pdf/govt-amendment package.pdf and 'Benn announces statutory review of 2050 climate targets', DEFRA News Release, 18 February 2008, http://www.defra.gov.uk/news/ 2008/080218a.htm (accessed 28/02/2008).

[95] Section 6 – Level of carbon budgets

(1) The carbon budget – (a) for the budgetary period including the year 2020, must be such that the annual equivalent of the carbon budget for the period is at least 26% lower than the 1990 baseline; (b) for the budgetary period including the year 2050, must be such that the annual equivalent of the carbon budget for the period is lower than the 1990 baseline by at least the percentage specified in section 2 (the target for 2050); (c) for the budgetary period including any later year specified by order of the Secretary of State, must be such that the annual equivalent of the carbon budget for the period is – (i) lower than the 1990 baseline by at least the percentage so specified, or (ii) at least the minimum percentage so specified, and not more than the maximum percentage so specified, lower than the 1990 baseline.

cally through domestic emissions reductions and domestic removal by sinks (Section 25(1). Controversially, emissions from international passenger travel or imports or exports of goods do not count as emissions from sources in the United Kingdom for the purposes of Part 1 of the Bill on Carbon Target and Budgeting (Section 30). Further, it is hoped that voluntary targets will also continue to be used, and perhaps the aviation and shipping sectors might be encouraged to engage in this.[96]

Part 2 (Sections 32–40) of the Bill legislates for the creation of a Committee on Climate Change which has been established in order to advise the Secretary of State on the issues contained within the Bill. This Committee is currently operating as a 'shadow' Committee without statutory basis as the Bill is not yet in force.[97]

Part 3 of the Bill (Sections 43–54) covers Trading Schemes as a means of helping the UK to meet its targets and commitments to reduce GHGs. The relevant national authority[98] will have powers to introduce new domestic emissions trading schemes (Section 43(1)) through secondary legislation subject to full public consultation, consultation by the Climate Change Committee and parliamentary scrutiny (Section 47).

Part 4 of the Bill concerns research into the impact of and adaptation in the UK to climate change, mandated through legal requirements to undertake reporting and to develop adaptation programmes. Interestingly, Part 5 of the Bill contains Other Provisions, most notably on Waste Reduction Schemes, with Section 69 calling for amendment of the 1990 Environmental Protection Act in this regard to provide for the creation of waste reduction schemes, largely through the promotion of recycling. The UK Government is commended here for explicitly linking the growing waste and related consumption and production issues to climate change and linking the Climate Change Bill in with the Government *Waste Strategy for England 2007*.[99] It is important to reiterate that landfilling, incineration, anaerobic digestion and composting all produce greenhouse gases, and though sequestration of gases

---

96 At the time of writing, this is being deliberated prior to the Bill attaining Royal Assent: http://www.defra.gov.uk/environment/climatechange/uk/legislation/pdf/govt-amendment-package.pdf (accessed 28/02/2008).

97 See DEFRA News Release, 22 February 2008: http://www.defra.gov.uk/news/2008/080222a.htm (accessed 28/02/2008).

98 Defined in Section 46.

99 http://www.defra.gov.uk/environment/waste/strategy/strategy07/pdf/waste07-strategy.pdf (accessed 28/03/2008). Under this strategy a key objective for the UK Government is to meet and exceed the Landfill Directive (Council Directive 1999/31/EC of 26 April 1999 on the landfill of waste) diversion targets for biodegradable municipal waste in 2010, 2013 and 2020.

and energy recovery are options, they are arguably not the only solution to safeguarding the planet.

We expect a series of other climate-related measures to be legislated for under the Climate Change Bill.

4.10 Other UK Initiatives: Now and Next

The UK is certainly taking an integrated approach towards GHG emissions reduction commitments. We have noted efforts largely related to the energy sector, touched on initiatives related to waste management and shall now briefly mention initiatives related to other areas.

4.10.1 EcoDesign for Energy-Using Products Regulations 2007

The EcoDesign for Energy-Using Products Regulations 2007 implement Directive 2005/32/EC of the European Parliament and of the Council establishing a framework for the setting of eco-design requirements for energy-using products,[100] which aims to encourage manufacturers to produce products which are designed to minimize their overall environmental impact, including the resources consumed in their production and disposal.

4.10.2 Energy Bill (2008)

The Energy Bill was introduced in the House of Commons on 10 January 2008 and finished in House of Commons committee on 11 March 2008. It is not yet in force. It contains the legislative provisions required to implement UK energy policy following the publication of the Energy Review 2006 and the Energy White Paper 2007, dealing with security of supply and climate change issues in the UK energy sector. Key aims and objectives include striving for: diverse, secure supply of electricity; reduced carbon dioxide emissions; strengthened regulatory framework to enable private sector investment in gas supply projects; creation of a regulatory framework to enable private sector investment in CCS; strengthening of the Renewables Obligation; dealing with the nuclear issues, especially in relation to waste and decommissioning issues; affordable energy.[101]

4.10.3 Carbon Reduction Commitment

On 23 May 2007 as part of the 2007 Energy White Paper, the Government

---

[100] Statutory Instrument 2007 No. 2037.

[101] Track the progress of the Bill: http://services.parliament.uk/bills/2007-08/energy.html (accessed 22/03/2008).

announced proposals for a Carbon Reduction Commitment[102] to cut carbon emissions from the large non-energy intensive commercial and public sectors[103] through mandatory emissions trading. This is driven by the Climate Change Bill and will be introduced through secondary implementing legislation. It is hoped that the scheme will commence in January 2010, with a three-year introductory phase, with the first capped phase expected to begin in January 2013.[104] The CRC will cover energy use emissions outside the EU Emissions Trading Scheme (EU ETS) and Climate Change Agreements (CCAs) – including both direct emissions and 'indirect emissions' so that there is no regulatory overlap. According to DEFRA, there will be an introductory three-year phase in which carbon allowances will be sold at a fixed price of £12/tCO_2. In the second phase, allowances will be auctioned.[105]

4.10.4 Environmental Transformation Fund

Also announced in the 2007 Energy White Paper, the Environmental Transformation Fund (ETF) is a new initiative that has been established to drive forward the development of new low-carbon energy and energy efficiency technologies in the UK. The fund was scheduled to begin operation in April 2008, and will be jointly administered by DEFRA and the Department for Business, Enterprise and Regulatory Reform (BERR).[106] It brings together DEFRA's and BERR's existing low-carbon technology funding programmes and will also promote new schemes. There will also be an international development and poverty reduction arm to the fund.[107]

4.10.5 Carbon Emissions Reduction Target[108]

The Carbon Emissions Reduction Target (CERT) came into effect on 1 April

---

102 Formerly the Energy Performance Commitment. The name of the scheme has been changed to prevent any confusion with Energy Performance Certificates. See DEFRA: http://www.defra.gov.uk/environment/climatechange/uk/business/crc/index.htm (accessed 29/03/2008).

103 Such as government departments, universities, retailers, banks, water companies, hotel chains and local authorities.

104 From 2013 there will be a Government imposed cap on the number of allowances, and all allowances will be sold each year via an auction – a world first for this type of scheme. See DEFRA: http://www.defra.gov.uk/environment/climatechange/uk/business/crc/qanda.htm (accessed 20/03/2008).

105 See DEFRA: http://www.defra.gov.uk/environment/climatechange/uk/business/crc/qanda.htm (accessed 20/03/2008).

106 http://www.defra.gov.uk/environment/climatechange/uk/energy/fund/index.htm.

107 See DEFRA for more information: http://www.defra.gov.uk/environment/climatechange/uk/energy/fund/index.htm (accessed 23/03/2008).

108 Previously referred to as the Energy Efficiency Commitment 2008-11, or EEC3.

2008 via the Electricity and Gas (Carbon Emissions Reduction) Order 2008 and will run until 2011. It requires energy suppliers to achieve targets for promoting reductions in carbon emissions in the household sector and is described by DEFRA as being 'the principal driver of energy efficiency improvements in existing homes in Great Britain'.[109]

4.11 UK Regional Initiatives: Are More Incentive-based Schemes Needed?

What is clear about the UK policy scene is that there appears to be a blend of both 'sticks' and 'carrots'. Decision-makers have adopted incentive-based, forward-thinking GHG emissions reductions policies, and have also adopted the application of punitive, reactive ones such as taxes and levies. Attempting to provide incentive-based schemes can be trialled at a local level until adopted as UK-wide policy. London offers a sensible test-bed for such initiatives, especially as the GLA[110] already has the Congestion Charge[111] and the Low Emissions Zone.[112] Though not specifically designed to reduce GHG emissions, there are related outputs. There are additional incentive-based initiatives within the Congestion Charge scheme that further help tackle the GHG emission issue: for example, if you drive a moped, electric car or hybrid vehicle, you are exempt from paying the charge. Taking this further, the UK competent authorities could consider, for example, offering Council Tax[113]

---

109 See legal basis: http://www.opsi.gov.uk/si/si2008/uksi_20080188_en_1 (accessed 02/04/2008) and http://www.defra.gov.uk/environment/climatechange/uk/household/supplier/index.htm (accessed 23/03/2008).

110 Greater London Authority http://www.london.gov.uk/ (accessed 16/02/2008).

111 Introduced under the Greater London Authority Act 1999 and the Transport Act 2000. See Consolidated Scheme Order of the Greater London (Central Zone) Congestion Charging Order 2004: http://www.tfl.gov.uk/assets/downloads/consolidated scheme-order-with-amendments-in-force.pdf (accessed 16/02/2008). Annex 4 to the scheme contains information on the application of the proceeds from congestion charging, including for environmental initiatives.

112 The LEZ is a specified area within which the most polluting diesel engine trucks, buses, coaches, large vans and minibuses will be required to meet specified EU emissions targets or pay a charge. The LEZ does not apply to cars or motorcycles. http://www.tfl.gov.uk/roadusers/lez/default.aspx (accessed 10/02/2008). Introduced under the Greater London Authority Act 1999 and the Transport Act 2000. See Greater London Low Emission Zone Charging Order 2006: http://www.tfl.gov.uk/assets/downloads/roadusers/lez/LEZ-scheme-order.pdf (accessed 20/02/2008).

113 Council tax is a local tax based upon the value of each domestic property (usually a home), at a fixed point in time. The income from council tax is collected by local councils in Wales to help pay for local services. See: http://www.voa.gov.uk/council_tax/index.htm (accessed 03/02/2008).

reductions for residents who do not own a car,[114] or for those who are registered moped or motorcycle users only. However, having to think of incentive-based GHG emissions schemes is more easily said than done. Operating and monitoring such schemes as proposed above would undoubtedly be complex. Elsewhere in the UK, there are currently incentive-based car-sharing 'fast lanes' (in Birmingham and Leeds, and soon-to-be in Hertfordshire – there are also such lanes in Greater London exclusively for Black Cabs ('taxi cabs'), coaches and motor bikes) though these are either underused, and cause traffic jams in the other lanes, or they are abused to such an extent that some cities, for example Leeds, are considering using heat-seeking cameras to detect the number of people in a vehicle.[115]

5. POLICY COHESION

Any assessment of the Climate Change Agreements would not be complete without dealing with the often intricate and complex relationship between the

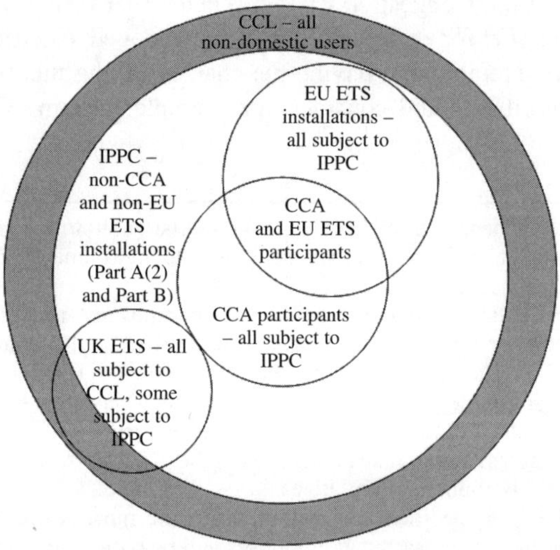

Figure 10.1 Interaction between CCL, IPPC, CCA, EU ETS and UK ETS

---

114 To regulate this, one idea is that the Driver and Vehicle Licensing Agency registers can be checked (as is the case for the LEZ and congestion charging) along with local authority parking permit registers.

115 See, *Car-sharing cameras being tested*, BBC News: http://news.bbc.co.uk/go/pr/fr/-/1/hi/uk/7260225.stm. Published: 2008/02/23 07:42:26 GMT.

Agreements and other environmental policies. We have already identified the role of the IPPC Directive in defining the eligibility of an installation to participate in the CCA process. Two other environmental policies that have a significant interaction with the Agreements are the UK Emissions Trading Scheme and the EU Emissions Trading Scheme. The interaction between IPPC, CCAs, CCL and other key policies such as EU ETS and UK ETS is demonstrated in Figure 10.1. IPPC forms the basis of qualification for both CCAs and EU ETS. Until 2007 installations covered by a CCA and EU ETS were permitted to opt out of Phase I. This exemption was withdrawn for Phase II. All installations covered by IPPC are subject to the CCL.

5.1 UK Emissions Trading Scheme

As we know, operators of target units are permitted to purchase allowances from the UK ETS emissions trading registry (ETR) in the event of a target unit not meeting the targets specified in the underlying agreement. The impact of CCA operators on the trading price of allowances has been significant. Allowance prices peaked in autumn 2002 at around £12.50 per tonne of carbon dioxide equivalent (CO_2e). This peak is linked to the rush from CCA signatories to purchase allowances prior to the end of the first target period in order to meet their energy consumption targets (National Audit Office, 2004). Since then the price of an allowance has decreased dramatically with the occasional elevation to around £5. In July 2007 Natsource Transaction Services were trading allowances for CCA participants below £2 per tonne of CO_2e (Natsource, 2007).

The ability of trade allowances via UK ETS has been utilized by many of CCA sectors. Thirty-eight per cent (n=6) of our sector respondents stated that less than 10% of their target units had purchased allowances from UK ETS in order to meet their targets. However, 56% (n=9) of respondents stated that between 11% and 50% of target units had purchased allowances. Only one respondent stated that between 71% and 90% of target units in their sector used the UK ETS facility to buy allowances. Purchase of allowances is clearly an important element of a firm's emissions reduction strategy. Conversely, very few sectors reported that more than 11% of their target units had sold excess allowances via the UK ETS, with only two sectors reporting between 11% and 50% of target units doing so.

Noting the co-appearance of the CCA, CCL and UK ETS, regulators might consider the extent to which instrument shopping will take place such that participation in more than one instrument will take place in order to minimize costs. Such a strategy may be acceptable to the extent that it does not undermine the objectives of any one instrument.

Several interview respondents raised the issue of UK ETS allowance prices and the impact this had on the credibility of the CCAs. One sector representative stated that there are

> some small sites that have had to buy credits from the sector. None of them have bought from UK ETS, but we produced a mechanism to allow small sites to buy carbon credits according to the market price on that day, It really is such a small amount that other companies [in the sector] are not bothered about it.

With regard to the cohesion between CCAs, UK ETS and other policy instruments, the same respondent stated that

> I would like to see CCAs and the CCL removed from the UK Climate Change Programme in favour of emissions trading. [The CCL/CCA] is unfair. [The CCL] is a tax and it makes us uncompetitive. While it [CCL/CCA] is there and is linked into UK ETS it is going to get horribly tangled up into EU ETS.

Regarding the price of UK ETS allowances, one interviewee stated that because UK ETS prices have been so low the CCAs do not provide the maximum incentive for energy efficiency improvements and 'people buy their way out of trouble'. One interviewee acknowledged that there may be a risk of CCA participants buying UK ETS allowances wholesale in order to avoid undertaking any energy efficiency improvements. However any risk associated with this is due to 'market failure of the UK ETS – it's not the fault of the CCAs'.

Another criticism of the interaction between UK ETS and CCAs concerned the impact of allowing companies to buy credits independently of the sector, thus disrupting the sense of 'collectivity' arising from the implementation of sector level CCAs:

> The CCAs were robust and in initial application generated a collective will to succeed [in the sector]. In this respect CCAs were hugely innovative and delivered a strong policy message. They represented a considerable achievement for DETR (now DEFRA). The introduction of UK ETS by another arm of DEFRA one year later in 2002 tended to disturb the collectiveness of CCAs.

Operators and sectors are also permitted to 'ring-fence' emission reductions. This means the operator/sector may 'preserve the over-achievement for possible future conversion [to allowances]' (DEFRA, 2002), commonly referred to as 'banking'. Studies show that allowance banking can prevent price volatility within trading schemes and subsequently increases certainty within the market (Ellerman, 2005; Cason and Gangadharan, 2006; Schleich et al., 2006; Tietenberg, 2006). Furthermore Schleich et al. (2006) state that 'theoretical and empirical analyses from existing programs suggest that banking (and borrowing) [of allowances] reduces overall compliance costs by allowing inter-temporal flexibility: cost savings can be traded over time'.

Herein, a brief comparison of two trading scheme in the United States: the sulphur dioxide (SO_2) trading scheme and the Regional Clean Air Incentives Market (RECLAIM). The RECLAIM programme is particularly rigid in terms of permitting banking and inter-temporal banking (where inter-temporal banking refers to the banking of emissions allowances for use in future commitment/compliance periods). As a result, the price fluctuations have been particularly severe (Ellerman, 2005). On the other hand, the more lenient approach to banking within the SO_2 programme has led to a significantly higher degree of price stability (Ellerman, 2005). In the case of CCAs, overachievement that is ring-fenced may be used in the instance of an operator or sector not meeting their interim targets. At the end of the third target period the majority of sectors, 41 out of 49, were able to ring-fence allowances for future use at the end of the third interim target period (Department for Environment, Food and Rural Affairs, 2007). The final quantity of allowances ring-fenced is equal to the difference between purchased allowances and the total number of excess allowances. The remaining ring-fenced allowances are available for subsequent sale via UK ETS or use during the emissions/energy use accounting prior to the interim target period. Evidence from the USA suggests that banking may have had a negative impact on the outcome of the CCAs. Scarcity of allowances would have led to high allowance prices and a fluid market. However, the perceived overachievement against unambitious targets led to high volumes of allowances entering the market and few being purchased. Problems associated with banking are neatly summarized by Carson and Gangadharan (2006): 'Price stability comes at a cost, however, since noncompliance and emissions are significantly greater when banking is allowed'. This issue was of particular concern during the period 2002–2006, when several key Direct Participants[116] in the UK ETS over-performed, generating a surplus of 3.5 million allowances (Smith and Swierzbinski, 2007).

5.2 EU Emissions Trading Scheme

The interaction between the EU ETS and CCAs is complex. The most pressing of issues relating to this interaction is double counting of emissions that are potentially covered by both the EU ETS and CCAs. Installations covered by both policies were given the opportunity to opt out of the EU ETS until the

[116] There were 32 Direct Participants in the UK Emissions Trading Scheme. These companies voluntarily agreemented to reduce their GHG emissions below a baseline (1998–2000) in return for a share of £215 million incentive fund supplied by HM Treasury. The UK ETS operated for Direct Participants from 2002 until the end of 2006. After the end of 2006, the ETS remained open to allow trading by CCA participants.

end of Phase I. Of the 500 installations eligible for EU ETS participation, 330 installations opted out of Phase I (DEFRA, 2007). In order to avoid companies receiving twice the benefit or twice the penalty for changes in emissions, DEFRA introduced a methodology to correct the CCA target of an installation to account for emissions included in the EU ETS. In the majority of cases the emissions associated with electricity consumption remain within the CCA. Direct emissions from processes fall within the EU ETS. The methodology is based on the correction of the CCA target. In some situations combustion emissions are covered by the EU ETS. These emissions may also be covered by a CCA. However some other process emissions may be included in the CCA but not in the EU ETS. Therefore, the CCA target is adjusted to remove EU ETS-covered emissions from the target. The performance of the installation is then compared against the revised CCA target (Department for Environment, Food and Rural Affairs, 2006b).

DEFRA guidance on double counting specifies that

> if an operator reduces emissions, then they may have a surplus of allowances for sale on EU ETS or banking for future use. The same reduction in emissions may also mean that the operator over-performs against their CCA target, which can be converted into allowances for sale on UK ETS. In other words, the operator gains allowances on both trading schemes for the same reduction in emissions. Conversely, if emissions increase, operators may find themselves forced to obtain allowances on both EU ETS and UK ETS to meet the requirements of different targets.

The National Audit Office (2007) reported that many sector associations and operators are concerned about the interaction between the two policies (National Audit Office, 2007). These concerns focus primarily on:

1. Data from different compliance periods being compared (e.g. 2006 CCA data compared with 2005 EU ETS data, 2008 CCA data compared with 2007 EU ETS data);
2. Difficulties in determining where the overlap between the two policies occurs on a given site, due in part to the coverage of EU ETS for entire installations and CCA coverage of only energy-intensive activities in some instances; and,
3. Operators experiencing difficulties during the accounting periods for both policies, which has on occasion resulted in an installation failing to meet adjusted targets for both the CCA and EU ETS despite having initially complied with both policies.

Several of the sector association interviewees expressed concern about the interaction between the EU ETS and CCAs due to the complex issue of double counting. There have been few investigations into the effects of double counting and the perceptions of double counting. DEFRA and DEFRA-appointed consultants remain the major sources of information and analysis. The Better Regulation Commission (BRC) recently highlighted the importance of Government ensuring that the appropriate level of intervention is applied within climate change policy (Better Regulation Commission, 2007). Furthermore the BRC stated that the Government should also ensure that 'double-banking' is avoided. In the late 1990s, when the CCAs were being developed the EU ETS was not even in the development phase. The potential problems arising from double counting between the two schemes could not have been accounted for. DEFRA was subsequently forced to implement a complicated methodology to minimize the impacts of double counting for companies participating in both policies.

One sector association highlighted the potential pitfalls of double counting with an illustration. One company within the sector reportedly passed its EU ETS target by a substantial amount. At the same time the company failed to meet its CCA target. This would suggest a lack of joined-up thinking in instrument design. The level of failure was increased further following adjustments made to the CCA target accounting for the EU ETS over-performance. Another sector association stated that the Government should decide upon the instrument that will act as the key climate change policy, such as the EU ETS, and then ensure that all other policies fit well within that framework as 'it is simpler and you avoid double regulation. Businesses want to minimise regulation and this is one area where you can do that easily'.

The methodology for calculating the EU ETS adjustments was described as 'tedious in the extreme, of doubtful legality and tedious to operate. Double counting is a significant problem. It is a significant administrative burden as well as an irritation.' The respondent responsible for this observation continued to question why DEFRA introduced the methodology to account for double counting before determining the extent of the problem, i.e. the amount of emissions that are incorporated in both policies and the potential financial reward/penalty for affected firms. The net environmental benefits of EU ETS Phase I have been low. Therefore installations participating in both EU ETS and CCAs may have achieved greater emission reduction by opting out of the EU ETS. However, as the EU ETS progresses, the case in favour of removing all EU ETS participants from CCAs becomes stronger. Certainly beyond Phase II (2012) there is little reason to support the continued inclusion of some installations in both schemes.

6. CONCLUSIONS

6.1 Next Steps: The 'Per Capita' Discussion in the UK

Personal carbon trading (PCT)[117] is one area which the UK Government is currently exploring in a bid to further address the ever-rising increase in carbon emissions.[118] There has been quite a concerted research effort in this area led by the Tyndall Centre for Climate Change,[119] the Environmental Change Institute[120] and the Royal Society for the encouragement of Arts, Manufactures and Commerce (RSA).[121] An initial scoping study[122] was commissioned by DEFRA in 2007 and undertaken by the Centre for Sustainable Energy (CSE).[123] According to the DEFRA website[124] the Government is now carrying out a study to assess whether personal carbon trading is a practical and feasible policy option, compared with other measures for constraining emissions, and if it is found to be so, a public debate will be held. This pre-feasibility study will examine how viable such a scheme is, what would be needed in order to administer it and how allocations would be made. The results of RSA's CarbonLimited project[125] recommend, *inter alia*, that PCT should initially be developed as a voluntary scheme in order for there to be sound public consultation and consideration of fairness issues, but that this should later evolve into a mandatory scheme to deliver a percentage of the UK's emissions reduction targets as set out in the (Draft) Climate Change Bill. PCT seems like a theoretically sound idea but in practice, devising the scheme and executing the scheme would likely raise moral issues (how do we decide who gets what quota and on what basis do we decide on the allocation?);

---

[117] Also described as personal carbon allowances, domestic tradable quotas, personal carbon rations, carbon credits – see Redgrove and Roberts (2007, p. 3).

[118] http://www.defra.gov.uk/environment/climatechange/uk/individual/pca/index.htm (accessed 02/03/2008).

[119] http://www.tyndall.ac.uk/ (accessed 02/03/2008).

[120] http://www.eci.ox.ac.uk/ (accessed 02/03/2008).

[121] http://www.thersa.org/index.asp (accessed 02/03/2008).

[122] Redgrove and Roberts (2007).

[123] http://www.cse.org.uk/ (accessed 02/03/2008).

[124] http://www.defra.gov.uk/environment/climatechange/uk/individual/pca/index.htm (accessed 29/03/2008).

[125] *Personal Carbon Trading The idea, its development and design*, RSA CarbonLimited Interim Recommendations September 2007: http://www.rsacarbonlimited.org/uploads/documents/CarbonLimited_InterimRecommendations_37.pdf (accessed 02/03/2008). CarbonLimited was established in 2006 as a 3-year RSA programme to analyse the effectiveness, feasibility and public acceptability of the concept of personal carbon trading. See: http://www.rsacarbonlimited.org/default.aspa (accessed 02/03/2008).

equity based arguments (should we in the UK have a right to continue to emit carbon in any amount?); and human rights-based dilemmas (would having a cap on my right to emit carbon impact on my right to life or right to property, for example?).

6.2 Far-Reaching Law and Policy – But a Bit Too Far?

UK domestic initiatives to address global warming and emissions reductions appear to be far ranging and integrating rather than general and non-specific. Policy and legal developments at the national level have resulted in some major ground-breaking legislative schemes, including the ETS and the proposed Climate Change Bill, and draws on a range of legal and policy instruments.

What is also interesting is that said policy initiatives span several sectors, such as horticulture, waste and industry, and are not solely focused on energy production and consumption.

UK policy and regulation in relation to driving down GHG emissions does not appear to be made lightly or undemocratically. There appear to be several layers of deliberation, scrutiny and accountability. A large number of policy initiatives appear to have been underpinned by sound policy-based and scientific research since the 1990s. It is interesting to note that regulation in the areas of climate change very much link economic and environmental concerns. This is evidenced by the wide number of stakeholders and also the variety of regulators, which are not just environmentally competent authorities, but financial and energy-related ones.

One criticism is that there may be too many climate change-related initiatives in operation throughout the UK, which may result in confusion or an over-burdening of some sectors. As our 'Policy Cohesion' analysis has indicated, there are issues of integration/joined-up thinking that need to be addressed.

It is quite difficult to keep track of all the schemes and the latest targets, though the Climate Change Bill does link things together somewhat. It is to be noted that DEFRA launched the Climate Change Simplification project consultation in December 2007[126] which not only addresses the inter-relationship of three climate change initiatives[127] but also their relationship with other

[126] *Consultation on the Recommendations of the Climate Change Simplification Project Climate Change Instruments Areas of overlap and options for simplification*, December 2007: http://www.defra.gov.uk/corporate/consult/cc-instruments/consultation.pdf (accessed 24/01/2008).

[127] EU Emissions Trading Scheme (EU ETS), Climate Change Agreements (CCAs), and the proposed Carbon Reduction Commitment.

areas of regulation such as integrated pollution prevention and control at the UK and wider levels. It is hoped that this will lead to greater overall policy cohesion in a jurisdiction which has few rivals in relation to the ubiquity of instruments in place for addressing climate change.

REFERENCES

MacDonald, K.E. (2006), 'Sustaining the Environmental Rights of Children: An Exploratory Critique', *Fordham Environmental Law Review*, Vol. 18, 2006.

Marshall, Lord (November 1998), *Economic Instruments and the Business use of Energy*, London, HM Treasury, http://www.hmtreasury.gov.uk/media/F/7/EconomicInstruments.pdf, (accessed 27/02/2008).

De Muizon, Gildas and Glachant, Matthieu (2004), 'The UK Climate Change Levy Agreements: Combining Negotiated Agreements with Tax and Emission Trading', in Baranzini, A. and Thalmann, P. (eds), *Voluntary Agreements to Climate Policies An Assessment*, Cheltenham, UK, Edward Elgar Publishers.

Nye, M. and Owens, S. (2008), 'Creating the UK Emission Trading Scheme: Motives and Symbolic Politics', *European Environment*, 18, 1–15.

Pearson, P.J. (2004), 'The UK Emissions Trading Scheme: Paying the Polluter – A Policy Experiment', *International Review for Environmental Strategies*, 5(1), 241–56.

Redgrove, Z. and Roberts, S. (June 2007), 'Making Carbon Personal? A Snapshot of Community Initiatives', Report to Defra, Centre for Sustainable Energy, http://www.cse.org.uk/pdf/pub1082.pdf (accessed 02/03/2008).

Sairinen, R. and Teittinen, O. (1999), 'Voluntary Agreements as a Policy Instrument in Finland', CAVA Working Paper No. 98/11/4.

ADDENDUM

Since this chapter was submitted there have been significant changes within the UK policy arena. As of October 3, 2008 the UK government created the new Department of Energy and Climate Change (DECC), indicating how seriously the UK government view climate change matters. The Department is headed by Ed Milliband, the new Secretary of State for Energy and Climate Change. Subsequently, the new Secretary of State has committed the UK to cutting greenhouse gas emissions by 80% on 1990 levels by 2050, an increase of 20% on the previously announced target. This target will not be binding law until it is formally introduced in the Climate Change Bill, due to receive Royal Assent in the autumn of 2008 though the Bill has had to be amended to accommodate the revised target.

11. Linking the EU ETS to other emissions trading schemes

Janneke Bazelmans[1]

1. INTRODUCTION

Emissions trading delivers a crucial tool for combating climate change. As a general trend, a number of domestic emissions trading schemes (dETSs) are emerging, each with their own characteristics. Efficiency would increase if these dETSs were linked to each other. The European Union's scheme (EU ETS)[2] offers the largest, broadest building block for developing a global network of systems. The EU ETS is already linked to domestic emissions trading systems in Norway, Iceland and Liechtenstein.[3] In addition, the current EU Trading Directive paves the way for linking the EU ETS to several dETSs in Kyoto Parties as well as in non-Kyoto countries. Such links can be created directly between different dETSs and indirectly through offsets, such as the Clean Development Mechanism (CDM) and Joint Implementation (JI).[4] In October 2007, the International Carbon Action Partnership (ICAP) was launched.[5] This consists of a coalition of European countries, some USA states,[6] Canadian

---

1 External PhD candidate, Centre for Environmental Law, University of Amsterdam. Janneke's PhD is focused on the future of the Clean Development Mechanism beyond 2012. This Chapter was finalized on 28 February 2008.

2 Based on Directive 2003/87/EC of the European Parliament and of the Council of 13 October 2003 establishing a scheme for greenhouse gas emission allowance trading within the Community (EU Trading Directive).

3 EEA agreement Decision of the EEA Joint Committee, no 146/2007 of 26 October 2007, amending Annex XX (Environment) to the EEA Agreement. This Decision was adopted by the EEA Joint Committee at its meeting of 26 October 2007 and will be published in the Official Journal of the European Union. It will not enter into force until the notification to the Joint Committee of the fulfilment of constitutional requirements in all three EEA EFTA States has taken place.

4 Article 6 resp. Article 12 of the Kyoto Protocol.

5 Political Declaration, International Carbon Action Partnership, Lisbon, 29 October 2007.

6 Arizona, British Columbia, California, Manitoba, New Mexico, Oregon, Washington (Western Climate Initiative Members) and Maine, Massachusetts, New Jersey and New York (Regional Greenhouse Gas Initiative Members).

provinces, New Zealand and Norway. It provides an international forum in which governments can share experiences and best practice on the design of their mandatory dETSs and discuss ways to link domestic schemes.

Linking also represents an important issue within the current review of the EU ETS. The European Commission (Commission) has proposed the possibility of connecting the EU ETS further with other dETSs, at national or regional levels (the Proposal).[7] Furthermore, the Commission proposed the harmonization of the use of credits from JI and CDM and the introduction of additional types of credits and mechanisms under the EU ETS. Because of its size and harmonization experience the EU ETS could be a starting point in linking dETSs. Linking requires sufficient institutional compatibility to establish the equivalence of allowances and to move them from one dETS to the EU ETS and vica versa. However, dETSs are often designed differently, which complicates direct linking. Participants in linked dETSs can take advantage of the compatibility rules between the different schemes. Emissions trading should be, first, an instrument for environmental protection, and, second, an instrument to reduce the costs to meet a given emission target. The environmental effectiveness of the EU ETS might be undermined if linking dETSs leads to smaller emission reductions than if they were operating independently or if environmental guarantees built into the EU ETS were to be bypassed. In addition, the economic efficiency of the EU ETS might be undermined if linking lead to higher compliance costs.

This chapter discusses the legal possibilities and prerequisites of linking the EU ETS to other dETSs. As connections between dETSs cannot be regarded in isolation from the international emissions trading framework as provided by the Kyoto Protocol, Section 2 describes the relationship between emissions trading under the present EU ETS scheme and under the Kyoto Protocol. It also briefly surveys possible linking partners. Section 3 assesses the legal possibilities of connecting the EU ETS with other dETSs. Section 4 looks at the issues and prerequisites to linking. Finally, Section 5 finishes with a few concluding remarks.

---

[7] COM(2008) 16 final; Proposal for a directive of the European Parliament and of the Council amending Directive 2003/87/EC so as to improve and extend the greenhouse gas emission allowance trading system of the Community.

2. EMISSIONS TRADING SCHEMES

2.1 Relationship of EU ETS with International Emissions Trading under the Kyoto Protocol

The Kyoto Protocol[8] affects the institutional setting for linking dETSs in two ways. In the first place, it provides a framework for international emissions trading, linking dETSs in Annex B Parties.[9] In the second place, it establishes internationally agreed procedures for generating emission reduction credits under CDM and JI.

The Kyoto Protocol provides emissions trading as a tool for Annex B Parties to meet their Kyoto target in a cost-effective way. The Kyoto Protocol established quantified targets for Annex B Parties in the form of an absolute emission cap for each Party for the 2008–2012 commitment period (a second such period has yet to be set). The emissions allowed to each Party are referred to as assigned amount units (AAUs), whereby one AAU equals one metric ton of CO_2-equivalent.[10] Each Annex B Party must cover its emissions by an equivalent amount of the Kyoto units. These are: (i) assigned amount units (AAUs); (ii) emission reduction units (ERUs); (iii) certified emission reductions (CERs), temporary CERs (tCER) and long-term CERs (lCERs); and (iv) removal units (RMUs).[11]

ETSs may be established as climate policy instruments at the national level (e.g. New Zealand, Japan) and the regional level (e.g. in the European Union). Each Annex B Party uses its own discretion in implementing a dETS for its domestic carbon emitting sources. Annex B Parties may set emissions obligations to be reached by the entities through a system of emissions trading. Depending on the rules of a scheme, the obligations may be fulfilled through holding either the Kyoto units or other units established specifically for those trading schemes, such as the European Union emission allowances (EUAs)

---

8 Kyoto Protocol to the United Nations Framework Convention on Climate Change was adopted at the third Conference of the Parties to the UNFCCC in Kyoto on 11 December 1997.

9 Article 17 of the Kyoto Protocol. Annex B Parties are Parties to the Kyoto Protocol with emission limitation and reduction commitments inscribed in Annex B to the Protocol.

10 The six main greenhouse gases are: carbon dioxide (CO_2), methane (CH_4), nitrous oxide (N_2O), hydrofluorocarbons (HFCs), perfluorocarbons (PFCs), and sulphur hexafluoride (SF6).

11 ERUs generated through Joint Implementation; CERs generated from Clean Development Mechanism whereby tCERs and lCERs are generated from forestry projects under the CDM; RMUs generated by land use, land-use change and forestry (LULUCF) activities under Articles 3.3 and 3.4 of the Kyoto Protocol.

under the EU ETS. Furthermore, the extent to which the Annex B Parties integrate the scheme into the international emissions trading structure and links it to other dETSs is up to them.

Emissions trading under the EU ETS is a domestic measure for the EU and it must be distinguished from international emissions trading under the Kyoto Protocol. The EU ETS started in 2005 and operates the largest scheme. The EU ETS aims to enable the European Union[12] and the Member States to achieve compliance with their Kyoto commitments in a cost-effective way. In fact, the EU ETS is a structure for linking trading programs of the Member States. The second phase of the EU ETS operates from 2008–2012, corresponding to the first commitment period of the Kyoto Protocol when the Member States and the EU must meet their Kyoto goals. The EU ETS has no final date.

Each year from 2008 on, based on their national allocation plans, the Member States provide EUAs to their entities which fall under the scope of the EU ETS. One EUA covers the emission of one metric ton of CO_2-equivalent. EUAs are valid for one trading period, although banking (i.e. the carry-over) of EUAs from the first to the second trading period is allowed.[13] In fact, EUAs are converted into that unit from the Member States' Kyoto budget, the AAUs. That makes EUAs specific Kyoto units designated as valid for trading under the EU ETS.

The participating entities are obliged to surrender after each calendar year an amount of EUAs to their government that is equivalent to their emissions in that calendar year. In addition, entities can also use CERs and ERUs through CDM and JI[14] to meet their targets. Furthermore, the EU ETS provides the possibility of covering emissions by 'mutually recognized third country allowances',[15] which is important for our focus on linking possibilities. Entities under the EU ETS are not allowed to use AAUs to cover their emissions. EUAs are not fully exchangeable with all Kyoto units, and will be tagged to keep their identity distinct. This allows the EU ETS to use a different definition of what is allowed to be traded within the scheme, and keeps the scheme as yet distinct from international emissions trading.

---

12 The EU is a distinct Party to the Kyoto Protocol, listed in Annex B. All Member States are Party to the Kyoto Protocol.

13 Article 13 EU Trading Directive.

14 Article 11 bis Directive 2004/01/EC of the European Parliament and of the Council of 27 October 2004 amending Directive 2003/87/EC establishing a scheme for greenhouse gas emission allowance trading within the Community, in respect of the Kyoto Protocol's project mechanisms.

15 Article 25 Directive 2003/87/EC.

The Community Independent Transaction Log (CITL), a central administrator on EU level, checks each transaction for any irregularities. For the start of the Kyoto commitment period in 2008, EU registries were to switch their connections from the CITL to the Kyoto registry, the International Transaction Log (ITL).[16] The ITL allows the trade and transfer of Kyoto units proposed by both EU and non-EU registries. In the case of transactions involving EU registries, the ITL will forward information to the CITL so that it can conduct supplementary checks defined under the EU ETS.

Although emissions trading under the EU ETS must be distinguished from international emissions trading under the Kyoto Protocol, it includes trading between Annex B Parties. This means that from 2008 on, those transactions also fit under the umbrella of Article 17 the Kyoto Protocol. Transactions in EUAs are therefore recorded automatically as transactions under the Kyoto Protocol. As a consequence, transfers of EUAs between entities in different Member States involve a corresponding adjustment of AAUs under the Kyoto Protocol.[17]

2.2 Other Emissions Trading Policy Schemes and Initiatives

There are more potential linking partners, countries which are developing a dETS, including the United States, that are not a Party to the Kyoto Protocol. This paragraph set out the main existing, announced and proposed dETSs.[18] Existing schemes are schemes for which the legislation has been passed. Announced dETSs are those that, although still under elaboration by the authorities, have been endorsed at the highest level of government. Proposed dETSs are schemes that have been suggested by parliamentarians and which are at a more explanatory stage.[19] Table 11.1 provides the main characteristics of the following dETSs:

ETSs in Kyoto Parties:

- Japanese Voluntary Emissions Trading Scheme (JV ETS); existing.[20]

[16] The EU was scheduled to connect its scheme to the ITL in 2007, but this has been delayed and will now take place by April 2009.

[17] Art. 45 and 59 EU Registration Regulation, 2216/2004/EC, 21 December 2004.

[18] For a comprehensive survey: Reinaud and Philibert (1997); Sterk and Braun et al. (2006).

[19] See also: Reinaud and Philibert (1997).

[20] The rules of the Japanese Voluntary Emissions Trading Scheme are only available in Japanese. See an explanation document: Ninomiya (2007) and www.et.chikyukankyo.com/english/.

Table 11.1 Main characteristics of dETSs

dETS	Start	Absolute/ relative targets	Allocation	Penalties/ Safety valve	Banking/ borrowing	Trading units	Opt in/ out
Existing dETSs							
EU ETS	2005	Absolute	Free and auctioning	EUR 100/t CO_2e plus surrender additional allowances next year.	Banking	EUAs, ERUs, CERs (no sinks, credits from nuclear projects)	limited
JVETS	2006	Absolute	Free (grandfathering)	No penalties; subsidies must be returned.	Banking	Allowances, CERs (incl. sinks)	No
NSW GGAS	2003	Relative	Free (benchmarking)	A$ 16/t CO2e (EUR 9)	Banking	Certificates from project-based emission reduction activities	X
RGGI	2009	Absolute	25% auctioning 75% discretion of states	3x market value /t CO_2e	Banking	Allowances, offset credits from RGGI states. Limited EUAs. CERs, ERUs	X
Announces dETSs							
Canadian ETS	2008	Relative	Free (benchmarking)	C$ 200/t CO_2e (EUR 126) Safety valve	Banking	Internal reduction credits, domestic offset credits, CERs (incl sinks), credits for contributing to a technology fund and early action credits	X
AUS ETS	2011	Absolute	Free and auctioning	Penalty level unclear Safety valve	X	Permits, early action credits, domestic offset credits, CERs	X

NZ ETS	2008 (in stages until 2013)	Absolute	Free for downstream industrial energy users (grandfathering). No free allocation to upstream points of obligation in some other sectors.	NZ$ 30/t CO_2e (EUR 15) or NZ$ 60 CO_2e (on purpose). Plus make good requirement	No	NZ units (NZUs) will be convertible to Kyoto units; AAUs, CERs, (no sinks and credits from nuclear projects), ERUs, RMUs	X
Swiss ETS	2008	Absolute	Free (ex post adjustment until 2010)	Payment of CO_2e tax plus interest	Banking and borrowing	Allowances, CERs (incl. sinks), ERUs, allowances from other dETSs	No
Proposed dETS							
CA ETS	2012	Absolute Possibility to suspend cap one year (emergency)	Free and auctioning	Penalty (amount not clear) Safety valve	X	Allowances, in phases: domestic offsets and external offsets	X
WRCAI	2008	X	X	X	X	X	X
Lieberman-Warner Climate Security Act	2012	Absolute	Free (grandfathering) and auctioning. Depends on sector.	X No safety valve	Banking and limited borrowing	Allowances, domestic and international offset credits, credits from recognized dETSs	Yes

X= not decided yet or unclear

- The New South Wales Greenhouse Gas Abatement Scheme (NSW GGAS), Australia; existing.[21]
- Canadian ETS; announced.[22]
- Australian Emissions Trading Scheme (AUS ETS) at the federal level; announced.[23]
- New Zealand Emissions Trading Scheme (NZ ETS); announced. [24]
- The Swiss ETS; announced.[25]

ETSs in non-Kyoto countries – USA ETSs:[26]

- Northeast Regional Greenhouse Gas Initiative (RGGI); existing;[27]
- Californian Emissisons Trading Scheme (CA ETS); proposed.[28]
- Western Regional Climate Action Initiative (WRCAI); proposed.[29]
- USA ETS at the federal level; proposed.[30]

---

[21] The NSW GGAS is underpinned by provisions in the Electricity Supply Act 1995 and will end when an ETS at the federal level commences, www.greenhousegas.nsw.gov.au.

[22] Regulatory Framework for Air Emissions, 26 April 2007, www.ec.gc.ca.

[23] Australia's Climate Change Policy, Our Economy, Our Environment, Our Future, 2007, www.pmc.gov.au; Prime Ministerial Report of the Task Group on Emissions Trading, 31/05/2007.

[24] Climate Change (Emissions Trading and Renewable Preference) Bill 2007, 187-1; See also: The Ministry of Environment and the Treasury (2007), www.climate-change.govt.nz.

[25] Emissionen nach CO_2-Gesetz und Kyoto-Protokoll, June 2007, www.bafu.admin.ch.

[26] See the contribution of Bluemel for an extensive description of USA dETSs.

[27] Northeast Regional Greenhouse Gas Initiative 'Memorandum of Understanding', dated 20 December, 2005, www.rggi.org.

[28] AB 32, the Global Warming Solutions Act.

[29] Western Climate Initiative, 'Western Climate Initiative Statement of Regional Goal', dated 22 August 2007, www.westernclimateinitiative.org. An agreement signed by seven Western States and two Canadian provinces..

[30] At the federal level the Senate has introduced several bills introducing a domestic ETS. They all call for adoption of some form of a cap-and-trade system, for example: Sen. Mc Cain-Liebermann Bill, 'the climate Stewardship and Innovation Act', S.280 (2003); Sen. Sanders-Boxter Bill, 'Global Warming Pollution Reduction Act,' S.309 (2007); Sen. Bingaman-Specter Bill, 'Low Carbon Economy Act,' H.R. 620 (2007).

In December 2007, the US Senate Environment and Public Works Committee approved the Lieberman-Warner Climate Security Act (America's Climate Security Act of 2007, S. 2191). This clears the way toward consideration of this legislation by the full Senate.

3. LINKING THE EU ETS TO OTHER dETSs

By linking dETSs a participant in a dETS in one country can pass its allowances directly or indirectly to a participant in another country's scheme for compliance purposes. Direct linking is the mutual recognition of allocated allowances in each dETS for compliance purpose. The allowances from one dETS are fully fungible and valid in another country's dETS.

Indirect linking means linking through offsets mechanisms with a common acceptance of credits generated under certain offsets. Offsets are project-based mechanisms outside a dETS, abroad or domestic. There is no mutual acceptance of allocated allowances under each dETS necessary, but only shared standards and acceptance of project-based credits. For example, if credits from CDM can be used toward compliance under two different schemes, such as the EU ETS and the Japanese ETS, these schemes are indirectly linked through each other. The EU Trading Directive and the Proposal allow direct as well as indirect linking.

3.1 Direct Linking

3.1.1 The current EU ETS Directive

Direct linking of the EU ETS is possible with dETSs in Kyoto Parties as well as in non-Kyoto countries. Article 25(1) of the EU Trading Directive provides a legal base for linking. According to this Article:

> agreements should be concluded with third countries listed in Annex B to the Kyoto Protocol which have ratified the Protocol to provide for the mutual recognition of allowances between the Community scheme and other greenhouse gas emissions trading schemes in accordance with the rules set out in Article 300 of the Treaty.

In addition the Commission should consider the possibility of concluding agreements with countries listed in Annex B to the Kyoto Protocol which have yet to ratify it (these are not Kyoto Parties) in order to provide for the recognition of allowances between the EU ETS and mandatory dETSs capping absolute emissions established within those countries.[31]

The EU Trading Directive does not mention further conditions that the third country scheme needs to meet. It could therefore be a similar trading system, but also a completely different system. Linking can be developed in a multilateral process involving simultaneously all governments that may have an interest in linking to the EU ETS. This type of process becomes a rather complex negotiating process, the greater the number of governments are

31 Recital 18 of the Linking Directive.

involved. Another possible and more realistic way to conclude a linking agreement can be through bilateral negotiations between two governments. An alternative approach is unilateral linking. Entities under a dETS can purchase allowances from the EU ETS and use them to meet their obligation under their dETS while the two schemes remain separate.

3.1.2 The Proposal to improve and extend the EU ETS

The Commission has proposed extending direct linking to any country or administrative entity (such as a state or group of states under a federal system) that has established a mandatory dETS capping absolute emissions.[32] This may open the door for linking to USA ETSs established on the federal level or on state level. However, according to the Commission the design elements of the linked ETSs may not undermine the environmental effectiveness of the EU ETS. Thus, differences in the designs of linked dETSs may not lead to smaller emission reductions in the EU ETS than if the schemes operate independently. Section 4 sets out the main design issues that may occur in case of linking.

3.2 Indirect Linking

3.2.1 The current EU ETS Directive

The Linking Directive is created to link CDM and JI to the EU ETS. It follows mainly the rules of the Kyoto Protocol and the Marrakesh Accords.[33] Under the EU ETS, entities can use credits generated from CDM and JI up to a certain limit (maximum of 10%) set in their national allocation plans to meet their reduction target. The EU ETS places some restriction on the types of credits which might be used for compliance. It excludes credits generated by project activities from land use, land use change, forestry (so called sinks) and credits from nuclear facilities.[34] Furthermore, the EU ETS does not allow credits from domestic offsets.

3.2.2 The Proposal to improve and extend the EU ETS

The Proposal aims to harmonize the use of credits for emission reductions by entities within the EU ETS. The Proposal sets out two scenarios which will extend the use of credits between 2013 and 2020.

---

32 Article 24a Proposal.

33 Report of the Conference of Parties on its seventh session, FCCC/CP/2001/13, 21 January 2002 (Marrakesh Accords), adopted by: Report of the Conference of Parties serving as the meeting of the Parties to the Kyoto Protocol on its first session, FCCC/KP/CMP/2005/8 (Montreal Accords).

34 Article 11bis sub 3 Linking Directive.

Scenario I

The first reflects only the EU's independent commitment[35] to reduce its emissions to at least 20% below 1990 levels by 2020.[36] In the absence of a satisfactory international agreement to combat climate change beyond 2012, entities will be able to use CERs and ERUs up to the remainder of the level which they were allowed in the period 2008–2012. These credits will be exchanged for allowances which will be valid from 2013 onwards. As the limit on these credits is generous it is expected that entities will be able to achieve more than one-third of the emission reductions required between 2013 and 2020 through their use.

CERs from CDM projects that were established before 2013 while issued after 2013 may be used from 2013 onwards.[37] ERUs cannot be created from 2013 onwards without new quantified emission targets being in place for host countries. As a consequence, JI projects which generated ERUs beforehand could continue to be recognized through bilateral or multilateral agreements with third countries.[38]

CERs from new CDM projects that started after 2013 would only be allowed from least developed countries without the need to conclude an agreement with these countries.[39] This is possible until 2020 unless those countries have ratified an agreement with the EU.

If the conclusion of an international agreement on climate change is delayed, entities may use credits from projects or other emission reducing activities which both promote technological transfer and sustainable development to comply with their obligations under the EU ETS in accordance with agreements concluded with third countries.[40]

There are two general conditions for the use of all credits beyond 2012 containing: (i) the extent that the levels of the credit use allowed by Member States for the period 2008 to 2012 have not been used; and (ii) only credits from project types which were accepted by all Member States during the second trading phase will be eligible for use.

---

35 COM(2008) 30 final Communication from the Commission to the European Parliament, the Council, the European Economic and Social Committee and the Committee of the Regions – 20 20 by 2020 – Europe's climate change opportunity; The EU emissions of greenhouse gases are to be reduced by 30% in 2020 provided that other developed countries will commit themselves to comparable emission reductions and economically more advanced developing countries contribute adequately according to their responsibilities and capabilities. Irrespective of any international agreement, the EU objective will be 20% in 2020.

36 Article 11a sub 1-6 Proposal.

37 Article 10a sub 3 Proposal.

38 Explanatory memorandum of the Proposal, section 5.

39 Article 11a sub 4 Proposal.

40 Article 11a sub 5 and 6 Proposal.

Scenario II

The second scenario is based on the EU's commitment to increase the emission reduction of 20% provided that a satisfactory international agreement to combat climate change beyond 2012 has been reached. In such cases, the limit on the use of credits will be automatically increased up to half of the additional reduction effort. This means that if the annual cap under the EU ETS were reduced by, e.g., 100 million tons following an international agreement, the limit on the use of credits would be increased automatically by 50 million credits.[41] These credits could be CERs, ERUs and other approved credits. The Commission will adopt measures to provide for the use of additional project types by entities in the EU ETS or the use by such entities of other mechanisms created under the international agreement.[42]

In addition to credits which are left over from 2008–2012, the EU ETS shall only accept CERs from third countries that have ratified the international agreement or from additional types of projects approved by the Commission.[43] However, the EU ETS will accept CERs from companies or administrative entities based in third countries which are linked to the EU ETS, although these countries haven't concluded an international agreement.[44]

It is doubtful whether the allowed amount of credits from projects outside the EU, which can be used in the EU ETS, will be in accordance with the supplementary principle as laid down in the Kyoto Protocol.[45] Under the current conditions for the second phase, a yearly average of 280 million tons of credits is allowed to enter the EU ETS. If the entities make full use of these credits, incentives for emission reductions and technological change within the EU ETS sectors might be taken away. The majority of emission reductions have to take place within EU ETS sectors.[46]

Furthermore, the Proposal requires that the use of CERs beyond 2012 must be in accordance with the EU's goal of generating 20% of energy from renewable sources by 2020, and promoting energy efficiency, innovation and technological development. In such cases, the possibility should be foreseen for concluding agreements with third countries in order to trigger investments in these countries, which would bring about real, additional reductions in greenhouse gas emissions while stimulating innovation in European companies and technological development in third countries. Such agreements may be ratified

41 Article 28 sub 3 Proposal.
42 Article 28 sub 4 Proposal.
43 Article 11a sub 7 Proposal.
44 Explanatory memorandum of the Proposal, under 5, p. 11.
45 Article 6b and 12 Kyoto Protocol.
46 Linking Directive, consideration 19.

by more than one country.[47] It is unclear whether the Proposal refers to agreements in relation to the use of CERs or acceptance of other credits, and which other countries are meant.

In addition, the Commission has proposed allowing domestic offset credits under the EU ETS. Projects in the Member States which reduce greenhouse gas emissions not covered by the EU ETS could generate credits. Such projects (such as climate neutral glasshouse horticulture) must comply with certain conditions necessary to safeguard the proper functioning of the EU ETS. These conditions must ensure that domestic credits do not result in double counting of emission reductions or impede other policy measures to reduce emissions not covered by the EU ETS and that they are based on simple, easily administered rules.[48]

This proposed possibility can be questioned. Domestic offsets do not reduce the net amount of greenhouse gas emissions, but they only allow a (capped) sector to emit more CO_2 emissions. Besides that, the rules in relation to emissions trading and other climate change aspects are always complicated and bureaucratic. Inclusion of domestic offsets would require the establishment of a new trading unit and would make it more complicated to determine the direct contribution of the ETS sectors to EU greenhouse gas emission reduction targets. In case domestic credits are indeed allowed under the EU ETS, these domestic offsets could probably take the JI road, avoiding the need to set up specific institutions.

4. ISSUES RAISED BY LINKING THE EU ETS TO OTHER dETSs

The designs of the dETSs vary significantly. Some schemes are voluntary, while others are mandatory. Some schemes are designed to be used for compliance with the Kyoto Protocol, while others are planned or in use in a non-Kyoto country. Differences also lie in compliance provisions and monitoring provisions etc. This complicates the task of linking and could undermine the environmental effectiveness and economic efficiency of the EU ETS. This section examines critical issues that could arise as a result of linking the EU ETS to other ETSs and identifies some minimum requirements for allowing linking.[49] Issues which will occur because of the mere existence of the ETSs,

47 Explanatory memorandum of the Proposal, under 5, p. 12.

48 Explanatory memorandum of the Proposal, under 5, p. 12.

49 See for extensive study's regarding linking implications: Reinaud and Philibert (1997); Ellis and Tirpak (2006); Fischer (2003, 3s2 s89–s103); Blyth and Bosi (2004); Anger (2006); Sterk and Braun et al. (2005); IEA (2005); Baron and Bygrave (2002); Haites and F Mullins (2001); Jepma (2003), August.

irrespective of linking them, won't be discussed. Not all the discussed issues are primarily legal issues, but rather economic and policy-related. However, as ETSs are entirely created by legal provisions on international and national levels, linking scenarios requires a legal design.

As there is no international agreement on climate change beyond 2012 yet and since most parts of the Proposal must be further developed, this section considers the Kyoto Protocol and the current EU ETS as a starting point for linking. However, the Proposal is taken into account to the extent possible.

4.1 Design Issues for Direct Linking

4.1.1 Mutual recognition of trading units

Mutual recognition of trading units is the main requisition for linking the EU ETSs to other ETSs. The Kyoto Protocol provides a framework for the recognition of trading units (AAUs, CERs, tCERs, lCERs, ERUs, RMUs) in the international emissions trading context. However, Kyoto Parties are free to decide on the definition of their trading units in their domestic trading schemes. The EU ETS allows EUAs, CERs, ERUs and mutually recognized third-country allowances. It excludes the use of credits from sinks and domestic offset projects. Other ETSs, for example the Japanese, the Swiss and the Canadian, allow CERs from sinks. The Canadian, the federal Australian and some USA ETSs allow credits from domestic offsets.

As long as the recognized units in linked dETSs can be used for compliance under the Kyoto Protocol, linking the EU ETS with a dETS which includes a broader range of Kyoto units, thus also AAUs, tCERs and lCERs, would not compromise meeting national emissions commitments under the Kyoto Protocol. However, linking schemes with different recognized units will affect the total supply of units in the combined scheme. The total amount of units in the combined scheme could be greater than if the schemes functioned independently.

Moreover, entities in a linked dETS could use the credits which are not covered by the EU ETS for domestic compliance purposes and sell their regular domestic allowances to European entities. In this way the decision of the EU not to include specific credits would be bypassed.

In case of linking the EU ETS to a dETS in a non-Kyoto country, such as the USA ETSs, the recognized units in linked ETSs cannot be used for compliance under the Kyoto Protocol. USA ETSs may generate trading units on a different basis which could not be recognized as environmentally friendly or with a broader definition of sinks. Therefore, linking the EU ETS to a USA ETS needs a system for the mutual recognition of the trading unit to maintain confidence in the environmental integrity of the linked scheme.

4.1.2 The currency of trading

Under the EU ETS as well as under the Kyoto Protocol, one EAU or AAU is equal to one metric ton of CO_2-equivalent. There is a different metric in, for example, the RGGI scheme where allowances would be denominated one short ton CO_2, which is less than a metric ton.

Linking the EU ETS to a scheme with a different trading unit would require an exchange rate. Each dETS involved must recognize the value of the trading units. In a most ideal situation dETSs (in Kyoto countries as well as in the USA) should have the same quantitative unit of trading based on the Kyoto Protocol: metric tons of CO_2.

4.1.3 Allocation method

The initial allocation method (auction or free)[50] in a cap-and-trade scheme has consequences for the distributions of costs and profits in a company.[51] Putting a price on emissions means that whoever is given the initial right to the remaining emissions has a valuable asset. However, the question is whether differences in allocation method are an obstacle for linking dETSs. These distortions will occur anyway due to the mere existence of the dETSs, irrespective of linking them. After the initial allocation, the price of an allowance will be determined by the supply and demand of allowances. Therefore, linking schemes with different initial allocation methods should not introduce any distortion.

In case of linking dETSs the subsequent allocation rules should be harmonized. In the second phase, Member States may take account of emission reductions that have occurred in the first period. The allocation can be based on an updated base-year instead of the same base – year as the first trading period (updating). The Commission proposes setting a single EU-wide cap and allocating allowances on the basis of fully harmonized auction rules. In that case national allocation plans and updating them will not be needed any more.[52] The Swiss ETS updates and adjusts the emission target each year to the companies' production growth (ex post adjustment). Linking two schemes where one uses updating and the other does not, could result in emissions being shifted to the scheme with updating for the purpose of receiving a more generous allocation.

50 In a cap-and-trade system, there are two different methods of allocating the allowances to the entities: by auction or free. The latter could be based on output (benchmarking) or based on historic activity (grandfathering).

51 Numerous economic studies have found that free allocation of CO_2 allowances typically overcompensates companies, see: Lans Bovenberg and Goulder (2001); Boemare and Quiron (2002, pp. 213–230); Burtraw et al. (2002, pp. 51–62); Burtraw (2001, pp. 13–16); Ahman et al. (2005).

52 Articles 9 and 10 Proposal.

In addition, different rules in the treatment of new entrants and plant closures can also lead to different incentives for behaviour during the first period. All Member States have provided new entrants with free allowances. The NZ ETS, on the other hand, does not allocate allowances free of charge to new entrants. The Proposal foresees the creation of a Community-wide new entrants' reserve. Allocations from this reserve should mirror the allocation rules for existing installations. Furthermore, the Proposal mentioned that installations that have closed shall no longer receive any allowances free of charge.[53] The question arises as to why closed installations should receive allowances anyway.

4.1.4 Stringency of targets

The stringency of targets refers to by how much emissions are reduced in comparison to historic or projected emissions. In the EU ETS most Member States have allocated for the first trading period (2005–2007) overgenerous amounts of allowances. The Commission proposes an EU-wide cap whereby the annual cap will decrease along a linear trend line, which will continue beyond the end of the third trading period (2013–2020).[54]

Linking the EU ETS to a scheme with very weak (or weaker) targets, such as a target above business-as-usual emissions level, will lead to higher emissions in the combined scheme than the emissions of the separate schemes. A company in one scheme with strict targets could largely meet its target by buying allowances from another scheme with lenient targets. This is particularly an issue in linking the EU ETS to USA ETSs. In case of linking the EU ETS to a scheme in a Kyoto Party, weak targets have to be compensated by additional reductions in order to meet the national Kyoto target.

4.1.5 Absolute versus relative targets

The EU ETS sets absolute targets of CO_2 emissions. Absolute targets limit the total emissions during a specific period (ex ante allocation). The total emissions should not exceed the target. Other ETSs such as the Canadian ETS and the NSW GGAS may be based on relative targets (ex post allocation), which are defined as emission per unit of output or activity. Emissions may increase as long as this is justified by an increase of production and the emissions have stayed below the relative target.

Linking the EU ETS to a dETS with relative targets could impair the liquidity of the combined scheme. Relative targets require that allocation takes place in two steps, an initial allocation based on production levels and adjustment ex

---

53 Explanatory memorandum of the Proposal, under 4.

54 Article 9 Proposal.

post when the actual production levels are known. This may lead to spikes in liquidity at the moment of adjustment and will also affect the EU ETS.[55] In a dETS with relative targets, emissions would typically be more or less linked to economic growth. Entities in a dETS with relative targets will receive more allowances the more they produce, provided that they will not exceed the relative target. Thus, in case of weak relative targets, entities may have an incentive to increase emissions. This may compromise the compatibility between the combined schemes because output increases will inflate the amount of allowances available in the EU ETS. This could lead to a smaller overall emission reduction.

Possible policy solutions to deal with this problem could be: (i) taxing trade between the two schemes, (ii) introducing an exchange rate to adjust for relative allowance values, (iii) adjusting allocation to the rate-based sectors to account for changes in expectations of growth levels resulting from linkage of the scheme and (iv) establishing a gateway.[56] However, all these options would render the system more complex and increase transaction costs.

As the Proposal requires that the EU ETS can only be linked with dETSs with absolute targets of CO_2 emissions, this issue could be solved.

4.1.6 Voluntary or mandatory participation

Participation in the EU ETS and most of the other schemes is mandatory. The JV ETS, the Swiss ETS and some USA ETSs, such as the RGGI are voluntary.

The environmental effectiveness of a voluntary scheme is likely lower than a mandatory scheme. When the EU ETS links to a voluntary scheme, an entity in the voluntary ETS may shift production and the attendant emissions to another entity that is not covered by the ETS (carbon leakage) in order to generate more allowances to sell. As voluntary dETSs typically achieve much lower coverage, the scope for leakage is greater unless non-participants are covered by other policies.

Linking requires that the targets of the participants in the voluntary scheme are guaranteed lower than business-as-usual emissions. Suitable penalties in case of non-compliance and a monitoring system are needed. However, as the Commission proposed that the EU ETS can be linked only with mandatory dETSs, this issue will no longer play any role.

4.1.7 Upstream versus downstream

The EU ETS is a downstream scheme, where allowances are allocated to entities based on their direct emissions at point of emission. It targets the end-

---

55 See also: Sterk and Braun (2005).

56 See also: Fischer (2003).

users of energy. Some ETSs are upstream schemes that allocate allowances at the point of entry of a fossil fuel into a country's energy system. The AUS ETS and the NZ ETS will combine an upstream regime for small sources, and a downstream regime for large sources. The Lieberman-Warner bill also mixes downstream with upstream features.

In case of linking two schemes with different rules of coverage, it is important to avoid any double counting that might arise. For example, if energy products are exported from an upstream scheme (producers and importers) to a downstream scheme (end-users), emissions will be accounted for in both schemes. Exports of these products should be excluded from the upstream system. The boundaries of the schemes must be clearly defined and there must be a proper accounting of emissions.

4.1.8 Non-compliance penalties

In the EU ETS the financial penalties in case of non-compliance are intended to be higher than the cost of EUAs.[57] There are no penalties in, for example, the Japanese ETS if targets are not met and none of the weaker targets in the NSW GGAS and the NZ ETS.

Linking schemes give all entities access to the price cap set by the lowest non-compliance penalty. A non-compliance penalty defined strictly might be lower than the market price if the scheme is linked with other schemes. Participants could sell allowances and credits to participants in linked dETSs and then fail to comply with their obligations. This situation can lead to non-compliance in the scheme with the lowest penalty, and lead to lower emission reductions and so compromise the environmental effectiveness of the EU ETS. Therefore, in case of linking, the penalties in the linked dETSs must be sufficient to ensure overall compliance.

4.1.9 Safety valve

Some dETSs, such as the Australian, the Canadian, and the Californian ETS have a safety valve. This is a pre-determined allowances price in case the market price rises above a certain level. Linking schemes give all entities access to this price cap. If the market price is higher than the safety valve, entities in the safety valve dETS would have an incentive to use the safety valve and sell their allowances and credits to entities in linked dETSs.

4.1.10 Opt-in and opt-out provisions

A further issue that needs to be considered is the existence of opt-in or opt-out provisions. Opt-in means that new sectors, gases or activities can be included

---

57 100 per ton of CO_2 for Phase II (2008–12).

in the scheme and opt-out means that entities can be excluded from the scheme. In the second trading phase, the EU ETS allows Member States to opt into activities, sectors and gases which are not covered yet by the scheme. Member States are not allowed to opt out of installations. However, the Proposal allows Member States to exclude certain combustion installations from the EU ETS.[58]

If an entity has the option to opt out and move to a less stringent compliance regime, it reduces the scope of the scheme and thus decreases its efficiency. In order to make a link between the EU ETS and other schemes, some restriction on opting-out is needed. The opting-out installations should be covered by other measures that would ensure the environmental effectiveness of the EU ETS.

4.1.11 Monitoring, reporting, verification and registration

For any dETS monitoring, reporting, verification and registration are fundamentally important to ensure confidence and underpinning value in the traded units.

Linking two schemes with differences in monitoring, reporting and verification should not pose any difficulties as long as the systems are transparent and robust enough to maintain confidence in the value of units. The question is whether these domestic monitoring, reporting, verification and registration systems are sufficiently robust to prevent fraud, such as underreporting of emissions. Therefore, efforts to develop international standards are very important for the establishment of an environmentally effective ETS. Linking dETSs also needs their registries to be sufficiently harmonized in order to allow for a smooth transfer of allowances between the schemes. The Kyoto Parties have set up national registries which have to abide by detailed guidelines in order to secure their compatibility. Domestic or regional ETSs that use Kyoto units also undertake their settlement through these registry systems. Linking with a USA ETS needs an agreement on connecting the registries to one other (see also in Section 4.2).

4.1.12 Banking

Banking is transferring entities' allowances from one trading period to the next. The EU ETS allows banking, although there are no standards for the banking rules between the Member States. Differences in banking rules could pose significant problems in linking. Entities in dETSs that do not allow banking would be able to bank through swaps with entities in dETSs that do allow banking. Harmonization of banking rules, or some limitations on banking, would be necessary to reduce concentration of banking in a few dETSs.

---

58 Article 27 sub 1 Proposal.

4.1.13 Borrowing

Borrowing allows an entity in a dETS to achieve compliance for the current compliance period by using allowances allocated to it for a future compliance period. The EU ETS does not allow borrowing. Under the Swiss ETS and the federal USA ETS borrowing is allowed. Linking the EU ETS to a scheme that allows borrowing can reduce environmental effectiveness if an entity ceases operation before the borrowed allowances are repaid. A solution: limit purchases from participants in the dETS with borrowing to ex post purchases from participants from the EU ETS that did not allow it.

4.2 Linking to a Non-Kyoto Country

A last, general obstacle is linking the EU ETS to a dETS in a non-Kyoto country, in fact a USA ETS.

In the first place, only Kyoto units can be used for compliance under the Kyoto Protocol. Thus, even if the EU ETS has recognized USA allowances, directly linking the EU ETS to a USA ETS will not help the Member States nor the EU to meet its Kyoto commitment.

In the second place, from 2008 on, transfers of EUAs are in fact transfers of AAUs. The EUAs will be backed by AAUs. In cases of transfers of USA allowances into the EU ETS, a complication arises: the USA allowances will not be backed up by AAUs. This situation would inflate emissions in the EU without corresponding acquisition of AAUs, with the possible consequence of non-compliance with the Kyoto Protocol.

In the third place, in cases of transactions from the EUAs to the USA ETSs another difficulty arises. Although an entity in a USA ETS can buy EUAs, such EUAs cannot be transferred to the non-Kyoto country. The Kyoto Protocol requires that transfers of Kyoto units may only occur between Kyoto Parties.[59] The main question here is whether an actual transfer is necessary in all circumstances. In cases where a USA ETS allows their entities to cancel the EUAs within the EU ETS and count this toward compliance in their own system, there will be no actual transfer of the EUAs needed. In such cases the buying entity (under an USA ETS) and selling entity (under the EU ETS) should arrange by contract that the selling entity will neither use nor sell the EUAs to a third entity and that the buying company has paid for the extinction of the EUAs under the EU ETS. However, if the EUAs have to be actually transferred, an option is to strip the EUAs of their AAU property. The AAUs

[59] Decision 11/CMP.1. Annex, paragraph 2, Modalities, rules and guidelines for emissions trading under Article 17 of the Kyoto Protocol. See also: De Witt Wijnen (2005, p. 409).

have to be cancelled. In both ways, the USA ETS can be unilaterally linked to the EU ETS. Thus, there will be no necessity for the EU ETS to recognize USA allowances.

In the fourth place, also in the case of a bilateral link between the EU ETS and the USA ETSs, the EUAs should be stripped of their AAU property. The AAUs should be put into a specific account and used to back up incoming USA allowances. The consequence is that acquisitions from the USA ETSs to the EU ETS could only be completed if there are sufficient AAUs available.[60]

The Commission proposed non-binding arrangements with third countries or with sub-federal or regional entities to provide for administrative and technical coordination in relation to allowances in the EU ETS or dETSs.[61]

4.3 Issues for Indirect Linking Through Offsets

Indirect linking requires less standardization than direct linking. There is no mutual acceptance of allocated allowances under each dETS necessary. This seems easy in the case of linking the EU ETS to schemes in Kyoto Parties because CDM and JI are already in place.

However, the EU ETS does not accept all CERs and ERUs. It excludes credits generated by project activities from sinks and credits from nuclear facilities currently. As mentioned in Section 4.1.1, the Japanese, the Swiss and the Canadian ETSs allow CERs from sinks. What does it mean for indirect linking when different criteria are used for the eligibility of credits from offsets in the different schemes?

Furthermore, the EU ETS and some other dETSs, such as the Canadian ETS, have restricted access to CERs. The question arises as to what the implications are for indirect linking of these dETSs to the EU ETS?

In the case of indirect linking of the EU ETS to USA ETSs there are no commonly accepted offset mechanisms at all. The current EU ETS only accepts the credits as generated under the CDM and JI (with limitations). Thus, for indirect linking of the EU ETS to USA ETSs, the latter must recognize these credits.

Also in this situation, a technical issue is how these credits can be transferred to accounts under USA ETSs. The registries in USA ETSs are not connected to the Kyoto registry, the ITL. A solution could be that USA ETSs establish an account in the national registry of an Annex B Party and the ITL. This requires an international agreement. A different option: participants under USA ETSs establish accounts in a national registry of an Annex B Party and

---

60 See also: Sterk and Braun (2006).

61 Article 25 sub 1bis Proposal.

the CDM registry.[62] However, only entities which are authorized to participate in a CDM project by Kyoto Parties may have an account in the CDM registry.

Although, CDM and JI are the established offsets at the multilateral level, there is no reason why they should be the exclusive vehicle for offsets trading. The Proposal provides for the possibility of recognizing a range of offsets other than CDM and JI. In the USA, a structure for offsets is established at the federal ETS. Environmental effectiveness can be protected by adopting common criteria and a common approval process for credits generated by such offsets.

5. CONCLUSION

An important goal of European climate change policy is to reach an international agreement about emission reductions beyond 2012. Even without such an international agreement, the EU made a commitment of at least a 20% reduction of GHG emissions by 2020. The EU is committed to reduce its greenhouse gas emissions by 30% in the event of an international agreement.

An international agreement would also play an important role in linking the EU ETS to other dETSs. An international agreement will most likely affect the combined number of allowances available in the EU ETS linked to other ETSs. Also, the amount of credits being used in the EU ETS depends on an international agreement. Once an international agreement has been reached, the use of credits from projects in third countries will be automatically increased up to half of the additional reduction effort. However, even without an international agreement credits generated from new projects are under certain conditions eligible for use under the EU ETS beyond 2012. The weak restrictions on the use of the amount of credits (CERs, ERUs and other approved credits) and the proposed use of credits from domestic offsets in the EU ETS may place a serious limit on the environmental effectiveness of the EU ETS.

Although the Commission has proposed the extension of direct linking to any country or administrative entity, direct linking to USA ETSs will remain complex. For this to happen, three keys to unlock the problem are required: (i) establish an international framework for the transfer of the Kyoto units; (ii) set emission reduction goals as well as (iii) set overall compliance standards.

Unilateral linking or indirect linking through offsets is an option in cases of linking with USA ETSs or where formal linkage between dETSs is not possible due to substantive differences in design.

62 Decision 1/CMP.1, Annex D.

However, to encourage bilateral linking of different dETSs, a further degree of harmonization is needed. It is important that the designs of the schemes do not hinder their linking. Such harmonization is less likely to be needed when establishing unilateral links between different dETSs.

Nevertheless, the proposed recognition of a range of other offsets created under the international agreement could be an important incentive to get the USA on board. Linking the EU ETS to schemes in the USA might present an important step toward integrating the USA into the international post-2012 process. Although the most ideal situation is linking the EU ETS to the federal scheme, even linkage between the EU ETS to regional schemes would be important. Such linkage would be a crucial sign of political support to the initiatives undertaken by States.

Although its system is far from perfect, the EU ETS has proved that trading in greenhouse gas emissions works.[63] The environmental effectiveness and economic efficiency of the EU ETS can only reach its maximum potential if the design of the EU ETS is optimized. Linking the EU ETS to other dETSs can improve this effectiveness. However, each possible link with the EU ETS has to be considered carefully and should be examined for environmental integrity, competitiveness and technical issues. The outcome of any linking negotiation may have important implications for the post-2012 climate change regime and will perhaps open the road for a global emissions trading system for greenhouse gas emitting sources.

REFERENCES

Ahman, M., Dallas Burtraw, Joseph A. Kruger, and Lars Zetterberg (2005), 'A Ten-Year Rule to Guide the Allocation of EU Emission Allowances Energy Policy', Discussion Paper, Resources for the future.

Anger, N. (2006), 'Emission Trading beyond Europe: Linking Schemes in a Post-Kyoto World', Discussion Paper, Centre for European Economic research.

Australia's Climate Change Policy, 'Our Economy, Our Environment, Our Future', 2007, www.pmc.gov.au.

Baron, R. and S. Bygrave (2002), 'Towards International Emission Trading: Design Implications for Linkages', Information paper, OECD/IEA.

Blyth, W. and M. Bosi (2004), 'Linking non-EU Domestic Emissions Trading Schemes with the EU Emission Trading Scheme', Information paper for the Annex I Expert Group on the UNFCCC OECD and IEA.

Boemare, C., and P. Quiron (2002), 'Implementing Greenhouse Gas Trading in Europe: Lessons from Economic Theory and International Experience', *Ecological Economics*, 43(2), 213–30.

63 See the contribution of Oosterhuis and Kuik.

Burtraw, D. (2001), 'Carbon Emissions Trading Costs and Allowance Allocations: Evaluating the Options', *Resources*, Issue 145, 13–16.

Burtraw, D., K. Palmer, R. Bharvirkar, and A. Paul (2002), 'The Effect on Asset Values of the Allocation of Carbon Dioxide Emission Allowances', *The Electricity Journal*, 15(5), 51–62.

Ellerman D. A., P. L. Joskow, and D. Harrison, (2003), *Emissions Trading in the United States – Experience, Lessons and Considerations for Greenhouse Gases*, paper prepared for Pew Center on Global Climate Change.

Ellis, J. and D. Tirpak (2006), *Linking GHG Emissions Trading Schemes and Markets*, Information paper for the Annex I Expert Group on the UNFCCC, OECD and IEA.

Fischer, C. (2003), *Combining Rate-based and Cap-and-Trade Emissions Policies*, Resources for the future.

Haites, E. (2006), *Possible Use of Kyoto Credits in a National Emissions Trading Scheme*, Report prepared for the Taskforce by Margaree Consultants Ltd.

Haites, E. and F. Mullins (2001), *Linking Domestic and Industry Greenhouse Gas Emission Trading Systems*, Report prepared for EPRI, International energy agency (IEA) and International Emissions Trading Association.

IEA (2005), *Act Locally, Trade Globally – Emissions Trading for Climate Policy*, OECD/IEA.

Jepma, C. (2003): 'KP and EU ETS – How Much to Link?', Editor's Note, *Joint Implementation Quarterly*.

Lans Bovenberg, A., and Lawrence H. Goulder (2001), 'Neutralizing the Adverse Industry Impacts of CO2 Abatement Policies: What Does It Cost?', in Carlo Carraro and Gilbert E. Metcalf (eds), *Behavioral and Distributional Effects of Environmental Policy*, Chicago: University of Chicago Press.

The Ministry of Environment and the Treasury, (2007), *The Framework for a New Zealand Emissions Trading Scheme*, www.climatechange.govt.nz.

Ninomiya Y. (2007), *Office of Market Mechanisms*, Ministry of the Environment Japan, 7 Dec.

Reinaud, J. and C. Philibert (1997), *Emissions Trading: Trends and Prospects*, Information paper for the Annex I Expert Group on the UNFCCC, OECD and IEA.

Sterk, W. and M. Braun et al. (2005), 'Ready to Link Up? The EU and the International Carbon Market', *Carbon Market Europe*.

Sterk, W. and M. Braun et al. (2006), *Implication of Design Difference for Linking Domestic Emissions Trading Schemes*, JET-SET.

De Witt Wijnen, R. (2005), 'Emissions Trading under Article 17 of the Kyoto Protocol', in David Freestone and Charlotte Streck (eds), *Legal Aspects of Implementing the Kyoto Protocol Mechanism: Making Kyoto Work*, Oxford, 399–411.

COM(2008) 16 final, Proposal for a directive of the European Parliament and of the Council amending Directive 2003/87/EC so as to improve and extend the greenhouse gas emission allowance trading system of the Community.

COM(2006) 676 final, Communication from the Commission to the Council, the European Parliament, The European Economic and Social Committee and the Committee of the regions, Building a global carbon market – Report pursuant to Article 30 of Directive 2003/87/EC.

COM(2008) 30 final Communication from the Commission to the European Parliament, the Council, the European Economic and Social Committee and

the Committee of the Regions – 2020 by 2020 – Europe's climate change opportunity.
SEC(2008) 52 Commission's Staff Working Document, Accompanying document to the Proposal for a directive of the European Parliament and of the Council amending Directive 2003/87/EC so as to improve and extend the greenhouse gas emission allowance trading system of the Community – Impact Assessment.

12. Expansion of the EU ETS: the case of emissions trading for aviation[1]

Giedre Kaminskaite-Salters

1. INTRODUCTION

The EU ETS represents the key instrument of the EU's climate change policy. The fact that only around 12 000 installations, accounting for about 40% of the EU's total carbon dioxide emissions, are covered by the EU Emissions Trading Scheme (EU ETS), inevitably means that any emission reductions achieved via the scheme are liable to be counterbalanced by growth in emissions in excluded sectors unless those sectors are carbon constrained via taxation, regulation or other means. Moreover, as only carbon dioxide (CO_2) is currently covered by the EU ETS, any rise in the emissions of other greenhouse gases[2] is bound to undermine the cuts achieved in the reductions of CO_2 ultimately hindering the efforts to prevent further climate change.

Article 30 of the EU ETS Directive[3] obliged the EU Commission to produce, by 30 June 2006, a report considering, inter alia, whether and how other sectors and activities should be included in the EU ETS, as well as the scope for extending its application to other greenhouse gases, with a view to further improving the economic efficiency of the scheme. Released in November 2006, the report, *Building a global carbon market*,[4] stated that legislative proposals would follow in late 2007, with regulatory changes taking effect in Phase III of the EU ETS (namely, 2013–2020). The report stressed the Commission's intention to expand the EU ETS Directive's current

---

1 Giedre Kaminskaite-Salters is a solicitor at the international law firm Norton Rose LLP. The views expressed in this chapter are those of the author and not the firm.

2 The term 'greenhouse gases' denotes carbon dioxide, methane, nitrous oxide, chlorofluorocarbons, and ozone, i.e. gases the concentration of which is thought to be increasing due to human activities, thus causing a 'greenhouse effect' on the earth's climate.

3 Directive 2003/87/EC of 13 October 2003 establishing a scheme for greenhouse gas emission allowance trading within the Community and amending Council Directive 96/61/EC.

4 COM(2006) 676 final.

scope to other sectors and activities (e.g. aviation, transportation and shipping) due to these sectors' increasing contribution to the overall levels of EU greenhouse gas emissions.[5] As regards inclusion of other greenhouse gases, the report noted the potential for including nitrous oxide from the production of nitric acid, methane released from coal mines, CO_2 and nitrous oxide from the production of chemicals, and others. While the draft Proposal amending the EU ETS Directive, issued on 23 January 2008,[6] does not deliver on all the areas reviewed in the report (for example, including shipping or transportation within the scope of the EU ETS was deemed premature without further analysis and pending international discussions), it does envisage a significantly expanded EU ETS that applies much greater greenhouse gas constraints and relies on substantially higher levels of harmonization across member states. It remains to be seen to what extent these ambitious plans for reviewing the EU ETS will survive the forthcoming negotiations in the European Parliament and the Council.

This chapter focuses on one sector which will almost certainly fall within the expanded scope of the EU ETS – namely, aviation. It is argued that the planned inclusion of aviation emissions into the EU ETS raises a number of important legal and policy issues, stemming largely from the unique nature of aviation as an economic activity. The chapter considers these issues in turn, suggesting that the majority of them can be resolved by building on the experience derived from the first years of the operation of the EU ETS and thereby drawing useful lessons for the Proposal to amend the EU ETS Directive, and any other future scheme extensions. However, the chapter also highlights the legal and policy issues arising from the technical and transboundary characteristics of aviation as an economic activity. Such issues are likely to be of relevance to the planned expansion of the EU ETS into sectors such as shipping and transportation; their resolution, therefore, is important not only for the effective operation of the aviation emissions trading scheme, but for the future of the ETS as a whole.

---

[5] The European Environment Agency indicate in their 2006 annual Environment Assessment of transport in Europe that emissions from transport have grown by approximately 34% in the period 1990–2004, with CO_2 emissions from shipping expected to grow by 75% in the next 20 years.

[6] Proposal for a Directive of the European Parliament and of the Council amending Directive 2003/87/EC so as to improve and extend the EU greenhouse gas emission allowance trading system of the Community, 2008/0013 (COD), published in January 2008 (further referred to as the 'Proposal to amend the EU ETS Directive').

2. THE PROPOSAL

Aviation emissions are one of the fastest growing climate change contributors in Europe. Although currently only approximately 3% of Europe's greenhouse gas emissions come from the aviation industry, emissions from the sector have increased by 87% in the period 1990–2004. If the trend continues, there is a risk that, by 2012, up to a quarter of the emissions reductions achieved with a view to meeting the EU's Kyoto commitments could be offset by aviation.[7] In recognition of this risk, in 2006 the EU Commission issued a proposal to include international aviation in the EU ETS.[8] The Proposal, which amends the EU ETS Directive, has now undergone the first reading by the European Parliament and the Council, and the final version of the Directive is not expected to be agreed before the end of 2008.

In short, the Proposal contains the following key provisions:

- The Proposal covers all flights arriving at or departing from an airport in the Community as of 1 January 2012, while intra-EU flights would be covered from 1 January 2011. The subsequent emissions trading periods would match those of the EU ETS. Aircraft operators (irrespective of nationality) would only be allowed to carry on aviation activities if, at the end of the relevant calendar year, they surrendered sufficient allowances equal to their total emissions in the calendar year.
- New aviation allowances would be allocated to operators. While they would be able to purchase and use for compliance EU emissions allowances (EUAs) as well as credits derived from Clean Development and Joint Implementation mechanisms (CERs and ERUs), operators currently covered by the EU ETS would not be able to use aviation emissions allowances for compliance purposes.
- By contrast with the current scheme, allocation of allowances would be harmonized across the Community, with a fixed percentage of the total quantity of allowances being allocated free of charge and the rest being auctioned.
- The cap to be allocated to the aviation sector would be determined by reference to average emissions from aviation in the period 2004–2006.

---

7 Commission Staff Working Document, Impact Assessment Report: Inclusion of Aviation in the EU Greenhouse Gas Emissions Trading Scheme (2006), p. 7, further referred to as the 'Impact Assessment'.

8 Proposal for a Directive of the European Parliament and of the Council amending Directive 2003/87/EC so as to include aviation activities in the scheme for greenhouse gas emission allowance trading within the Community, COM (2006) 818, further referred to as the 'Proposal'.

- Only CO_2 emissions would be covered, although the Commission would put forward a proposal by the end of 2008 as to how nitrogen oxide emissions from aviation are to be addressed.
- Certain aviation activities and aircraft would be exempt from the proposal, including aircraft with a maximum take-off weight of less than 5700kg, flights by state aircraft,[9] and rescue flights.
- The scheme would be administered by member states (each operator being administered by just one state), and verification guidelines would be harmonized.

The following paragraphs will consider some of the key aspects of the Proposal, focusing on the legal and policy issues arising. It will do so by reviewing both the original Proposal and the subsequent legislative amendments by the European Parliament and Council, as well as examining, where appropriate, the interaction of the Proposal with other legislation, including the current EU ETS Directive, the Proposal to amend the EU ETS Directive, the Kyoto Protocol, and others.

3. ALLOCATION

The issue of allocation – i.e. the method by which allowances that are due to operators are calculated and allocated – lies at the heart of any emissions trading scheme. A surplus allocation may result in a low market price for the allowances which in turn may undermine any incentives to reduce emissions. By contrast, a substantial under-allocation of allowances may have a negative economic impact on the industries affected, undermining their competitiveness.

Prior to examining the approach to allocation taken in the Proposal, it is worth reviewing the experience acquired in the course of the first trading period of the EU ETS.

Under the EU ETS Directive, National Allocation Plans (NAPs) are drawn up by individual member states, and approved by the Commission. 'Grandfathering' (namely, free allocation of allowances to operators based on historical emissions) was the most commonly used method of calculating the original allocations of allowances for Phase I of trading (2005–2008), with a total of around 6.57 billion emission allowances having been allocated for

[9] This exemption proposed by the Commission did not have a clear justification and was later rejected by the European Parliament.

free, resulting in an annual cap of 2.19 billion tonnes of carbon dioxide.[10] Release of verified emissions data in 2005, however, dealt a serious blow to the confidence about the future of the EU ETS, both as a market-based instrument aimed at reducing carbon dioxide emissions, and as the EU's key means of attaining its Kyoto targets. The data – which resulted in a sudden crash in the price of EUAs – confirmed that there had been substantial (approximately 100m units) over-allocation of allowances such that, rather than experiencing operational carbon constraints, some industries had in fact enjoyed windfall profits. Verified emissions data released in 2006 confirmed that although emissions had grown slightly, they remained lower than the allocated levels. The price of EUAs consequently fell substantially below 1, and remained at that level until the end of their validity period.[11]

One of the main explanations for the disappointing performance of the scheme, which became apparent very early on in the first Phase of trading, stemmed from the different approaches to allocation adopted by the member states, which made estimates of scarcity of allowances (fundamental for any cap-and-trade system) almost impossible to arrive at accurately.[12] This, in turn, meant that the affected industries were not given a clear signal as to whether cuts in emissions were necessary, with many operators correctly guessing that purchasing allowances in the market made greater economic sense. Admittedly, Phase I of the EU ETS was always intended to act as a trial phase, and NAPs for Phase II of the EU ETS trading (2008–2012) have been subjected to much greater scrutiny by the Commission, whose insistence on more consistent allocation methodologies and stricter allocation caps is expected to ensure that greater scarcity of allowances is achieved. Nevertheless, it is clear that unless further harmonization in the allocation methodologies is achieved, the EU ETS will continue to suffer from unpredictability and inefficiency.

In light of the above, it is not surprising that the Proposal to amend the EU ETS Directive should have recognized that 'a more harmonized emission trading system is imperative, in order to better exploit the benefits of emission trading'.[13] Thus, going forward, it will be the Commission that will carry out

---

[10] Questions and Answers on Emissions Trading and National Allocation Plans: Commission MEMO/05/84, 8 March 2005.

[11] S. Long, and G.K. Salters (2007), 'The EU ETS – Latest Developments and the Way Forward', *Climate Change Law Review*. For information on prices of EUAs, see www.pointcarbon.com.

[12] The Commission has recognized the problems associated with different allocation methodologies; see, for example, Commission's guidance, COM (2005) 703 final: Further Guidance on Allocation Plans for the 2008 to 2012 Trading Period of the EU ETS.

[13] Proposal to amend the EU ETS Directive, Recital 7.

allocation to the sectors covered by the EU ETS, although member states will retain the right to decide individual allocations to installations, subject to auctioning requirements.

Secondly, initial experience from the operation of the EU ETS has confirmed that grandfathering is an unsatisfactory method of allocating allowances. Because calculations are based on historic data, the method effectively rewards previously high emissions, thus incentivizing operators to maintain high emissions in order to receive more allowances. Conversely, efforts to reduce emissions through technological innovation or greater efficiencies are inevitably discouraged, as this carries the risk of receiving a lower allocation of EUAs.[14] The failure of grandfathering as an allocation method has been enhanced by the fact that to date the majority of EUAs have been allocated to the operators free of charge. Free allocation of allowances based on historical emissions is unlikely to result in a sufficiently high price for carbon, necessary for an effective emissions market to emerge. Again, the Proposal to amend the EU ETS Directive recognizes this, and notes that the new Community-wide allocation rules must 'take account of the most greenhouse gas and energy efficient techniques' and 'not give incentives to increase emissions'.[15]

As regards allocation, the Proposal mirrors the approach taken by the Proposal to amend the EU ETS Directive, thus avoiding the shortcomings of the current scheme. Firstly, it proposes that allocation of aviation allowances would be managed by the EU Commission, rather than the individual member states.[16] The Commission would utilize a single method of allocation,[17] and may in the future issue detailed provisions for harmonized verification of emission reports.[18]

Furthermore, the Proposal rejects grandfathering as an inefficient means of allocating carbon allowances to the aviation sector, favouring benchmarking instead.[19] Benchmarking, which involves determining the level of allocation by reference to a certain technical benchmark or an activity indicator, allows the allocation to reward those operators who utilize best-available techniques and maximize their operational efficiencies, and to penalize operators who have failed to take measures to improve their emissions performance. As the

14 See, for example, G. Stuart and A.M. Fisher (2007), 'One World? International Aviation and the EU Emission Trading Scheme', *Environmental Law and Management*, 19. Also see DEFRA, Consultation on the Commission's Proposal to Include Aviation in the European Union Emissions Trading Scheme, March 2007.

15 Id., Recital 18.

16 Proposal, Article 3(d)(3).

17 Id.

18 Proposal to amend the EU ETS Directive, Article 15.

19 Proposal, Article 3(d)(3).

Impact Assessment accompanying the Proposal points out, whereas the extensive data needed for implementing benchmarking in other sectors of the EU ETS has been lacking, for the aviation sector the data requirements are fewer due to the relative mechanical homogeneity of carrying passengers or freight between airports, and therefore benchmarking becomes a straightforward option.[20]

Thirdly, the Proposal envisages a greater role for auctioning[21] as a means of ensuring that those allowances that would have been allocated free of charge are actually paid for by the operators; this, in turn, reduces the potential to experience windfall profits from passing on to the customers the cost of having to obtain the allowances, even where they were in fact obtained for free. The precise share of auctioning remains to be determined. Notably, the Proposal envisaged that the percentage of allowances to be auctioned in 2011–2012 would correspond to the average percentage proposed by the member states.[22] The Parliament, however, proposed that at least 25% of allowances should be auctioned initially, with the level of emissions auctioned in subsequent periods corresponding to the maximum level of auctioning in other sectors.[23] The Council adopted a more conservative approach, preferring a 10% share for auctioning in 2011–2012,[24] with scope for further increases in the future. It is doubtful that this comparatively low share of auctioning favoured by the Council will be retained, however, in light of the ambitious plans endorsed by the Proposal to amend the EU ETS Directive.[25] Indeed, the Proposal to amend the EU ETS Directive explicitly states that 'as regards the approach to allocation, aviation should be treated as other industries which receive transitional free allocation' so that such allocations would gradually decrease, resulting in 'no free allocation by 2020'.[26]

The cumulative effect of these measures is likely to be that of facilitating the development of an aviation emissions trading scheme that would be considerably more harmonized and transparent than the current EU ETS, creating an element of predictability necessary for the effective assessment of the 'investment risk-reward', an essential feature of any market-based regula-

---

20 Impact Assessment, p. 20.

21 Proposal, Article 3(c)(1).

22 An average of approximately 3% of EUAs will be auctioned in Phase II of the EU ETS.

23 European Parliament legislative resolution of 13 November 2007 on the Proposal (further referred to as the 'Parliament Proposal'), Article 3(c)(1).

24 Council of the European Union, political agreement on the Proposal adopted in December 2007 (further referred to as the 'Council Proposal'), Article 3(c)(1).

25 Under the proposed plans, around 60% of EUAs will be auctioned, with the power sector being required to auction 100% of its allowances.

26 Proposal to amend the EU ETS Directive, Recital 33.

tory mechanism. Moreover, by rejecting grandfathering in favour of benchmarking and placing greater reliance on auctioning, the scheme is likely to create the necessary level of scarcity of emission allowances, triggering the emergence of a stable, and growing, price for carbon, which in turn would encourage measures to reduce emissions.

However, a few issues connected to allocation remain to be resolved. Firstly, the overall cap on emissions which the aviation sector would be allowed to emit every year during 2011–2012 and in Phase III of trading is set at 100% of the mean average of the 2004–2005 emissions,[27] 2005 being the year for which the most recent emissions data was available to the Commission at the time of drafting the Proposal. The Council Proposal broadly endorses the recommendations of the Proposal (albeit stating that the scheme would commence in 2012 – please see below).[28] However, as the European Economic and Social Committee[29] points out, there seems little justification for not bringing the baseline date of the scheme into closer alignment with the Kyoto commitments (based on the 1990 emission levels) or the EU's unilateral target of achieving a 20% reduction in CO_2 emissions by 2020, as compared to the 1990 levels (or 30%, if other countries follow suit). Indeed, any other arrangement would, arguably, give preferential treatment to the aviation sector, as compared to other sectors covered by the EU ETS. By way of a counter-argument, the so-called 'Kyoto Protocol freeze scenario' would have a disproportionately negative effect on the industry which experienced its greatest economic boom (and therefore rise in emissions) in the past decade. It is also arguable that capping the aviation sector's emissions at the 1990 levels would create an artificial and unjustified link with the Kyoto Protocol, from which aviation emissions are currently excluded.

The alternative would be to cap the emission allowances at less than 100% of the 2004–2006 emissions. Thus, for example, the Environment Committee Report recommends setting the cap at 75% of the 2004–2006 emissions, with further reductions in the cap to be determined in light of the EU's international or unilateral commitments.[30] The Parliament, less ambitiously, has opted for 90% of the 2004–2006 emissions with further reductions envisaged for the future.[31]

27 Proposal, Article 3(b).

28 Council Proposal, Article 3(b).

29 Opinion of the European Economic and Social Committee on the Proposal, Brussels, 31 May 2007, p. 7.

30 Committee on the Environment, Public Health and Food Safety report on the Proposal, 19 October 2007, Article 3(b), further referred to as the 'Environment Committee report'.

31 Parliament Proposal, Article 3(b).

The key arguments for setting the cap at lower levels than those stipulated in the Proposal are the maximization of the environmental benefits of the scheme, as well as providing for a possibility to adjust the cap in line with the EU's increasingly stringent commitments to reduce its CO_2 emissions. Regardless of whether the caps proposed by the Environment Committee report and the Parliament Proposal are considered to be the appropriate ones, more detailed justification for the choice of the 2004–2006 baseline is required if the scheme is to be put on a sound policy footing and the environmental rewards it is intended to bring are to be maximized.

Furthermore, the treatment of new entrants remains to be clarified. Under the EU ETS Directive, member states are required to maintain a new entrant reserve (namely, a reserve of allowances to be allocated to new members of the scheme as and when they fall under its scope).[32] Whereas the EU ETS Directive defines a new operator as a new installation, or a substantially changed/extended installation, the same concept cannot be easily applied to the aviation sector, where aircraft (the equivalent of an installation) is frequently transferred into and out of the scheme; the same fluidity defines the opening and closing of new routes. While defining a new entrant as a new commercial operator may appear an attractive alternative, the Impact Assessment points out that the link between new aircraft activity and new aircraft operators is not automatic, as businesses may be restructured into multiple operators without increasing aviation activity. As a result, no new entrants' reserve for aviation emissions is envisaged in the Proposal, with the Impact Assessment recommending that 100% of the allowances be purchased by new operators via auctioning or the free market. Regardless of its attractive straightforwardness, the solution clearly creates considerable competitive disadvantages for the potential new entrants, who may be a key source of technological innovation and yet who would be competing with existing operators auctioning only a portion of their allowances. Not surprisingly, therefore, the Parliament Proposal requires the Commission to implement measures in respect of the allocation of allowances to make provision for new aviation entrants (defined as aircraft operators who have been issued with Air Operators' Certificates for the first time after the commencement of the scheme, with specific provisions being envisaged for mergers and takeovers as well as cessation of operations).[33] The Council Proposal also contains certain provisions relating to new entrants. The Proposal to amend the EU ETS Directive provides that 5% of the total quantity of allowances will be placed in the new entrants' reserve to be allocated to new installations and 'airlines' in a manner which mirrors the allocations to the

---

[32] Article 11(3) of the EU ETS Directive.

[33] Parliament Proposal, Article 3(d), paragraph 5a.

existing installations;[34] however, the definition of 'airlines' is not included. It seems justified to suggest that defining 'new entrants' along the lines proposed by the Parliament, and allocating the allowances from the new entrants' reserve on the same basis that they are allocated to the existing operators (namely, via a mixture of benchmarking and auctioning) would help to ensure a level playing field and encourage innovation and introduction of best-available techniques into the industry.

In summary, the approach to allocation adopted in the Proposal is one which should ensure that the scheme avoids the key shortcomings of the current EU ETS. However, the choice of the baseline for setting the allocation cap, as well as provisions relating to new entrants raise a number of issues that to date remain insufficiently addressed. If the new scheme is to operate successfully, the said issues ought to be prioritized.

4. INTERPLAY WITH THE OTHER EU ETS SECTORS AND THE KYOTO PROTOCOL

The second phase of trading of the EU ETS will coincide with the Kyoto Commitment Period (namely, the period during which signatories to the Kyoto Protocol will have to meet their respective emission reduction targets). As a result, the European Registries Regulation[35] provides that any Phase II EUAs will have to be created from the Assigned Amount Units (or AAUs, being the tradable units derived from the Assigned Amount of emissions which a signatory of the Kyoto Protocol is allowed to emit during the Commitment Period). Effectively, therefore, every time EUAs are transferred within two accounts in the registries system, the equivalent number of AAUs are also transferred, in order to keep an accurate track of the reductions (or otherwise) in the EU's emissions and the effect this has on its Assigned Amount.

Given that international aviation falls outside the scope of the Kyoto Protocol targets[36] (which means that any allowances representing the sector's emissions cannot be backed up by AAUs), this poses a problem for the inclusion of aviation emissions into the EU ETS, as transfers of aviation emissions allowances from and into the EU ETS could have a distorting effect on the member states' Assigned Amounts. The following examples will illustrate this point.[37]

---

[34] Questions and Answers on the Commission's proposal to revise the EU missions Trading System, MEMO/08/35, 23 January 2008.

[35] Regulation 2216/2004/EC.

[36] By contrast, domestic aviation is included within the Kyoto targets.

[37] See G. Schwarze, 'Including Aviation into the European Union's Emissions Trading Scheme', *European Environmental Law Review*, January 2007.

If an entity currently covered by the EU ETS successfully reduces its emissions by one tonne of CO_2, it can sell one EUA to an aircraft operator, who would now acquire the right to emit one ton of CO_2. However, because aviation falls outside the scope of the Kyoto Protocol, an AAU cannot be transferred alongside the EUA, which would leave the member state of the installation that sold the EUA with one spare AAU. The AAU could be allocated to a non-ETS entity, which could now emit one ton of CO_2. As a result of the transaction, a reduction in emissions of CO_2 by one ton by the original installation would have the effect of allowing the emission of two tons of CO_2 – one by the aircraft operator, and one by a non-ETS entity within the member state.

Conversely, if an aircraft operator achieved a one ton reduction in its CO_2 emissions and then sold an aviation emission allowance to an installation covered by the EU ETS (thereby giving that installation the right to emit one ton of CO_2), there would now be an additional allowance being traded within the ETS, with no AAU to back it up. The member state would have to deal with this discrepancy by achieving a one ton reduction in CO_2 emissions in a non-ETS sector. Thus, a one ton saving in CO_2 emissions by the aircraft operator would require a further saving by a non-ETS entity within the member state if the shortage in the number of AAUs were to be eliminated.

The most obvious solution to the problems outlined above would be to establish a closed aviation emissions trading scheme, whereby aircraft operators would only be allowed to buy and sell aviation emission allowances between themselves, with no access to EUAs. While this would ensure that impacts on, and distortions of, the EU ETS are minimized, the viability of a closed scheme would be doubtful. According to the Impact Assessment, it is not certain whether an emissions trading scheme for the aviation sector in isolation would be large enough to ensure a viable market, given the relatively small number of players and the very limited size of the market (8% of the EU ETS).[38] Moreover, as pointed out by the European Economic and Social Committee, including aviation would have a positive impact on the ETS and the environment in general: due to its limited capacity to achieve emissions reduction, and the almost inevitable growth, aviation is likely to be the net buyer of EUAs, which would result in a significant influx of new funds into the ETS, providing investment for carbon savings in other sectors.[39] The Commission's conclusion – namely, that 'the best way forward, from an economic and environmental point of view, lies in including the climate

---

38 Impact Assessment, p. 39.

39 Opinion of the European Economic and Social Committee on the Proposal, Brussels, 31 May 2007, p. 5.

impact of the aviation sector in the EU emissions trading scheme',[40] whereby all sectors would collectively meet the emission reduction goal, with the average market price for CO_2 reflecting the marginal abatement costs of attaining that goal – seems justified.

However, how does one avoid the problems of interplay with the Kyoto Protocol outlined above which would inevitably arise if an open scheme for aviation emissions were adopted? In the absence of a solution, serious distortions to the member states' Assigned Amounts could lead to difficulties in compliance with the Kyoto targets. Two broad options have been considered by the Commission in this regard, namely: allocating no allowances to the aviation sector, or imposing trading restrictions.

The first method (whereby aircraft operators would be required to purchase all of their allowances from the EU ETS), while ensuring that the allowances held by aircraft operators would be EUAs and therefore fully backed by AAUs, has been dismissed on the grounds that it would place an excessive burden on the sectors currently covered by the EU ETS, as they would be the only source of allowances accessible to the aviation sector. In addition, this would impose a high economic burden on the aviation sector, creating a potential competition distortion between EU and non-EU carriers.[41]

Several options involving the imposition of trading restrictions have also been considered. A 'gateway' system, for example, would involve aircraft operators being allowed to buy only as many EUAs as the number of aviation emission allowances sold by the aviation sector to the rest of the EU ETS. The AAUs that accompany the EUAs being transferred into the aviation sector would be held in a separate 'gateway' account to be used to underpin the aviation allowances entering the scheme. As soon as the last AAUs are used, the gates would 'shut', prohibiting any further influx of aviation allowances until surplus AAUs become available. While this method has the advantage of eliminating any inconsistencies between the EU ETS and the EU's Assigned Amount, the uncertainty of trading transactions would inevitably increase the volatility of the price of carbon. Another alternative – namely, 'borrowing' AAUs from sectors not currently covered by the EU ETS to temporarily underpin aviation allowances – has also been dismissed. Although this method would facilitate free emissions trading between the aviation sector and other sectors, in the absence of a gateway system the risk of there being a net influx of aviation allowances and therefore insufficient AAUs at the end of the trading period, requiring the relevant member state to bridge the shortage by purchas-

40 Communication on Reducing Climate Change Impact of Aviation, COM(2005) 459.

41 Aviation Working Group, European Climate Change Programme II, Final report, Annex 3a, Minutes of Meeting 3 – day 1, April 2006, p. 4.

ing additional AAUs – has been deemed unacceptable by the member states. Finally, the 'baseline' option – namely, the requirement for the aviation operators to surrender allowances only above a certain baseline (thus reducing the demand for EUAs) – has been rejected on the grounds that it provided no incentives to reduce emissions below the baseline.[42]

The arrangement contained in the Proposal at first glance appears to avoid the problems of the other options outlined here. Namely, by providing that aircraft operators can sell their allowances, purchase EUAs, and use both types of allowances for compliance purposes without any restrictions, the Proposal gives aircraft operators full access to the EU ETS, and avoids imposing trading restrictions. Moreover, by prohibiting stationary installations currently covered by the EU ETS from using aviation allowances for compliance purposes,[43] and providing that aviation allowances can be exchanged by aircraft operators into AAU-backed EUAs,[44] the Proposal eliminates the risk of discrepancies occurring between the EU ETS and the member states' Assigned Amounts.

However, two important issues remain unaddressed in the Proposal. To begin with, providing the aviation sector with an unrestricted access to EUAs creates a real risk of 'carbon leakage'[45] in the energy-intensive sectors (such as the cement, lime or steel sectors). The Parliament Proposal recommends two ways for tackling this: firstly, it provides that access to EUAs be restricted in line with the quotas for the use of credits derived from Clean Development and Joint Implementation mechanisms envisaged by the Proposal (i.e. it is proposed that the aviation sector can use EUAs, CERs and ERUs only up to the average of the percentages specified by the member states for the use of CERs and ERUs for the relevant trading period). While inevitably increasing the trading restrictions on the sector, these measures would go some way towards minimizing the 'carbon leakage' from the most affected industries, while placing a greater burden on the aviation sector to invest in emission reducing measures. Secondly, it requires certain efficiency goals to be met by aircraft operators before they are allowed to surrender allowances other than aviation allowances for the purposes of compliance.[46] Disappointingly, the Council Proposal supports the original drafting of the Proposal, whereby there

42 Impact Assessment, pp. 16–17.

43 Proposal to amend the EU ETS Directive, Article 12.

44 Proposal to amend the EU ETS Directive, Article 19(3).

45 'Carbon leakage' arises when industries most affected by carbon constraints are forced to relocate production to countries outside the EU; this can then increase global emissions without any environmental benefits.

46 Parliament Proposal, Article 12(2)(e).

is no limit on the aviation sector's access to EUAs,[47] and no provision for an efficiency indicator, with the problem of 'carbon leakage' remaining unresolved. It seems imperative that alternative solutions should be developed as the Proposal undergoes further legislative scrutiny if the issue is to be properly addressed.

Secondly, it remains unclear how the provision regarding the swapping of aviation allowances for EUAs would work in practice. Like other similar options considered by the Commission, this requires the member states to provide aircraft operators with access to the AAUs which have not been earmarked for EUAs. This requirement to accept non-Kyoto allowances in exchange for AAUs is further complicated by the absence of any 'gateway' system. If its AAU reserves were exhausted, the member state would be forced to acquire additional AAUs externally, rather than shut down the gates on further swaps. Not surprisingly, the Parliament Proposal has deleted this provision from the Proposal. This, however, does not of itself offer a complete solution. For, if the aviation sector is unable to convert its allowances into EUAs, can only purchase EUAs up to a certain amount, and is unable to sell its allowances to other sectors due to the prohibition against the use of aviation allowances for compliance purposes, then the trading system created is a semi-closed one. As outlined above, such a system would be of limited environmental and economic benefit. It remains to be seen how this important issue will be addressed as the Proposal is further debated; the compromise reached in the Council Proposal (namely, that the Commission should make proposals by 1 June 2015 as to whether a gateway system should be included to facilitate the trading of allowances between aircraft operators and other EU ETS sectors[48]) indicates that substantial further investigation and analysis will be required in this area.

In summary, the linkages between the aviation sector, other sectors of the EU ETS, and the Kyoto Protocol, raise complex legal and policy issues. Having considered a number of potential solutions, the Commission has proposed to include aviation emissions within the EU ETS (rather than opting for a closed aviation emissions trading scheme). Further, apart from the prohibition against stationary installations using aviation allowances for compliance, the Proposal has not imposed trading restrictions on the aviation sector. While in broad policy terms this is to be welcomed, two key issues – namely, that of carbon leakage from the energy intensive sectors, and the ability of member states to back up aviation emissions with AAUs– must be addressed in greater detail if the scheme is to be put on a sound footing.

47 Council Proposal, Article 11a.

48 Council Proposal, Article 30(4)(g).

5. NON-CO_2 GREENHOUSE GAS EMISSIONS

A clear limitation of the EU ETS, as it currently stands, is that its application is limited to CO_2 emissions only, although it does anticipate that other greenhouse gases may be included, either unilaterally by the member states (subject to the Commission's approval), or following a review of the EU ETS by the Commission. The Proposal to amend the EU ETS Directive goes some way towards addressing this problem by expanding the scope of the EU ETS to include, inter alia, the emissions of petrofluorocarbons from the production of aluminium, and nitrous oxide from the production of nitric acid and other activities. The Proposal, by contrast, covers only CO_2 emissions from aviation, although the Commission has indicated that it will put forward a proposal to address the impacts of nitrogen oxide (NO_x) emissions from aviation by the end of 2008.[49]

In addition to NO_x, other non-CO_2 emissions from the aviation sector include water vapour, sulphate and soot particles. Although it has been established that, when released at high altitudes, water vapour leads to the formation of condensation trails which contribute to global warming, the exact effect of these emissions has not been sufficiently understood, which has led some experts to argue that the overall climate impact of aviation emissions is significantly more substantial than the impact of its CO_2 emissions alone; indeed, the IPCC estimate that such impact may be two to four times greater.[50]

The Commission has considered two potential ways of addressing this issue pending scientific progress:[51] firstly, a multiplier could be applied to the calculation of the number of allowances that an aircraft operator must surrender, as a precautionary means of taking into account the impact of non-CO_2 emissions; alternatively, ancillary instruments (such as airport charges) could be applied specifically to the non-CO_2 emissions. Given that production of NO_x is not directly related to CO_2 production, the case for implementing separate measures for tackling at least the NO_x emissions seems convincing. Moreover, while being easy to administer, the multiplier option lacks any scientific basis. Not surprisingly, therefore, the Parliament's attempt to introduce an 'impact factor' of 2 to be applied to aviation emissions, pending Community measures to incentivize the reduction of NO_x,[52] has been dismissed by the Council Proposal. While this should be welcomed, in that it ensures that the Proposal does not embrace scientifically unsubstantiated

---

49 NO_x produces ozone, a greenhouse gas.

50 IPCC, 'Aviation and the Global Atmosphere' (Cambridge University Press, 1999).

51 Impact Assessment, p. 15.

52 Parliament Proposal, Article 12.2(b).

measures, the deletion of the multiplier provisions also means that the issue of additional impacts of non-CO_2 emissions remains unaddressed. In light of the potential significance of such impacts, as well as bearing in mind developments in other areas of the EU ETS, where the trend towards expanding the scope of the scheme to include other harmful greenhouse gases has been evident, further policy efforts in this area seem a priority.

6. SCOPE

The preceding sections of this chapter have shown that the Proposal has to a large extent succeeded in overcoming the problems of the current EU ETS Directive; issues that still need addressing (such as the interaction of aviation emission allowances with those issued under the Kyoto Protocol, the treatment of new entrants, or the need to 'index' aviation-specific emissions to take account of their additional environmental impacts) stem from the specific characteristics of aviation as an economic activity, rather than any limitations of the EU ETS as a vehicle for combating climate change. Similar issues are likely to arise in the context of the anticipated expansion of the EU ETS into shipping and transportation, and therefore their resolution in the context of aviation emissions is a necessary (if difficult) process. By contrast, the final issue considered in this chapter (namely, the scope of the Proposal) arises largely from the international legal framework within which aviation operates and as such is unlikely to be particularly relevant to other EU ETS sectors. However, its resolution will determine whether or not an aviation emissions trading scheme can take off.

The Proposal covers all intra-EU flights from 2011 and all arrivals at, and departures from, EU airports from 2012 – an option rejected in the Parliament Proposal on the grounds that if a level playing field between airports and between aircraft operators is to be ensured and the effects on the reduction of greenhouse gas emissions maximized, then a common date (2010) for the commencement of aviation emissions trading must be adopted.[53] The Council has accepted the 'level playing field' position, but has shifted the commencement date to 2012.[54]

The timing considerations outlined here conceal a much greater issue underlying the Proposal – namely, the possibility that, because of its geographical scope (i.e. the fact that it extends to both European and non-European operators who operate flights to and from the EU), the Proposal potentially

---

53 Parliament Proposal, Annex I, 1(b).
54 Council Proposal, Annex I, 1(b).

contravenes the current international legal regime. The fundamental point of dispute that arises here is whether the European Union can integrate international aviation emissions from aircraft operators of non-European states in its emissions trading schemes without mutual agreement. The resolution of this fundamental issue will determine whether or not the Proposal can be implemented in practice.

Although international aviation emissions fall outside the scope of the Kyoto Protocol,[55] Article 2(1) of the Protocol requires parties to 'pursue limitation or reduction of emissions of greenhouse gases [...] from aviation [...], working through the International Civil Aviation Organisation' (ICAO).

As a specialized United Nations Agency, ICAO adopts standards and recommended practices in relation to international civil aviation. In 2004, it endorsed the recommendation of the ICAO Committee on Aviation Environmental Protection not to pursue an aviation-specific emissions trading system based on a new legal instrument under ICAO. Instead, it committed itself to the development of an open voluntary aviation emissions trading system, while also agreeing to provide guidance to contracting states on the incorporation of emissions from international aviation into their emissions trading schemes consistent with the UNFCCC process.[56]

It is in response to ICAO's reluctance to require its members to make any binding commitments to reduce emissions from aviation that the Commission issued the Proposal in 2006. Moreover, ICAO's endorsement of an open emissions trading scheme and readiness to offer guidance as to the integration of aviation emissions into parties' trading schemes was interpreted by the Commission as an indication that the Proposal would be consistent with the international legal framework. However, ICAO's non-European members have viewed the Proposal as a one-sided measure imposed by the EU on other countries without their mutual agreement which undermines the international framework for tackling aviation emissions via ICAO envisaged in the Kyoto Protocol. ICAO's 36th Assembly, for example, noted that the EU's Proposal was 'viewed by many non-EU States as unilateral imposition which should be reconsidered', and required ICAO to 'oppose any unilateral action and non-differentiated coercive emissions reduction measures', as well taking on a 'leadership role' in tackling global warming.[57]

The issue here is clearly centred on the interpretation of the obligations imposed on the parties by the Kyoto Protocol. The proponents of the ICAO

---

[55] The Kyoto Protocol covers domestic aviation emissions only.

[56] ICAO Assembly Resolution A35-5, Consolidated Statement of continuing ICAO policies and practices related to environmental protection, October 2004.

[57] ICAO Assembly – 36th Session, report of the Executive Committee on Agenda Item 17, paras.17.4.2.8–12.

route have argued that the requirement to work through ICAO in devising measures to tackle emissions from international aviation is at the exclusion of unilateral measures such as those contained in the Proposal; this is especially so given that the membership of ICAO does not coincide with the list of signatories of the Kyoto Protocol; requiring the non-Kyoto countries to participate in an emissions trading scheme established under the auspices of the Kyoto Protocol potentially subjects those countries to the Kyoto targets 'through the back door'.[58] Another, very convincing, argument in favour of those who see the Proposal as undermining the role intended for ICAO by the Kyoto Protocol is that, by setting national emissions reduction targets, the Kyoto Protocol makes its parties responsible for all emission sources located in their territory, regardless of nationality; conversely, parties cannot, under the Kyoto Protocol, take on responsibility for emission sources outside their territory. Hence the decision to exclude aviation, as a transboundary activity, from the scope of the Kyoto Protocol and the recommendation that parties work through ICAO on developing a separate solution for aviation emissions.[59]

Opponents of the position outlined above argue that the obligation to 'work' through ICAO contained in Article 2 of the Kyoto Protocol arguably does not of itself prohibit unilateral measures from being taken by individual states, provided this does not contradict a position endorsed by ICAO. As stated above, the EU Commission has viewed the Proposal as consistent with ICAO's resolution 35-5, which endorsed emissions trading in principle. While EU member states continue to cooperate with other states within the framework of ICAO as required under the Protocol, they have considered it necessary to supplement their efforts via the Proposal, in light of the limited progress achieved by ICAO on the one hand, and the rapid growth in air traffic and the resultant aviation emissions on the other. The fact that such unilateral measures are not prohibited under the Protocol is evidenced by the general requirement (also contained in Article 2(1) of the Kyoto Protocol) for the parties to implement further 'policies and measures in accordance with [their] national circumstances' in limiting emissions from the transport sector. The Proposal, arguably, is precisely such a measure designed to tackle emissions from one branch of the transport sector.

It remains to be seen to what extent, if at all, these two different positions can be reconciled.

The geographical scope of the Proposal has also been argued to be inconsistent with the 1944 Convention on International Civil Aviation (the Chicago Convention), which regulates international airspace, airplane registration,

---

58 Id., para.17.4.2.13.

59 See A. Hardeman, 'A Common Approach to Aviation Emissions Trading', *Air & Space Law*, Vol. 32 (Feb. 2006).

safety and rights of the parties in relation to air travel. The relevant provisions of the Convention can be summarized as follows:

> Article 1 stresses states' sovereignty over the airspace above their territories. Thus, states are free to regulate such airspace as they see fit, provided this is done in conformity with international law (namely, the Chicago Convention and any bilateral agreements).
>
> Article 11 requires states to apply laws as to the admission to or departure of aircraft from their territory or operation of aircraft while in their territory without distinction as to nationality.
>
> Article 15 regulates the fees, dues and other charges for the use of airports and air navigation facilities, and prohibits the imposition of fees, dues or other charges in respect solely of the right of transit over or entry into or exit from a state's territory of any aircraft.
>
> Article 24 prohibits the charging of customs duties.
>
> Annex 16 (Environmental Protection) sets standards which limit the emissions of aircraft engines and aircraft noise.

Provided the emissions trading obligations are applied to all aircraft operators in a non-discriminatory manner regardless of nationality, the Proposal is unlikely to fall foul of the requirements of Article 11. However, it has been argued that, since it is not linked to the use of airport facilities or services,[60] the costs arising from participating in an emissions trading scheme effectively amount to a charge prohibited under Article 15 under the Chicago Convention. Alternatively, it has been proposed that the requirement to purchase allowances effectively amounts to a custom duty prohibited by Article 24.

It is difficult to sustain either of the above arguments. Firstly, it is hard to see how the purchase of allowances would fit the definition of a charge, given that allowances are either allocated free of charge or purchased via a commercial transaction (whereas charges are financial costs imposed by the regulatory authority), as well as considering that the sanctions for failure to comply with

[60] ICAO states that charges must be based on the cost of providing an airport or air navigation service or facility and the related cost must be directly attributable to the operator; ICAO Doc 90827 Policies on Charges for Airports and Air Navigation Services. See A. Hardeman, 'A Common Approach to Aviation Emissions Trading', *Air and Space Law*, Vol. 32 (Feb. 2006); see also K. Dunseath and R. Macrory, 'Time to Act? Should the UK be Taxing Aviation Fuel?', *New Law Journal*, 27 April 2007.

the EU ETS include financial fines for non-compliance with regulations, rather than the refusal of airport services for non-payment of charges, for example. Similarly, though it is not debated that the acquisition of allowances will result in additional costs for aircraft operators, thus having an economic effect similar to that of a customs duty, it is also clear that the legal nature of the measure cannot be equated with a custom duty prohibited by Article 24.[61] Lastly, while Annex 16 sets maximum levels of emissions for certain aircraft engines, the EU ETS does not interfere with such standards, as it does not target the design of aircraft engines, leaving it up to the operators to decide how they would achieve their share of the required emission reductions.

Similarly to Article 11 of the Chicago Convention, bilateral air transport agreements (such as the US–EU Open Skies Agreement) into which a number of EU member states have entered with their non-EU counterparts, invariably contain a provision against discrimination and require parties to give operators 'fair and equal opportunity' to compete in providing aviation services. Again, the EU ETS, if applied in a non-discriminatory fashion to all aircraft operators, will not infringe this requirement. Other requirements contained in the Open Skies Agreement restrict user charges to the full cost of providing airport services, and require parties to abstain from limiting the volume of traffic, or aircraft type, except as may be required for customs, technical and operational reasons in compliance with Article 15 of the Chicago Convention. As stated above, it is difficult to see how the obligations imposed by the EU ETS would amount to an airport charge. Moreover, as indicated by the Impact Assessment, the scheme is projected to have only a modest impact on service or aircraft types.[62]

In summary, therefore, there are sound arguments to support the position which sees the Proposal as consistent with the international legal framework and an effective means of dealing with the increasingly pressing problem of emissions from international aviation. At the same time, a letter written to the European Commission by several countries stating that the Proposal, if implemented, would violate international law and undercut international efforts to better manage the impact of aviation emissions[63] sends a stark warning that the building of international policy consensus around this issue ought to be prioritized if the ambitious goals set by the Proposal are not to be undermined by legal disputes and disagreements.

---

61 Hardeman, supra.

62 Impact Assessment, p. 9.

63 Letter to the European Commission and the European Member States from Australia, Canada, China, Japan, South Korea and the US, 6 April 2007, cited in D. Crespo and M. Crompton, 'The European Approach to Aviation and Emissions Trading', *The Air and Space Lawyer*, Vol. 21, Number 3, 2007.

7. CONCLUSION

This chapter has surveyed the key legal and policy issues arising from the EU Commission's proposal to include aviation emissions within the scope of the EU Emissions Trading Scheme. It has shown that, to a large extent, the Proposal has successfully avoided the pitfalls of the original EU ETS Directive, creating a framework for aviation emissions trading that ought to be more effective than the EU ETS in the first two phases of its operation. Indeed, a number of provisions included in the Proposal (such as harmonized allocation, reliance on auctioning and benchmarking, tighter allocation caps and others) have also been incorporated in the Proposal to amend the EU ETS Directive, and will almost certainly form the basis of any future expansion of the EU ETS. At the same time, the Proposal undeniably requires further analysis and development in key areas such as allocation, interplay with the Kyoto Protocol, inclusion of aviation-specific emissions and geographical scope. The issues that remain to be addressed in these areas arise largely from the characteristics of aviation as an economic activity, rather than stemming from inadequacies or limitations of the EU ETS as a market-based mechanism. Nevertheless, such issues must be resolved as a matter of priority if the Proposal is to achieve the ambitious goal it has set itself, namely, to minimize the greenhouse gas emissions from the aviation sector, and if it is to set a valuable precedent for future expansion of the EU ETS.

REFERENCES

Crespo, D. and M. Crompton (2007), 'The European Approach to Aviation and Emissions Trading', *The Air and Space Lawyer*, 21(3).

Dunseath, K. and R. Macrory (2007), 'Time to Act? Should the UK be Taxing Aviation Fuel?', *New Law Journal*, April.

Hansjürgens, B. (ed) (2005), *Emissions Trading for Climate Policy: US and European Perspectives*, Cambridge University Press, July.

Hardeman, A. (2006), 'A Common Approach to Aviation Emissions Trading', *Air & Space Law*, Vol. 32.

IPCC (1999), *Aviation and the Global Atmosphere*, Cambridge University Press.

Long S. and G. K. Salters (2007), 'The EU ETS – Latest Developments and the Way Forward', *Climate Change Law Review*, 1.

Robinson J. (2007), 'Climate Change Law: Emissions Trading in the EU and the UK', Cameron May.

Schwarze, G. (2007), 'Including Aviation into the European Union's Emissions Trading Scheme', *European Environmental Law Review*, January.

Skjaerseth J.B. and J. Wettestad (2008), 'EU Emissions Trading: Initiation, Decision-making and Implementation', Ashgate Publishing Group.

Stuart, G. and A. M. Fisher (2007), 'One World? International Aviation and the EU Emissions Trading Scheme', *Environmental Law and Management*, 19.

13. The European emissions trading system: auctions and their challenges

Stefan Weishaar

1. INTRODUCTION

On 25 October 2003, Directive 2003/87/EC governing the European Emissions Trading System (EU ETS) for greenhouse gas emission allowances for energy intensive installations entered into force. Thereby all Member States were obliged to establish an emissions trading scheme as of 1 January 2005. Around 5000 operators with approximately 12 000 installations participate in this multi-jurisdictional attempt to reduce CO_2 emissions from four broad sectors: energy, ferrous metals, minerals, pulp and paper.[1] The programme is implemented in multiple phases: the first ranging from 2005–2007; the second from 2008–2012, which resembles the Kyoto Protocol compliance period; and subsequent phases are consecutive five-year periods.[2]

The operation of the Emissions Trading System required the introduction of emission allowances into the market. Member States were obliged to draft so-called National Allocation Plans (NAPs) in order to indicate how to allocate the allowances to operators. During the first trading period 95% of all emission allowances had to be allocated free of charge and in the second trading phase this number only marginally increased to 90%.[3] In January 2008 the European Commission delivered a proposal to amend Directive 2003/87/EC in order to improve and to extend the greenhouse gas emission allowance trading system of the Community. At the heart of the proposed amendment lies a fundamental change of the prescribed allocation methods. While free allocation mechanisms, in particular grandfathering, have been employed by the Member States, auctioning is to become the most prominent allocation format in future trading periods.

---

1 See Directive 2003/87/EC of 13 October 2003 on the establishment of an Greenhouse Gas Emissions Trading System, [2003] OJ L 275/32, Annex I.

2 Article 11(1) and (2) Directive 2003/87/EC. The present Commission proposal extends the trading period to 8 years. See Article 13 COM (2008) 16 final of 23.01.2008.

3 Article 10 Directive 2003/87/EC.

This represents an important and crucial change in approach to allocations of tradable rights in the context of environmental law. Allocation formats such as, for example, grandfathering are more easily implemented politically since they are better able to protect the vested interests of existing industries and induce fewer direct financial costs upon them[4] than auction systems. Notable examples of auctions that have been successfully employed to reduce emissions, albeit to a varying degree, include the Acid Rain Programme in the US[5] and the Virginia NO_x auctioning system.[6] Auctioning systems for greenhouse gas emissions that have been employed in the European context prior to the introduction of the EU ETS were, however, circumventing the deficiency of stakeholder support by creating voluntary participation and essentially reversing the auction. In the UK Emissions Trading Scheme that ran from 2002 to 2006 the State offered incentive money to buy emission allowance reductions from undertakings.[7]

Following the successful implementation of large-scale auctioning in the US Regional Greenhouse Gas Initiative (RGGI)[8] the European Union has now proposed the introduction of auctions as well. This is remarkable, in particular since the lobbying resistance of European stake holders may be more intense than the resistance encountered for the setup of RGGI where the economic structure is not characterized by heavy industry. In light of an increasing number of international conferences in this field and an increased receptiveness of decision makers to use auction proceeds in order to enhance the energy efficiency of the economy and to increase overall social welfare, it appears that auctioning is an upcoming method.

In emphasizing auctions as the main allocation rule, the European Commission follows suggestions from scholars who have long been identifying the superiority of auctions with regard to allocative efficiency.[9] While in theory the method of auctioning appears to be the most appealing solution to the problem at hand, detailed consideration reveals that there are a number of issues that may limit the positive effects expected to stem from auctions. This chapter identifies challenges the legislator faces in implementing the currently proposed auctioning scheme without reviewing the multitude of auction mechanism designs that the European Commission could employ within the framework of the EU ETS and without identifying the specific legal aspects of an auctioning scheme within the EU legal order. It focuses on basic elements of

---

4 Cramton P. and Kerr, S. (2002).
5 Evans and Peck (2007), p. 20.
6 Ibid., (2007), p. 25.
7 See Defra, Enviros (2006), p. 28 following.
8 Bogdonoff, S. and Rubin, J. (2007).
9 Cramton, P. and Kerr, S. (1999).

auctioning tradable rights as being defined in economic theory, and following these basic insights, identifies the main challenges the European and/or national legislator faces in order to effectively implement auctioning within the framework of the EU ETS. In doing so it first presents relevant background information regarding the proposed amendment of the Directive (Section 2). Section 3 concisely introduces auctions in general and highlights their appealing properties. Section 4 examines the decisions the European and/or the national legislator will have to make and their implications and complexities for the effective operation of the EU ETS allowance market. A conclusion summarizes the main findings of the paper.

2. THE PROPOSED AMENDMENT

The proposed amendment of Directive 2003/87/EC[10] differs considerably from the current legislation. The present one uses free allocation – in particular the so-called grandfathering method has been employed by Member States – and only up to 10% of emission allowances were allowed to be sold or auctioned.[11] In contrast to this the default allocation rule of emission allowances proposed for the period post 2013 is auctioning with derogations based on benchmarking.

Pursuant to its proposal the European Commission shall adopt by 30 June 2011 Community-wide and fully harmonized allocation rules for free allocation relating to the electricity sector, new entrants, and sectors exposed to strong international competition and that give rise to a significant risk of carbon leakage.[12] The Commission will regularly review which sectors are exposed to a significant risk of carbon leakage, and will take issues such as substantial increases in production costs, emission reduction potentials, market structure, ability to pass on costs (price elasticities) and effects of climate change and energy policies into account.[13] Sectors other than electricity – electricity will in principle not benefit from any free allocation[14] – will be subject to ever more auctioning.[15] A transitory rule envisages a partially benchmark-based free allocation of 80% of the average emissions measured

---

[10] COM (2008) 16 final of 23.01.2008.

[11] Article 10(3) Directive 2003/87/EC.

[12] See recital 19 and Article 10(a)(1), COM (2008) 16 final of 23.01.2008.

[13] See Article 10(a)(8) and (9), COM (2008) 16 final of 23.01.2008. This is done at the latest by 30 June 2010 and then every 3 years.

[14] See the derogation for cogeneration installations, Article 10(a)(3), COM (2008) 16 final of 23.01.2008.

[15] Recital 16–19 of the Explanatory memorandum and Article 10(a)(3), COM (2008) 16 final of 23.01.2008.

during the period 2005–2007. The allocation under the transitional Community-wide rules for harmonized free allocation is reduced annually by equal amounts so that by 2020 free allocation is completely faded out in these sectors.[16]

In light of its independent EU commitment of reducing Member States' greenhouse gas emissions by at least 20% by 2020 compared to 1990[17] Article 9 of the proposed amendment stipulates a linear annual decrease of emission allowances. Beginning from the mid-point of the second allocation period the Community-wide quantity of allowances decreases by 1.74%. Envisaging an emission reduction of 21% below reported 2005 levels, the EU ETS contributes cost-effectively to the attainment of the EU's self-induced commitment.[18]

With regard to auctioning, the European Commission would like to be the key decision maker. It wants to oblige Member States to auction allowances that are not allocated free of charge[19] in accordance with its own procedures. The Commission wants to adopt a Regulation on the timing, administration and other aspects of auctioning by 31 December 2010 in order to ensure that it is conducted in an open, transparent and non-discriminatory way.[20] In particular the Commission wants to ensure that all operators do have full access and that participants cannot undermine the auction. Furthermore the Commission seeks to influence the distribution of allowances by obliging Member States to give up 10% of their total quantity of allowances to be auctioned so that they can be distributed for the purpose of Community growth and solidarity.[21] Besides this redistribution it proposes that Member States[22] should – not shall – use at least 20% of the auction proceeds for climate change related measures. These measures include contributions to a Global Energy Efficiency and Renewable Energy Fund, development of new energies, carbon storage and capture, avoidance of deforestation, adaptation strategies, mitigation of social climate change impacts and administrative expenses.[23] It may, however, be questioned whether Member States are indeed prepared to follow, in particular, this element of the Commission's proposal that appears to be closely related to such politically sensitive areas as taxation and fiscal policies. It

---

16 Recital 17 and 18, and Article 10(7) COM (2008) 16 final of 23.01.2008. The adaptation pressure for installations benefiting from free allocation is intensified through the annual reduction of the emission allowances issued under the EU ETS. See Article 9 COM (2008) 16 final of 23.01.2008.

17 Council of the European Union (2007), recital 32.

18 See COM (2008) 16 final of 23.01.2008, p. 7.

19 See Article 10(1) COM (2008) 16 final of 23.01.2008.

20 See Article 10(5), COM (2008) 16 final of 23.01.2008.

21 See Article 10(2)(b), COM (2008) 16 final of 23.01.2008.

22 Article 10(3), COM (2008) 16 final of 23.01.2008 employs the word 'should' not 'shall'.

23 Article 10(3) COM (2008) 16 final of 23.01.2008.

should be noticed in this context, that while the EC Treaty does not grant the Commission power to adopt harmonizing legislation in the area of taxation in the absence of unanimous consent of Member States,[24] the legal concept of a tax is more restrictive than its economic counterpart and that emissions trading in the EU ET may perhaps not qualify a a tax.[25]

The Commission proposal thus takes a highly centralized approach in which it seeks to determine the permissible auctioning rules and also which sectors are allowed to fall under special exemptions. These far-reaching plans turn the Commission into the core decision maker regarding the application and design of the EU ETS, whose powers even emanate into areas of national climate change as well as energy policy.

3. AUCTION FORMATS AND UNDERLYING ASSUMPTIONS

After having introduced important elements of the proposed amendment to the EU ETS Directive, this section presents relevant information in order to understand the working of auctions in general and within the EU ETS in particular. It thereby reviews what auctions actually are, why they are attractive as an allocation mechanism and which elements auction designers need to consider in order to fully realize its capabilities. Section 4 will build upon this section in order to examine which challenges the Commission faces.

Even though the term 'auction' is certainly familiar to everybody, it is still worthwhile to review how auctions are understood on a more conceptual level. An auction can be described as a set of rules that translates information revealed by bidders, by means of an allocation rule and a payment rule, into efficient outcomes. The challenge of auction theory is to develop auction rules which are tailored to the preferences of bidders in such a way as to provide 'Pareto optimal'[26] allocations. This means that the bidder that has the highest valuation for an emission allowance will be able to attain it through the working of an auctioning process. There is, however, no 'one size fits all' auction but the mechanism design has to be fine-tuned to the potential bidders to avoid losses to the auctioneer and/or the bidders.[27] Indeed the auction mechanism

---

24 See Article 93 EC Treaty.

25 Heij, G. (2001), 'The Definition of Tax', *Asia-Pacific Bulletin*, April, 74–79.

26 Pareto optimality describes situations in which it is impossible to make one person better off without making at least someone worse off. In the absence of side payments between bidders, Pareto efficient but sub-optimal allocations can occur.

27 If an auction mechanism is not tailored appropriately bidders may pay too much for the goods they wish to acquire or on the contrary behave in a non-competitive fashion and pay only very little.

design can give rise to auctions in which bidders compete so fiercely with each other that they end up paying such high prices that sizable amounts of money are 'left on the table'. Milgrom (2004) present a powerful example of the 1997 Brazil wireless telephone services auction in which a consortium, also including Bellsouth and Splice, won the licence for the Sao Paulo area. It paid for it with 2.45 billion US Dollars, about 60% more than any of its competitors had been bidding.

Auctions do not only differ with regard to allocation and payment rules but also with respect to the amount of information they require bidders to reveal. Before examining how this translates in practice into actual auctions, it is good to review why auctions are beneficial.

There are four reasons that make auctions attractive as an allocation mechanism. Firstly, an auction is designed to lead to self-revelation of the bidder's private values. In the presence of inherent information asymmetry, in which a potential seller is unable to determine the market value of a particular object, an auction mechanism can yield higher revenues than by simply quoting a price or by repeated negotiations with potential buyers. While this is very desirable from a theoretical point of view, it should be noted that bidders are generally reluctant to reveal their preferences because they fear that competitors could take advantage of them – protection of such information is crucial for firms. Secondly, auctions can be designed in such a way as to ensure allocative efficiency. It should be noted that efficiency here is to be understood as to award the bidder with the highest valuation for an object with the tender.[28] Thirdly, auctions legitimize transfers which would otherwise be suspect. Prior knowledge of the auction rules provides bidders with a transparent framework of how their bids will be assessed, while at the same time assuring bidders that selling agents have clear and indiscrete tender selection criteria.[29] Fourthly, since no time-consuming negotiation has to take place, auctions are fast allocation mechanisms, though it should be noted that the development of an auction mechanism depends on the object being auctioned as well as its potential market, and that it can be a non-trivial, time-consuming process.

There are numerous possibilities for the design of auctions. These models fall into several categories, or formats. After introducing four generic types of

28 Implicitly assuming away the possibility of credit rationed bidders. See Milgrom, P. (2004), p. 57. Maximization of social welfare, defined as the maximization of the sum of producer surplus and consumer surplus can be reached if side-payments (in the presence of budget balance constraints) are possible. In such cases Pareto optimal allocations are feasible in which one person is better off without someone else being worse off.

29 See Rothkopf, M. and Harstad, R. (1994), p. 368.

auctions, relevant dimensions influencing bidders' preferences will be reviewed. These are the number, the homogeneity and substitutability of the objects that are being auctioned and the market structure. Furthermore the information available to bidders, the independence of bidders' values, the distributive symmetry and the risk averseness of participants will be examined.

A standard auction is an auction in which the highest bidder among potential buyers, or the lowest bidder among potential sellers, wins. Since there is an almost perfect correspondence in results,[30] it is quite unimportant to distinguish between each form.[31] Standard auctions are commonly distinguished as 'open' and 'closed' auctions. In open auctions bidders are aware of their competitors' bids while in closed ones they are not. Two examples of open auctions are the ascending price auction, also called the English auction, and the descending price auction, also known as the Dutch auction. Two examples of closed auctions are the second-price sealed-bid auction, frequently referred to as the Vickrey auction,[32] and the first-price sealed-bid auction. The four standard auction types are presented in Table 13.1.

In an (open) ascending price (English) auction, the price is raised by the auctioneer or by bidders themselves until only one bidder remains. At any particular point in time bidders know the level of the current best bid.[32] Such auctions are often used by auction houses like Sotheby's. In the (open) descending price (Dutch) auction the price decreases continuously until one bidder accepts the current price. A well known example where (sequential) (open) descending price (Dutch) auctions are used is the flower auction in Aalsmeer (the Netherlands). In a closed sealed-bid auction bidders are only

Table 13.1 Standard auction types

Open auctions	Closed auctions
Ascending price auctions	
(English auction)	Second-price sealed bid auction
Descending price auction	(Vickrey auction)
(Dutch auction)	First-price sealed-bid auction

[30] With the possible exception of the invalidity of reserve prices and treating zero as an implicit limit to acceptable bids. Despite the intuitive appeal of the later argument, Shubik, M. (1983), p. 39ff., cites Herodotus reporting on Babylonian marriage markets which did include auctions starting at negative bidding values.

[31] Rothkopf, M. and Harstad R. (1994), p. 366.

[32] Named after Nobel laureate William Vickrey, who first presented this auction in his seminal paper on auctions. Vickrey, W. (1961).

[33] McAfee, P. and McMillan, J. (1987), p. 702.

allowed to enter one bid, thus they are unable to react *ex post* to their rivals. In the (closed) first-price sealed-bid auction the highest bid wins, while in the closed second-price sealed-bid (Vickrey) auction, the highest bidder is only required to pay a price equal to the second-highest bid.

Since the winning bidder alone determines the amount to be paid in (open) descending price (Dutch) auctions and in (closed) first-price sealed-bid auctions, both models are equivalent. Similarly, since the price the winning bidder has to pay is determined by the second-highest bidder, (open) ascending price (English) auctions and closed second-price sealed-bid (Vickrey) auction outcomes are equivalent.[34]

After having reviewed the above-mentioned standard auction types, elements influencing bidders' preferences are discussed. A good understanding of the products, market structure and bidders' preferences enables auctioneers to set behavioural incentives in such a way as to maximize their benefits. Adequately taking into consideration these factors will enable the European Commission to maximize allocative efficiency and internalize negative externalities without creating windfall profits. Windfall profits could, for instance, be created by auctions if bidders pay less for emission allowances than what they would be prepared to pay. If bidders were always motivated to pay amounts that reflect their maximum willingness to pay,[35] the likelihood of windfall profits would be mitigated.[36]

The number and nature of the goods that are being auctioned have important implications. Generally one distinguishes between single-unit auctions and multiple-unit auctions, which are designed to address bidder's preferences. In the former, the good is assumed to be indivisible, while in the latter, any number of goods can be auctioned.[37] Modelling of multiple-unit auctions or several sequential single-unit auctions is complicated by the introduction of two important dimensions which have a bearing on bidders' valuation and willingness to participate in a tender.

Goods can be complements or substitutes and they can be homogeneous or heterogeneous. If obtaining one good makes the bidder willing to pay more for a second good, the goods are complements, if the bidder is willing to pay less, they are substitutes. Single-unit auctions do not impair allocative efficiency

---

34 Wolfstetter, E. (1999), p. 187. The best strategy in these models is for bidders to bid their true value.

35 Economists call such prices 'indifference prices' as the benefit attained from holding the product equals its costs.

36 Though it should be noticed that the cost burden on bidder would naturally be higher and set European production sites at a comparative disadvantage vis-à-vis non-European production sites and non-covered sectors.

37 See Börgers, T. and Van Damme, E. (2004), pp. 28ff., for auction models addressing such issues.

considerations irrespective of goods being complements or substitutes. In the case of multiple-unit auctions, particularly if one allows for heterogeneous goods, concerns regarding allocative efficiency can be particularly severe if complements are auctioned in inefficiency-inducing packages.[38] Given the exponentially large number of possible allocations, auction algorithms and bidder strategies of models of multiple heterogeneous goods are very complex.

The Commission's legislative proposal extends the scope of the current Directive to greenhouse gases other than carbon dioxide.[39] The legislator's decision to employ the concept of carbon dioxide equivalents[40] in Directive 2003/87/EC appears to suggest that the various greenhouse gases that may be included within the framework of the EU ETS[41] will be integrated on the basis of their CO_2 emission reduction potential. This is particularly relevant in this context of heterogeneous goods since, from an auction-theoretical point of view, the establishment of a common denominator would suggest that those gases would be perceived by the market as functional substitutes to the extent that their conversion rate would be fixed. It is, however, noticeable that the issue of fixing conversion factors for greenhouse gases has not been addressed in the legislative proposal.

Besides the more structural elements discussed above, auctions can be designed to address such elements as information availability, independence, distributive symmetry and risk averseness, which will have a bearing on bidders' preferences. Each will be discussed in turn.

There are three different assumptions regarding information that is available to bidders. The literature distinguishes between private-value models, common-value models and correlated-value models. In private-value models bidders have knowledge, which is strictly private to them such as, for example, the price they would be willing to pay in a tender. Common-value models, in contrast, assume that the actual value is identical to all participants, but bidders do have diverging private information about this value.[42] It is assumed that bidders' values depend on other bidders' signals.[43] Correlated-value models assume that bidders' valuations are positively correlated and hence dependent upon each other.

---

[38] For a recent review see Milgrom, P. (2004), chapter 8.

[39] See explanatory memorandum at p. 4 and Annex I, COM (2008) 16 final of 23.01.2008.

[40] See for example Article 3(a), Directive 2003/87/EC.

[41] See recital 15, Directive 2003/87/EC.

[42] For example, bidders would have different geological predictions about gold reserves when bidding for mining rights, though all participants' valuation of the existing gold is uniformly determined by the quantity to be mined and the current market price.

[43] See Klemperer, P. (1999), pp. 6 ff., for an elaboration of hybrid models.

In much of the auction literature it is assumed that each bidder's private information is independent of other bidders' private information. Myerson (1981)[44] uses the term 'preference uncertainty' to describe private information that, even if known by other bidder's, would not induce them to reassess their personal utilities.[45] Information that would induce competitors to re-evaluate their own value is referred to as 'quality uncertainty'.[46] Assumptions about independence are important to anticipate bidder behaviour. If private information is correlated and auctioneers successfully induce bidders to reveal their preferences, information and beliefs, auctioneers are able to reap substantial parts of the bidder's surplus.[47]

A large body of auction theory literature assumes symmetry of beliefs. Buyers are modelled to have common underlying preferences and draw their information from a symmetric probability distribution.[48] Long after auctions addressed asymmetric information of bidders,[49] did scholars examine the complexity of equilibria under truly asymmetric circumstances.[50]

Another bidder preference, which auction models have to take into consideration, is the degree of risk aversion. Participants can be risk-taking, risk neutral or risk averse. Risk takers are willing to take unfair bets, risk neutral participants are prepared to take fair bets and risk averse bidders are unwilling to take fair bets.[51] Bidding systems with large variability in prices or uncer-

---

44 Myerson, R. (1981), p. 60.

45 Though, of course, they may well alter their bidding strategies. Matthews, S. (1995), p. 3, cites art auctions as an example where assumptions of independence may hold. Ashenfelter, O. (1989), p. 27, postulates that bidders do alter their valuations in art auctions once other bidder's valuation is known and that artwork which is not being sold at an auction can 'get burned' (i.e. substantially loses in value).

46 Myerson, R. (1981), p. 60.

47 See Myerson, R. (1981), pp. 70ff.

48 For examples of private-value models see Maskin, E. and Riley, J. (1984), Matthews, S. (1983), Milgrom, P. and Weber, R. (1982a), Riley, J. and Samuelson, W. (1981). Early examples of symmetric common-value models include Wilson, R. (1977).

49 Early examples of asymmetric common-value models include Milgrom, P. and Weber, R. (1982b), and Engelbrecht-Wiggans et al. (1983).

50 Maskin, E. and Riley, J. (2000a), and Maskin, E. and Riley, J. (2000b), construct models where bidders' independent and private values are drawn from heterogeneous distributions. Marshall, R. et al. (1994), p. 194, point out that even though Myerson R. (1981), introduced distributional heterogeneity in his revenue equivalence model, he did not establish that Nash equilibrium bidding strategies existed for such distributions.

51 For a model of risk averse sellers and a discussion of risk in independent-value models see Maskin, E. and Riley, J. (1984). For a discussion of risk in common-value models see Milgrom, P. and Weber, R. (1982a). For risk averse buyers see Matthews, S. (1983) and Matthews, S. (1987).

tainty about the actual price to be paid are evaluated negatively by risk averse participants.

In light of the above parameters that auction theory has identified as being important to tailor auction mechanism designs to the bidder population, it can be expected that realization of the superior allocation properties commonly attributed to auctions may not be a self-fulfilling prophecy.

4. DESIGN CHALLENGES WITH RESPECT TO AUCTIONING GREENHOUSE GAS ALLOWANCES

In order to examine the design challenges of auction mechanism and their implications, as faced by the European Commission, this section builds on the issues raised in the previous one. The implications of the eventuality that the European Commission does not provide a sufficient degree of harmonization in the design of its auctioning procedures is examined. In such cases Member State discretion could lead to the adoption of differing auction mechanism designs and auctioning schedules. It is, however, not the objective of this chapter to present a technical review of a wide array of auction mechanism designs that the Commission (or any legislator establishing a greenhouse gas auction scheme) could be selecting and to review their respective advantages and disadvantages. Instead of focusing on the specificities of auction mechanism design, this section will focus on the decisions to be taken when introducing an auctioning system and its associated implications and complexities.

The starting point for examining the difficulties of employing large-scale auctioning under the EU ETS is to define the product that is being auctioned. Under Directive 2003/87/EC each tradable allowance is defined as 1 ton of CO_2 equivalent over a designated period of time.[52] The Commission proposal also extends the trading scheme to other greenhouse gases with an equivalent global warming potential of 1 CO_2 ton.[53] Because the benefit derived by operators from attaining 1 ton of CO_2 equivalent is the same irrespective of the greenhouse gas they purchase, the good is clearly a substitute[54] and can thus be treated as a homogeneous, i.e. a similar good, from an auction design point of view.

Since participants to an emission trading auction will generally acquire larger quantities of allowances – particularly when one considers institutional

52 Defined in Article 3(a) of Directive 2003/87/EC.

53 See Annex II, COM (2008) 16 final of 23.01.2008.

54 If obtaining one good makes the bidder willing to pay more for a second good, the goods are complements, if the bidder is willing to pay less, they are substitutes.

investors such as brokers or banks[55] – multiple-unit auctions are of core interest. As in the case for single-unit auctions, auction designers strive to reach two goals that may well be assumed to be close to the objectives of the government. They try to ensure an efficient outcome and to maximize government revenues. It is, however, also entirely possible that the European Commission is not seeking to maximize auction revenues. Reasons for this can be that high auction prices lead to carbon leakage or to fierce resistance of stake holders. While the latter does not appear to be evidenced in the Commission proposal, the former consideration of carbon leakage associated with a higher cost burden of auctions seems to be reflected in Article 10(a)(9)(a) of the Proposal. The extent to which auctions lead to substantial increases in production costs and to loss of market share of carbon-efficient Community installations to less carbon-efficient non-community installations is expressly mentioned as a relevant factor to be considered for awarding free allocation and also suggested by the literature.[56]

Independent of the preferences the Commission may have, it has to be observed that surprisingly little is known about efficiency properties of multiple-unit auctions. A lot of the conventional wisdom comes by analogy from single-unit auctions[57] but a sound understanding of how equilibria respond when assumptions about values and information change has not yet been reached at any level of generality.[58] This dilemma may be exacerbated in the presence of an inherent uncertainty about the bidders and their preferences that characterizes the EU ETS market. In light of the severe over-allocation that characterized the first trading period, the ability of administrators to correctly estimate the underlying parameters should not be overestimated.

Related to the task to allocate a large number of emission allowances is the choice between two general multiple auction methods that could be used to allocate greenhouse gas emission allowances. Firstly, sequential auctions in which one allowance would be sold after the other and secondly, simultaneous auction models in which multiple greenhouse gas emission allowances are sold at the same time. In light of the long trading period (eight years) and the large number of emission allowances that has to be introduced into the market it can be expected that the Commission will employ a combination of both types. It may allow in its regulation the organization of repeated auctions in which larger and smaller bundles of allowances will be auctioned.

55 Any person within the Community is allowed to hold emission allowances. See Article 3(g) and 12, Directive 2003/87/EC as well as recital 29 of COM (2008) 16 final of 23.01.2008.

56 See Kuik, O. (2005), pp. 14ff.

57 Ausubel, L. and Cramton, P. (2002), p. 1.

58 See Börgers, T. and Van Damme, E. (2004), p. 43.

In this context the Commission, in its desired capacity as the 'drafting authority' for auctioning rules, and Member States as the relevant authorities who have to apply these rules, must consider three elements upon which also bidders' preferences are expected to have important influence. These are the bundling of the allowances, the timing of the auctions and the market structure; each will be addressed in turn.

Single-unit auctions do not impair allocative efficiency considerations irrespective of goods being complements or substitutes. In the case of multiple-unit auctions, particularly if one allows for heterogeneous goods, concerns regarding allocative efficiency can be particularly severe if complements are auctioned in inefficiency-inducing packages.[59] Given the exponentially large number of possible allocations, auction algorithms and bidder strategies of models of multiple heterogeneous goods are very complex. With regard to the EU ETS the theoretical complexity of bundling of emission allowances is limited to strike a balance between sufficiently small lots that allow a large number of natural or legal persons[60] to participate in auctions and increased costs from having to organize a large number of auctions. Cost-effectiveness will thus have to be balanced against the self-imposed Commission's duty to draft an auction regulation that allows small and medium sized enterprises covered by the Community scheme full market access.[61] While the Commission will have to take decisions in this area, it is not expected that this will be particularly problematic if sufficiently harmonized auctioning schemes are employed.

With regard to the second issue, the timing of auctions, it can be stated that sequential auctions are in general easy to implement for auctioneers since they merely have to repeat the (electronic) auction over and over again without adaptations of the auction design mechanism. Such auctions can be organized consecutively or with identical or dissimilar time periods between them. One could imagine for example that auctions could be organized every day, or during the course of the last week of every month or that auctions are in particular concentrated in the months prior to the submission of emission allowances. Despite the simplicity of organizing (electronic) repeat auctions for auctioneers, sequential auctions are, however, not very much favoured by bidders. One reason is the strategic complexity of bidding decisions and the inevitable price variation of sequential auctions of homogeneous goods.[62]

---

59 For a recent review see Milgrom, P. (2004), chapter 8.

60 Any person within the Community is allowed to hold emission allowances. See Article 3(g) and 12, Directive 2003/87/EC as well as recital 29 of COM (2008) 16 final of 23.01.2008.

61 See Article 10(5), COM (2008) 16 final of 23.01.2008.

62 Inefficiencies from synergies and complementariness of goods appear to be smaller if goods are homogeneous.

Ashenfelter (1989) has termed this 'declining price anomaly'[63] and explained a falling price in subsequent auctions[64] in terms of bidders' necessity to acquire particular quantities, risk aversion[65] and uncertainty.[66] In order to allow for a well-placed scheduling of auctions during the eight-year trading period starting in 2013, the European Commission will have to estimate the degree of risk aversion of the various bidders and the degree to which parties' information on the actual EU ETS markets differs. It could for example be expected that the large energy producers, as prominent actors on the EU ETS market, will have a better understanding of the market dynamics than will a medium-sized paper producer or a small installation in any other sector. In such situations prices paid at auctions could differ considerably.

From an allocative efficiency point of view, auctions ensuring a single market price, as for example uniform-price auctions, are preferred over sequential auctions because they mitigate the 'price risk' of paying too much for the same good.[67] Yet this does not appear to be an option that is easily reconciled with the Commission's objective to allow new entrants access to the EU ETS market. According to the current proposal new entrants may benefit from free allocation under the transitional Community-wide rules.[68]

Also related to the difficulty of measuring bidders' risk aversion and associated price anomaly are the varying amounts of revenues that are paid per emission allowance. The amount of revenues does not only depend on the type of auctioning mechanisms that the Commission would be permitting under the Regulation that it would like to adopt by 31 December 2010, but also crucially on the timing of the auction. If the Commission decided that Member States were allowed to auction at differing points in time, average revenues could differ considerably. Similarly, if the Commission were to prescribe for each Member State a particular point in time for it to organize an auction, the Commission would have to find a way to reimburse Member States in such a way as to pay an equal amount per ton of CO_2 emission allowances to each or face likely criticism from the Member States.

Two issues that are very closely linked to both the amount of revenue generation and the decision the Commission has to take to provide for centrally publicized auctions or fully decentralized ones are market structure

63 Ashenfelter, O. (1989), pp. 29ff.

64 See Ashenfelter, O. and Graddy, K. (2002), pp. 34–36 and Table 8 for a review of subsequent research.

65 Risk aversion implies that bidders dislike taking fair bids. McAfee, R. and Vincent, D. (1993) show that risk aversion can create declining prices.

66 See Neugebauer, T. and Pezanis-Christou, P. (2005).

67 Milgrom, P. (2004), p. 256.

68 See Article 10(a)(6), COM (2008) 16 final of 23.01.2008.

and transparency. In the EU ETS market for emission allowances large emitters appear to be relatively strong in terms of market size. The largest 7% of installations account for 60% of the total emissions while the 1400 smallest installations only account for 0.14%.[69] To the extent that these large installations belong to a limited number of undertakings or that their operators are particularly active in auctions because they are continuously adjusting their stocks to their predicted needs – such as energy producers may be doing – such operators may become influential. They might even be able to recognize their interdependence and thereby set small and medium sized bidders at a comparative disadvantage in the sense that large operators pay a relatively lower price than they otherwise would have been paying.

If a few undertakings would recognize their interdependence this may give rise to inefficiency[70] inducing 'demand reduction' strategies.[71] In multiple-unit auctions dominant players recognize the interdependence of their bidding strategy and competitors' bidding behaviour. A strategy of self-restricting the quantity demanded while bidding the minimum price to indicate interest in a number of units can generate large consumer surpluses. The inherent inefficiency stems from the fact that users with the highest value for a good do in fact prefer not to attain it: large bidders win too little and small bidders win too much. Salmon (2003)[72] points out that if such behaviour is strictly unilateral, this does not amount to collusion. However, if it does involve strategic considerations exemplified by trigger strategies, such behaviour would amount to (tacit) collusion.[73]

Bidders' ability to use backward induction[74] to limit their capital investments will depend on the degree of interdependence that they are recognizing they can employ. If bidders fail to immediately reach a low price outcome, signalling[75] can be employed to 'negotiate' a mutually acceptable allocation.

---

69 COM (2008) 16 final of 23.01.2008, p. 5.

70 Inefficiency is created by 'differential bid shading', i.e. when bidders with identical marginal values reduce their bids by different amounts so that awarding the bidder who values the item most is impossible. See Ausubel, L. and Cramton, P. (2002), p. 4.

71 For examples see Weber, R. (1997), and Ausubel, L. and Cramton, P. (2002).

72 Salmon, T. (2003), p. 5.

73 Distinguishing between tacit collusion and pure strategic firm behaviour is complicated if not impossible. In a multiple-unit auction both bidders could, for example, independently decide to pursue a 'demand reducing' strategy. The outcome would be identical to tacit and indeed outright collusion.

74 Bidders are assumed to imagine how the auction will be developing and to derive from this a mutually acceptable offer at the beginning of the auction.

75 By for example using the financially inconsequential digits of their bids, parties can signal their identity or indicate the market for which they are retaliating.

In a simultaneous auction environment with a limited number of participants and known limit prices of fringe firms, Grimm, Riedel and Wolfstetter (2001)[76] cite the German GSM Spectrum Auctions as a powerful example of how effectively signalling can be used to reach an almost immediate mutually acceptable strategic demand reduction. While signalling and demand reduction certainly are difficult points of some (multiple-unit) auctions, there is one factor[77] that is relevant in the context of the EU ETS that complicates effective collusion: a large number of bidders.[78] In light of the above, it appears desirable to not only limit the information that bidders are able to exchange through the auctioning process but also to allow for processes that enable a large number of bidders to participate in the auction. Both transparency and sufficiently low transaction costs can be expected to have a positive impact on the number of bidders that can credibly be expected to participate in an auction.

In setting up a Community-wide Regulation on the application and implementation of auctions, the Commission also has to consider how it ensures market transparency. Enhanced transparency can influence the amount of auction proceeds that depend on the degree of actual and potential competition. Bidders may be more willing to fiercely compete with each other if they cannot determine the identity and the quantity of bidders during or before the auction. If participation in an auction does not entail high transaction costs, such as for example expenses for designing a bidding strategy, the wide publication of an auction can effectively attract a larger number of potential bidders, including small and medium sized undertakings. This in turn will affect the bidding behaviour of participants. Since transparency will only lead to behavioural change if costs of participation in an auction are low enough, and also in particular to small and medium sized undertakings, the Commission has to weigh transparency and transaction cost considerations against benefits stemming from mechanism design variety. In particular in decentralized auctions in which auctioneers can more easily determine bidder preferences and fine-tune the auction mechanism design to the expected bidders, auctions may be expected to realize their full potential.

---

[76] Grimm, V. et al. (2001).

[77] Other factors that are relevant particularly in the context of heterogeneous products are synergies and externalities. For an analysis of externalities in single-unit auctions, the interested reader is referred to Caillaud, B. and Jehiel, P. (1998).

[78] Brusco, S. and Lopomo, G. (1999) show that the collusion becomes more difficult as the number of bidders relative to the number of items rises. Weber, R. (1997) presents a vivid example of how difficult it can be to reach a mutually acceptable allocation of heterogeneous permits when the number of participants is large.

There is thus a trade off between the Commission prescribing a 'one size fits all' fully harmonized and community-wide publicized auction system through its regulation and granting Member States discretion to tailor auction mechanisms to the assumed preferences of the expected bidders that sets different behavioural incentives. It is in the context of tailor-made solutions in which auctions are developing their full potential. Besides this important consideration it should be borne in mind that the Commission's regulation on auctioning has to be compatible with the principle of subsidiarity and that the regulation may bring the Commission in conflict with the conferral of power principle.

5. CONCLUSION

The Commission proposal for amending Directive 2003/87/EC attributes a strong role to auctioning. It is to become the core allocation format to bring emission allowances onto the market. Auctions are not only expected to successfully address the problems of windfall profits and new entrants but also lead to the generation of auction revenues. They shall be used to enhance economic growth and solidarity and – according to the wishes of the Commission – also used for environmental purposes. While the Commission has given itself time until 31 December 2010 to adopt a regulation on the timing, administration and other aspects of auctioning to ensure that it is conducted in an open, transparent and non-discriminatory way, a number of problematic issues relating to the introduction of large-scale auctioning in the EU ETS can be identified.

The design of auction mechanisms takes relevant properties of the bidders into account in order to design efficient auctions that yield the desired objectives. Auction theory is best developed with regard to single-unit objects that are sold, and insights that are gained may not be directly transferable to multiple-unit auctions as likely to be employed under the EU ETS. Independent of any theoretical challenges that may still have to be addressed, it appears clear that the European Commission will have to make assumptions about bidders' characteristics and preferences. Given the inherent uncertainty relating to the characteristics of the market operators, it can be expected to be difficult for the Commission to design an optimal auctioning system, and that the realization of the superior allocation properties commonly attributed to auctions may not be a self-fulfilling prophecy.

In particular the Commission will have to decide whether it wishes to set up regulation to create strongly harmonized and simple auctioning rules that reduce costs associated with comprehending auctioning mechanisms and limited strategic complexity or whether it wants to grant Member States a larger degree of discretion. A very substantial harmonization of auctioning rules can reduce transaction costs sufficiently and allow for a large number of bidders to be credibly expected to participate in any given auction. This in turn will enable in particular small bidders to be better able to engage in auctions without being confined to conduct standard market operations via exchange houses or other traders. In contrast to a strong degree of harmonization, the decentralized approach allows for more appropriate auction mechanism designs and to realize the full potential of auctions. It also allows Member States to raise the funds they wish to raise through auctioning and to avoid problems related to subsidiarity and competences of the Commission. Depending on the degree of discretion afforded to Member States and the diversity of auction mechanism designs that would actually be employed, it could effectively exclude small firms from participating in certain auctions without incurring substantial costs. Whether this is decisive does, however, also depend on the frequency with which small installations are expected to participate in auctions. Operators who are already weary about participating in the EU ETS market may be even more reserved when it comes to participating in auctions. Such operators may then predominantly rely on secondary markets.

REFERENCES

Ashenfelter, O. (1989), 'How Auctions Work for Wine and Art', *The Journal of Economic Perspectives*, Vol. 3, No. 3, pp. 23–36.

Ashenfelter, O. and K. Graddy, (2002), 'Art Auctions: A Survey of Empirical Studies', *NBER Working Paper Series*, Working Paper No. 8997, pp. 2–43.

Ausubel, L. and P. Cramton (2002), *Demand Reduction and Inefficiency in Multi-Unit Auctions*, University of Maryland, 12 July 2002, pp. 1–30.

Bogdonoff, S. and J. Rubin (2007), 'The Regional Greenhouse Gas Initiative, Taking Action in Maine', *Environment*, Vol. 49, No. 2, pp. 9–16.

Börgers, T. and E. Van Damme (2004), 'Auction Theory for Auction Design', in Janssen, M. C. W. (eds.) *Auctioning Public Assets, Analysis and Alternatives*, Cambridge University Press, pp. 19–63.

Brusco, S. and G. Lopomo (1999), *Collusion via Signalling in Open Ascending Auctions with Multiple Objects and Complementarities*, New York University, Leonard N. Stern School of Business, Department of Economics Working Paper Series, 99-05, pp. 1–27.

Caillaud, B. and P. Jehiel, (1998), 'Collusion in Auctions with Externalities', *The RAND Journal of Economics*, Vol. 29, No. 4, pp. 680–702.

COM (2000) 87, Commission Communication of 8 March 2000, Green Paper on greenhouse gas emissions trading within the European Union, COM (2000) 87 final, p. 28.

COM (2001) 579 final, *Proposal for a Council Decision concerning the approval, on behalf of the European Community, of the Kyoto Protocol to the United Nations Framework Convention on Climate Change and the joint fulfilment of commitments there under*, OJ 075 E, 26/03/2002 P. 0017 – 0032.

COM (2008) 16 final, *Proposal for a Directive of the European Parliament and of the Council amending Directive 2003/87/EC so as to improve and extend the greenhouse gas emission allowance trading system of the Community*, p. 51, Brussels, 23.01.2008.

Council of the European Union (2007), Presidency conclusions of the Brussels European Council, 8/9 March 2007, 7224/1/07 REV 1 CONCL 1, Brussles, 2 May 2007, available at http://register.consilium.europa.eu/pdf/en/07/st07/st07224-re01.en07.pdf.

Cramton, P. and S. Kerr (1999), *Tradable Carbon Permit Auctions, How and Why to Auction not Grandfather*, Wharton, Financial Institutions Center, Implications of Auction Theory for New Issues Markets, No. 02-19, pp. 1–19.

Cramton, P. and S. Kerr (2002), *Tradable Carbon Permit Auctions, How and Why to Auction not Grandfather*, Energy Policy, Vol. 30, pp. 333–45.

Defra, Enviros (2006), 'Appraisal of Years 1–4 of the UK Emissions Trading Scheme', available at http://www.defra.gov.uk/Environment/climatechange/trading/uk/pdf/ukets1-4yr-appraisal.pdf.

Engelbrecht-Wiggans, R., P. Milgrom and R. Weber (1983), 'Competitive Bidding and Proprietary Information', *Journal of Mathematical Economy*, Vol. 11, pp. 161–69.

Evans & Peck (2007), *National Emissions Trading Taskforce, Possible Design for a Greenhouse Gas Trading System*, p. 74.

Grimm, V., F. Riedel and E. Wolfstetter (2001), *Low Price Equilibrium in Multi-Unit Auctions: The GSM Spectrum Auction in Germany*, Institut fuer Wirtschaftstheorie I, Humboldt Universitaet zu Berlin, p. 18.

Klemperer, P. (1999), 'Auction Theory: A Guide to the Literature', Introductory chapter to *The Economic Theory of Auctions*, P. Klemperer (ed.), Cheltenham, UK, Edward Elgar, 2000. (Also published in *Journal of Economic Surveys*, 1999, and reprinted in *The Current State of Economic Science* S. Dahiya (ed.), 1999.) Paul Klemperer, see http://www.nuff.ox.ac.uk/users/klemperer/papers.html, p. 94.

Kuik, O. (2005), *Climate change policies, international trade and carbon leakage: An applied general equilibrium analysis*, dissertation, Vrije Universiteit Amsterdam, p. 192.

Marshall, R., M. Meurer, J.F. Richard and W. Stromquist (1994), 'Numerical Analysis of Asymmetric First Price Auctions', *Games and Economic Behaviour*, Vol. 7, pp. 193–220.

Maskin, E. and J. Riley (1984), 'Optimal Auctions with Risk Averse Buyers', *Econometrica*, Vol. 52, pp. 1473–1518.

Maskin, E. and J. Riley (2000a), 'Asymmetric Auctions', *The Review of Economic Studies*, Vol. 67, No. 3, pp. 413–38.

Maskin, E. and J. Riley (2000b), 'Equilibrium in Sealed High Bid Auctions', *The Review of Economic Studies*, Vol. 67, No. 3, pp. 439–54.

Matthews, S. (1983), 'Selling to Risk Averse Buyers with Unobservable Tastes', *Journal of Economic Theory*, Vol. 30, pp. 370–400.

Matthews, S. (1987), 'Comparing Auctions for Risk Averse Buyers: A Buyer's Point of View', *Econometrica*, Vol. 55, No. 3, pp. 633–46.

McAfee, R. and J. McMillan (1987), 'Auctions and Bidding', *Journal of Economic Literature*, Vol. 25, pp. 699–738.

McAfee, R. and D. Vincent (1993), 'The Declining Price Anomaly', *Journal of Economic Theory*, Vol. 60, pp. 191–212.

Milgrom, P. (2004), *Putting Auction Theory to Work*, Cambridge University Press, p. 368.

Milgrom P. and R. Weber (1982a), 'A Theory of Auctions and Competitive Bidding', *Econometrica*, Vol. 50, No. 5, pp. 1089–1122.
Milgrom, P. and R. Weber (1982b), 'The Value of Information in a Sealed-Bid Auction', *Journal of Mathematical Economy*, Vol. 10, pp. 105–14.
Myerson R. (1981), 'Optimal Auction Design', *Mathematics of Operations Research*, Vol. 6, No. 1, pp. 58–73.
Neugebauer, T. and P. Pezanis-Christou (2005), *Bidding Behavior at Sequential First-Price Auctions with(out) Supply Uncertainty: A Laboratory Analysis*, 13 January 2005, p. 23, republished in *Journal of Economic Behavior & Organization*, 2007, Vol. 63, issue 1, pp. 55-72.
Peeters, M. (2003), 'Internationale klimaatafspraken en Europese emissiehandel. Een onwaarschijnlijk kort implementatietraject voor een niew fenomeen' 10 *Nederlands tijdschrift voor Europees recht*, p. 280.
Peeters, M., J. De Cendra De Larragán and S. Weishaar (2007), 'Perspectives on the Fundamental Choice between "Cap and Trade" and "Credit and Trade"', *EELR*, Vol. 16, No. 7, p. 191.
Riley, J. and W. Samuelson (1981), 'Optimal Auctions', *The American Economic Review*, Vol. 71, No. 3, pp. 381–92.
Rothkopf, M. and R. Harstad (1994), 'Modeling Competitive Bidding: A Critical Essay', *Management Science*, Vol. 40, No. 3, pp. 364–84.
Salmon, T. (2003), *Preventing Collusion between Firms in Auctions, Department of Economics* (5 February), Florida State University, pp. 1–25.
Shubik, M. (1983), 'Auctions, bidding, and markets: An historical sketch', in Engelbrecht-Wiggans, R., Shubik, M. and Stark, J., eds., *Auctions, Bidding, and Contracting*, New York, New York University Press, pp. 33–52.
Vickrey, W. (1961), 'Counter-speculation, Auctions, and Competitive Sealed Tenders', *The Journal of Finance*, Vol. 16, No. 1, pp. 8–37.
Weber, R. (1997), 'Making More from Less: Strategic Demand Reduction in the FCC Spectrum Auctions', *Journal of Economics & Management Strategy*, Vol. 6, No. 3, pp. 529–48.
Weishaar, S. (2007), 'The EU ETS: Current Problems and Possible Ways to Move Forward', invited paper prepared for the climate change conference 'Comparing North American and European approaches to climate change', held by the Viessmann European Research Centre, Wilfrid Laurier University, Waterloo, Canada, available at http://www.wlu.ca/viessmann/html_pages/Climate.htm.
Wilson, R. (1977), 'A Bidding Model of Perfect Competition', *Review of Economic Studies*, Vol. 11, pp. 511–18.
Wolfstetter, E. (1999), *Topics in Microeconomics: Industrial Organization, Auctions, and Incentives*, Cambridge University Press, p. 370.

PART IV

Conclusions: Future look

14. Concluding remarks

Michael Faure and Marjan Peeters

1. THE EU ETS AT THE CORE

Emissions trading is probably the topic best suited for a combination of a legal and economic analysis. At this time, we have a fascinating opportunity to review practical applications of the emissions trading instrument. Indeed, after many environmental economists have already for a long time been advocating emissions trading as an efficient system for achieving an internalization of externalities caused by environmental pollution, some interesting large-scale experiments have finally taken place. These occurred first in the US, notably to deal with a range of air pollution problems, and later in Europe through the establishment of an EU-wide greenhouse gas emissions trading scheme. Most of the contributions to this book are devoted to Directive 2003/87/EC of 13 October 2003 establishing a scheme for greenhouse gas emission allowance trading, commonly referred to as the EU ETS. However, most of the contributions provide a look to the future by discussing the proposal of the European Commission delivered on 23 January 2008 for a major revision of this scheme (COM(2008)16).

Emissions trading seems particularly fit for the climate change problem, especially for the greenhouse gases that lack local effects as is the case with carbon dioxide. After some initial experiments or initiatives at the national level, notably in the UK (see the contribution of MacDonald and Makuch in chapter 10), Denmark and The Netherlands, Europe decided to implement its Kyoto obligations via an EU-wide emissions trading scheme, thereby preventing the emergence of different national greenhouse gas emissions trading applications. The idea of using emissions trading had already been accepted at the international level through the Kyoto Protocol, specifically in the form of three different options: Interstate emissions trading, which is a cap-and-trade mechanism, and two project-based trading options with basically a credit-and-trade design, which are the Clean Development Mechanism and Joint Implementation. However, the design of the instrument within a domestic or regional context needs to be done by taking account of the specific legal culture and legal framework. In this vein, the choice for and design of greenhouse gas emissions trading within the EU also needs to be understood particularly within the context of the competences as being provided by the EC Treaty. One very legalistic circumstance is that the classic instrument to internalize environmental externalities (the Pigovian tax) could only be adopted within the European Community through a unanimous position by the Council, which is different for emissions trading with free allocation as, then, only a qualified majority is required (Rodi, 2005). Moreover, the emissions trading instrument did in fact not come as a surprise since emissions trading with grandfathering obviously serves the interests of industry better than costly taxation measures. To a large extent, the current shape of the emissions trading directive can thus be considered as the result of effective lobbying by industry.

Even the strongest opponents of the so-called law and economics approach have probably realized that when a typical 'economic' instrument like emissions trading is introduced in legislation, economic analysis of law is the obvious tool to examine whether the particular legal rules chosen to implement this economic instrument are in fact effective to achieve the economic goals set. Hence many contributors to the book have, in some cases implicitly, adopted an economic approach to emissions trading in order to analyse several aspects of the directive.

On the other hand a careful analysis of the EU ETS is not only an invitation to lawyers to take economic analysis seriously. Many contributions to this book equally make clear that the economic efficiency and effectiveness of an emissions trading scheme may well to a large extent depend on what some non-lawyers would refer to as 'legal details'. It is more particularly these 'legal details' or, to put it more politely, legal design issues, that will have an important bearing upon the economic effects of an emission trading scheme. Hence we hope to have convinced those interested in climate change instruments, but traditionally disregarding legal issues, that the legal design of a particular (also economic) instrument is of course crucial for the effectiveness of that particular policy tool.[1]

Of course with this book we did not attempt to provide a traditional handbook which would comment on every possible legalistic aspect of the EU ETS. We have rather attempted to bring together lawyers and economists from different jurisdictions within and outside Europe to look at various major questions regarding the European emissions trading scheme, but also more generally regarding the effectiveness of emissions trading as such. As a result many aspects were unavoidably, even in a rather lengthy book like this, not discussed. For example we did not discuss at length

1 Also T.H. Tietenberg refers to the importance of the design issues regarding the effectiveness of the emissions trading instrument, and to the fact that the practical experiments, concluded through political negotiations, differ from the theoretical models, see his concluding remarks in Tietenberg (2006).

particular legal aspects of how trading between different partners exactly takes place and which different contracts might be used. Nevertheless we have the impression that the chapters brought together in this book permit a few generalizations. In these concluding remarks we would like to attempt to look at a few similarities and differences in the research results of the various chapters, thereby of course focusing on the main ideas. Striking issues on which we would like to focus for the simple reason that they are discussed in many contributions include *inter alia* the importance of the legal design of an emissions trading scheme, the effectiveness of the EU ETS, proposals for improvement and the well-known subsidiarity debate, meaning to what extent a further centralization (more power to Europe) or decentralization (towards the member states) is indicated in this area of emissions trading.

2. LEGAL DESIGN: ALLOCATION METHOD

Many contributors stress that a first crucial design issue of an emissions trading scheme is of course how allowances are going to be allocated. Economic literature has always stressed the superiority of auctioning as an allocation mechanism for the simple reason that one then can avoid the tendency of industry to simply inflate its own emissions projections in order to maximize the number of free allowances it will receive. Under auctioning, these distorting effects will not take place and polluting enterprises will calculate their abatement costs as accurately as they can (Baldwin, 2008, p. 11). The European Commission, however, initially clearly chose not to follow this advice from economists and laid down the rule in article 10 of the original directive that for the three-year period beginning 1 January 2005 member states shall allocate at least 95% of the allowances free of charge. For the second period (starting 1 January 2008 for five years) this free allocation had to be at least 90%.

This free allocation (in the literature also referred to as grandfathering) has given rise to a number of interesting legal and economic questions addressed in many of the chapters in this book. One obvious question is whether giving away the allocations for free is not violating the polluter-pays principle. In a well-known paper Jonathan Nash had argued that grandfathering does violate the polluter-pays principle (Nash, 2000). De Cendra De Larragán holds in chapter 3 that free allocation, if coupled with updating of historical emissions, violates the polluter-pays principle, since it provides an incentive to companies to increase their emissions in order to receive more allowances. A different line of argument is, however, presented in chapter 5 by Woerdman, Clò and Arcuri. They first of all make a distinction between an efficiency interpretation on the one hand and an equity interpretation on the other hand. As far as the first criterion is concerned they hold that it is not correct to argue that firms with grandfathered emission rights would not have any costs. They argue that those firms have an opportunity cost which is equal to the price at which the allowance can be sold. Instead of using allowances to cover the emissions the firm could have sold emission rights. Hence they argue that firms 'must' pass on these opportunity costs in the product price. This 'obligation' is not a legal obligation, though an economic rationale. Their point of view is also followed in chapter 8 by Kuik and Oosterhuis, who show that there equally has been empirical evidence that many power producers did indeed pass on the opportunity costs of their allowances to consumers. Woerdman, Clò and Arcuri do, however, argue that whereas free allocation as such does not violate the polluter-pays-principle, the conclusion is different for the over-allocation which has taken place for the ETS sectors in the EU. Their straightforward economic reasoning, supported with empirical evidence, is that this over-allocation has led to a near to zero price of allowances as a result of which no cost internalization has taken place. Thus they argue that the free allocation as such did not, but over-allocation did, violate the polluter-pays principle.

An equally interesting legal aspect resulting from the free allocation of allowances is whether this constitutes state aid. Again the contributions show that a straightforward answer to this question is not easy to give. De Cendra de Larragán shows in chapter 3 that on the basis of the community guidelines on state aid for environmental protection an ETS can lead to state aid when a member state grants allowances to companies below their market value. He equally shows that although the Commission considers that free allocation is state aid, the academic literature is much more divided on the issue. Chapter 6 by Weishaar is to a large extent devoted to this question. He concludes, on the basis of a careful examination of the state aid criteria, that grandfathering systems can be liable to constitute state aid within the meaning of article 87(1) of the EC Treaty. However, he equally argues that whether this is the case depends upon a number of characteristics of the particular market on which the specific entity is operating. General answers therefore remain dangerous. Such a specific examination of the relationship between the directive and state aid was also put forward in the *EnBW Energy Baden Württemberg v Commission* case before the European Court of First Instance, discussed in chapter 4 by Van Aken. The court, however, did not review the claim on the merits as it stated that the Commission does not take a formal state aid decision when deciding about a national allocation plan.

The Commission proposed, however, that auctioning will replace free allocation as the principal allocation method, to start in 2013 with the power sector, gradually including other sectors.[2] Both Peeters in chapter 2 and Weishaar in chapter 13 discuss the consequences of moving to auctioning as a main allocation mechanism, since this equally leads to interesting and important design questions. Weishaar identifies a number of problematic issues relating to the introduction of large-scale auctioning in the EU ETS. He argues that the European Commission will have to make certain complicated assumptions about bidders' characteristics and preferences. Given the inherent uncertainty relating to the characteristics of market operators, difficulties for the Commission in designing an optimal auctioning system can be expected.

Moreover, following the proposal from the Commission the sectors and sub-sectors facing significant international competition will be subject to another approach, in order to avoid carbon leakage from the EU. Here, important questions appear, to begin with, the question of how the determination of these exposed sectors will be done, and, subsequently, what the alternative approach will look like.

In addition, many discuss yet another allocation mechanism than free allocation or auctioning, which both fall within the traditional 'cap-and-trade' approach. An alternative system, discussed in detail in chapter 2 by Peeters, is referred to as 'credit-and-trade', sometimes also called 'benchmark and trade' or 'performance standard rate trading' (PSR). In this system allowances take place on the basis of emissions per unit of production, or per unit of fuel. If industry produces fewer emissions than indicated by the relative standard it can sell these credits (or when allowed reserve these); when industry exceeds the relative standard it has to obtain additional allowances. As Peeters shows, the major advantage of PSR is that it is more attractive within carbon policies where a threat of so-called carbon leakage (on which more below) would be a real concern. In addition, she discusses a proposal that originated from the private sector, which entails a combination of credit and trade with a cap. This would be a new design option, which of course needs to be considered carefully, especially regarding the question of whether the cap indeed ultimately will be ensured. Another interesting view on credit and trade is provided by Weishaar, who powerfully shows in chapter 6 that, differently from a grandfathering system, PSR systems are not liable to constitute state aid. Weishaar therefore criticizes the European Commission, which held

2 Of course the reader has to be careful here: we had to rely on the draft of a revised directive as proposed by the European Commission in the beginning of 2008. Some later changes which we could no longer consider in the publication phase of this book may thus well have occurred.

that the Dutch NO_X system (as an example of the PSR system) constitutes state aid. He argues that it can be questioned if state resources are involved in a PSR system and that this would constitute state aid. The Court of First Instance has recently confirmed the view that this Dutch PSR system does not constitute state aid.

3. LEGAL DESIGN: ROLE OF COURTS AND PRINCIPLES

A crucial issue in any environmental policy introducing new instruments is how a fair balance can be struck between, on the one hand, the social welfare goal of internalizing the externality caused by CO_2 emissions and, on the other hand, the individual rights of companies involved not to be overburdened as a result of the implementation of a new instrument in an unreasonable way. In this respect the question always arises whether the judiciary is able to fulfil the important task of providing the necessary legal protection to the regulated in case administrative authorities pursue policy goals in an unreasonable way in individual cases. This, of course, also raises an important question with respect to the relationship between the regulatory authorities executing the ETS on the one hand (more particularly the European Commission and national authorities deciding on national allocation plans) and the judiciary on the other hand. A question that more particularly plays in this respect is what legal rules in addition obviously to the text of the directive and guidelines issued by the Commission can constitute the basis for addressing these issues by the judiciary. De Cendra De Larragán shows in chapter 3 that to some important extent legal principles like the polluter-pays principle (also discussed by Woerdman, Clò and Arcuri in chapter 5), equality and other principles can play an important role, for example, in critically reviewing specific aspects of the emissions trading scheme (like the setting of the cap, scope of the ETS and allocation rules).

A particular characteristic to be taken into account when discussing the EU ETS is of course that the role of the court system is totally different at the national level than at the EU level. Many contributors, more particularly Backes, Deketelaere, Peeters and Schurmans addressing *ex post* adjustments in chapter 7 discuss some national case law in member states showing how the judiciary attempts to provide (with modest success) protection to enterprises who argue that decisions of national authorities (for example with respect to national allocation plans) may violate legal rules, principles and interests in the individual case. The task of the European Court of Justice is, as is for example shown in the detailed discussion of the case law by Van Aken in chapter 4, obviously different since the ECJ reflects on the task of what necessarily should be regulated at the EC level and what could be left to the discretion of member states. This as such quite policy-oriented principle of subsidiarity (further interpreted by De Cendra De Larragán in chapter 3) is the main criterion for deciding whether the particular division of labour between the EU level and the member states in this case of emissions trading would be justified. Within EC law, the national courts are furthermore the principal courts to address allocation decisions of the national governments, with of course the possibility that the national courts could refer a preliminary question to the ECJ.

The overview of case law provided by Van Aken shows that industries directly addressing the European Court of First Instance regarding their concerns towards allocation issues have not been successful. Van Aken shows that the Court of First Instance, followed by the European Court of Justice provided a very restrictive interpretation of the requirement that one has to be 'individually concerned' to be able to lodge an appeal against a decision of the Commission regarding the national allocation plan. This jurisprudence is however in line with the case law of the ECJ. Formally the Commission's decision is of course only affecting the member state which drafted the NAP. However, it can be clear that in practice the enterprises involved will have a large interest at stake in the decision of the Commission concerning the NAP of their particular member state. Still Van Aken shows that the CFI and ECJ apparently assume that it has to be the member state that acts as 'advocate' of the interests of industry in their state, or, to put it differently, to assess the legality of the consequences from a national allocation plan or decision to the specific positions of industries and other sources. As a result only the member states and not individuals (like industry) are allowed to bring a case to the Court of First Instance challenging a decision of the Commission concerning a NAP. Consequently, the industries need to address national courts when they want to fight the national allocation plan and the subsequent national allocation decision. However, as we will show below, in some cases national courts (more particularly in The Netherlands) all too quickly seem to follow the agenda of the policy maker (thus refusing to examine the possibility of *ex post* adjustments seriously) whereas at European level the Court of First Instance adopts a more flexible approach (thus allowing *ex post* adjustments).

Van Aken calls in his concluding remarks for an increasing importance of the role of the courts in the struggle against global warming and in this respect he refers *inter alia* to the well-known US Supreme Court case of *Massachusetts v Environmental Protection Agency*,[3] even though the context

3 For a discussion see e.g. Smith (2007) and Reitze (2007).

of that decision was of course a totally different one. It is indeed interesting to ask the question to what extent national and European courts can play an important role not only in deciding issues of subsidiarity, but also on distributional issues thereby discussing proportionality and the principle of equality, and for example, also promoting public participation during the allocation process. Indeed, an aspect stressed by some contributors as well is that there is serious doubt on the democratic nature of the emissions trading instrument in general and of the way it has been implemented within the EU in particular. The analysis provided by Peeters in chapter 2 shows that individuals concerned (in this respect referring to, for example, NGOs or potential victims) have in fact less possibility to intervene in an emissions trading scheme, which follows from the nature of the instrument combined with the nature of the pollution problem at hand. Especially when compared to other more traditional (command and control) type of instruments for traditional (local) problems, public participation is usually provided case-by-case, connected to the permit procedure for a specific industrial site. Furthermore, when using a market-based instrument, the decision about how much will be polluted by a certain industry is deliberately left to the market, which naturally excludes public participation. Given the specific nature of emissions trading, public participation is only made available, and, importantly enough, is even more obliged by the EU directive regarding the national procedures concerning the national allocation plan and the national allocation decision. However, strikingly enough, the present assessments of emissions trading thus far do not give serious attention to the use and effectiveness of public participation. It thus seems that there is a one-sided attention to economic and environmental effectiveness, and less to other values like democratic decision-making which, taken in a broad sense, also includes public participation. However, as soon as public participation requirements are breached, courts can play a meaningful role by enforcing them. A more fundamental research question, to be taken in hand in future research, is however to what extent public participation indeed could lead to improvement of the climate change policies, especially in the case of the emissions trading instrument.

4. LEGAL DESIGN: *EX POST* ADJUSTMENTS

The question whether the allocation system within an emissions trading scheme is either final or can be subject to adjustments *ex post* is of course largely connected to the allocation mechanism chosen. The main question is for what reason *ex post* adjustments would be justified. Clearly enough, when emissions decrease because of technological investments stimulated by the market price of the tradable rights, the operator should be rewarded by being able to sell allowances. However, in case of a change of production level, the situation is different. Especially when the allocation of allowances has been based on a forecast of production levels, there is a clear incentive for industries to present higher production levels than actually will occur. For such situations, a dynamic perspective would fit better than a static approach. When we take such a dynamic perspective, it is not hard to argue that when factual circumstances change, in particular when it appears that production levels are much lower than predicted, an *ex post* adjustment could be necessary. Such *ex post* adjustments could either be downward or upward. A downward adjustment then refers to the fact that allowances would be reduced (for example if production levels fall dramatically below the predicted levels) or upward (thus granting industry additional allowances given a change in factual circumstances).

The chapters dealing with this *ex post* adjustment (and more particularly chapter 7 which is entirely devoted to this phenomenon) make clear that, perhaps rather surprisingly, the European Commission is strongly opposed to these *ex post* adjustments. The Commission argued that when the operator is aware that any fall in production diverging from his own forecasts will be penalized with the application of *ex post* adjustments his incentives to reduce production freely would be negatively affected (see the arguments of the Commission summarized in chapter 4 by Van Aken). Backes, Deketelaere, Peeters and Schurmans discuss in detail in chapter 7 how national law in more particularly Belgium, Germany and The Netherlands deals with *ex post* adjustments and also show that this is in fact a broad notion which can cover diverging situations. They discuss an interesting case before the Dutch Council of State where industry pleaded in favour of *ex post* adjustments. A request to ask for a preliminary ruling of the European Court of Justice was rejected. This is rather remarkable in light of the fact that, as is shown both in chapter 4 by Van Aken and in chapter 7 that the European Court of First Instance on 7 November 2007 held, against the Commission, that downwards *ex post* adjustments (as these were proposed by Germany) could not have been rejected by the Commission. Interestingly, in the original German NAP not only downward but also upward (favouring industry) adjustments were foreseen. Thus one can speculate that to the extent a cap-and-trade regime with free allocation remains in place *ex post* adjustments related to actual production levels would (differently from what the Commission argued) be allowed in the future. Interestingly Bazelmans discusses in chapter 11 a few other emissions trading schemes, among which is one applicable in Switzerland which allows an adjustment of emission target each year according to the companies production growth.

5. EFFECTIVENESS OF THE EU ETS?

One always has to be very careful when attempting to assess the effectiveness of a regulatory instrument or measure since opinions on the definition of an effectiveness test already could easily diverge. Moreover, a distinction should of course be made between on the one hand the effectiveness of emissions trading in general and of the EU ETS in particular. It may well be that, indeed, given the design of the EU ETS (more particularly cap and trade and free allocation), criticism would be possible about the particular way in which ETS has been shaped in the EU directive, but that should not necessarily be an argument against the instrument as such. Generally, theoretical studies and empirical analysis (of course based on other case studies) is rather enthusiastic concerning emissions trading as an environmental policy tool, albeit that sometimes questions are asked with respect to the suitability of emissions trading for the particular case of developing countries (Faure, Peeters and Wibisana, 2006). Also on the particular design of emissions trading in the EU directive much criticism has been formulated both by economists (Endres and Ohl, 2005) and by lawyers.[4] If one were to narrow down the broad effectiveness question somewhat one could ask to what extent the EU ETS has contributed to achieving a behavioural change in industry as a result of which CO_2 emissions would be reduced. Here, we need to note that the first period of the EU ETS did not take place under the firm condition of an international emissions reduction obligation, as is the case in the first Kyoto emissions reduction period that runs from 2008 till 2013.[5]

Many contributions provide at least some indication of the general effectiveness of the EU ETS (in addition to more detailed criticism with respect to the legal design referred to above). Kuik and Oosterhuis make clear in chapter 8 that *ex ante* many predictions were formulated based on a variety of models and assumptions concerning the economic effects of implementing the ETS in Europe. Based on an evolution of the price of allowances it is not hard to argue that member states have (of course as a result of exaggerated predictions by industry) largely over-allocated allowances. Kuik and Oosterhuis show the evolution of the price of allowances that, from 2007 on, resulted in a price lower than one Euro. Indeed, many stress that caps were probably set

---

4 See for a critical discussion Mehling (2005). See furthermore different contributions discussing complexities of the EU ETS in Peeters and Deketelaere (2006).

5 Note that one could of course question the effectiveness (or efficiency) of the Kyoto Protocol as such, as has, for example, been done by Victor (2001). Most contributors to this book have, however, accepted the Kyoto goals as given and have thus examined to what extent the EU ETS can help in achieving the reduction commitments agreed to in the Kyoto Protocol.

in relation to business-as-usual projections so that over-allocation should not necessarily come as a surprise. Also Woerdman, Clò and Arcuri come to the same conclusion as Kuik and Oosterhuis, that there has been a substantial over-allocation, as they show convincingly with an empirical analysis.

Less clear is, however, whether the over-allocation as such means that the EU ETS therefore necessarily has been totally ineffective in providing incentives for emissions reductions. The 'over-allocation' could also be (partly) the result of investments in technological and other innovations causing emissions reductions. In that respect the introduction of the EU ETS would have had a desired incentive effect of reducing emissions. That emissions were lower than predicted should then not necessarily be considered as a negative outcome, but may well have been the result of the introduction of the EU ETS. Kuik and Oosterhuis refer indeed to studies showing that the EU ETS led to an additional abatement of between 50 and 200 million tonnes. They also refer to studies showing that the EU ETS played a key role in long-term decisions of companies to develop innovative technologies, with more particularly a strong impact on the steel industry. Hence, these considerations show that one can not, from the simple fact that the price of an allowance dropped below a dramatically low one Euro level at the beginning of 2007, conclude that the EU ETS would have had no incentive effect on innovation and thus would in that sense be ineffective.

Still, notwithstanding these relatively enthusiastic evaluations concerning the effectiveness of the EU ETS one has to remain cautious in generalizing the conclusion that the EU ETS would have been an effective instrument in reaching the overall emissions reduction targets. One problem is that by merely analysing whether there has been any effect on industrial behaviour as far as reducing emissions is concerned one does not take into account whether additional and further-going reductions could equally have been achieved with other instruments (like, for example, taxation). The conclusion that the EU ETS is effective (in the sense of having some effect) therefore does not imply that it is optimal. Moreover, the fact that the EU ETS has had any effect does not take into account the cost at which the EU ETS has been implemented. Peeters mentions in chapter 2 the costs of the national allocation procedures and recently Baldwin also argued that the transaction costs of trading systems can be high, referring to the EU ETS as 'an administrative nightmare'. This expression was actually also used by one of the most important civil servants of the office of the European Commission dealing with the assessment of the national allocation plans.[6] Without taking into account the relative costs of EU

[6] Baldwin (2008, p. 15) and the presentation of Mr Jos Delbeke on the conference 'The challenge of implementing new regulatory initiatives: state of affairs and critical issues of EU Climate Change Policy', 17 & 18 September 2004, Leuven.

ETS one can therefore not argue that the system, notably when *a free allocation* has been used, would be cost-effective as well.

Finally, as many contributors show, even though there are indications that the EU ETS has been effective this does not at all mean that there is no room for improvement, thus increasing the effectiveness. Many issues were already mentioned as far as the legal design is concerned and will be further discussed when analysing alternatives and the way forward. See in this respect, for example, the concluding remarks in chapter 11 by Bazelmans arguing that the cost-effectiveness of the EU ETS can be improved by linking it to other ETSs.

6. WHY EUROPE?

Many chapters in the book devote attention to the crucial question of why an emissions trading system needs to be completely regulated at the EU level and if so, which issues can still be left to member states. It is the classic issue which by lawyers is referred to as the subsidiarity question and which economists classify as issues of 'environmental federalism'. Both refer to the question of what particular legal design issues have to be taken care of at either central or decentralized level within federal systems like the EU. There are many quite diverging arguments both criticizing centralization and decentralization in the EU ETS in the various contributions to this book, of course depending upon the different perspectives one takes. It is striking that whereas in general economists show themselves to favour a differentiation of legal standards (respecting differing preferences) and lawyers seem to favour more harmonization (arguing that harmonized rules lower transaction costs) the division seems to be the other way around as far as the EU ETS is concerned.

Indeed, when addressing traditional economic arguments in the so-called environmental federalism debate it is not difficult to argue that climate change constitutes undoubtedly a transboundary externality which therefore calls for regulation at a centralized level to avoid free riding problems resulting from externalization of harm to other legal systems. However, at the same time economists usually argue that some legal issues can to a large extent be left to the member states, thus reducing the need for centralization as much as possible.

However, as far as the division of labour between member states and the EU in the EU ETS is concerned economists seem to be remarkably critical. This concerns more particularly the complicated division of labour whereby member states according to article 11 of the directive decide upon the total quantity of allowances and draft national allocation plans on the basis of article 9, but where the European Commission subsequently has been given the power to reject the plan or any aspect thereof. The fact that member states have received the power to decide on the allocation of allowances to the operator of each installation and to draft national allocation plans has of course provided ample scope for lobbying by national industry, subsequently leading to a generous allocation of allowances to national industries. This is why, so Kuik and Oosterhuis *inter alia* argue in chapter 8, one should not be surprised that over-allocation followed. A final conclusion about the optimal choice between decentralization and centralization of free allocation of allowances should of course only be formulated by taking into account the effects of lobbying at the EU level, which is also abundantly present. The latter phenomenon is even more important to follow because of the proposed shift of decision-making to the EU level, with some important choices left to the Commission, like the determination of sectors exposed to international competition and the adoption of a regulation on auction.

Furthermore, other economists have also held that the fact that the EU directive is unspecific in many respects and leaves many decisions defining the rules of the game to the individual member state has resulted in uncertainty and heterogeneity which increases transaction costs and thereby hampers the effectiveness of the system (Endres and Ohl, 2005). However, some of the more legally oriented contributions are remarkably critical of the proposal of the Commission to award large powers to the EU. Peeters, for example, discusses the emissions trading instrument from a viewpoint of democratic accountability, where she distinguishes between the core question 'what may be polluted during a certain period in a certain region?' and the complementary question 'which sector needs to reduce how much?'. Within the specific legal system of the EU, with the concept of shared competences and the principle of subsidiarity, it needs to be considered on what level those distinctive questions should be answered. The initial EU ETS provides a quite clear arrangement regarding both questions, and it would as such have been possible to proceed with that for some years: on the EU level there would then be a discussion of the commitments of states contributing to the overall reduction target of, for instance, minus 20% or 30% in 2020, while the national governments and parliaments then would decide about the distribution of efforts between the EU ETS sector and other sectors. In this form, the democratic accountability seems transparently arranged: the EU level will then focus on commitments among states, while the states themselves focus on commitments among sectors. The proposal of the Commission to revise the EU ETS implies however an important shift of decision-making from member states to the EU level, which means that the distributional question cannot be comprehensively addressed any longer by one legislator: both the European as well as the national legislators will be responsible, each for their part, for the distributional choices made. The push for a level playing field thus seems to win from getting a broader experience with different national approaches for allocation issues, and, moreover, from leaving important distributional choices to national democratic processes. Moreover, as far as industries are covered by an EU-wide cap, the adoption by a member state of further-going measures to the covered industries would be meaningless as the saved emissions could be simply used by other industries, under the same cap. As it seems realistic that a shift of decision-making to the EU level indeed will be adopted following the proposal of the Commission, Peeters raises the question whether an EU-wide greenhouse gas emissions trading scheme should be complemented by other means that would give citizens an alternative say in climate change policies, for instance through labelling.

Also De Cendra De Larragán is using a principles approach to the harmonization issue that is quite critical in chapter 3 concerning the large allocation of powers to the EU level. He argues that setting both the cap and the trading rules at EU level seems justified on the basis of legal principles. On the other hand, he argues that harmonization cannot as such ensure equal treatment within sectors, as the Commission suggests. He therefore concludes that, on the basis of the principles of subsidiarity and equality, it is not so clear that a completely harmonized approach to allocation rules in this respect is really necessary.

Interestingly, to some extent the case law, more particularly of the EU Court of First Instance, has relied upon the subsidiarity principle more particularly to argue that the member state may keep a large discretion when it transposes the directive into its national legal order and can thus for example also argue that *ex post* adjustments are necessary. Both Van Aken in chapter 4 and Backes, Deketelaere, Peeters and Schurmans in chapter 7 see this decision of 7 November 2007 of the Court of First Instance as a confirmation of the subsidiarity principle and of the particular nature of a directive, which is binding as to the result to be achieved, but not as far as the choice of forms and methods is concerned. The latter is explicitly left to the member states.

Notwithstanding the criticism of the lawyers on the large EU involvement and shift to the central level it seems that their voices will not be heard since the proposal of the Commission to amend the directive even suggests that the decision about the total amount of allowances available for each sector would be decided at EU level. Peeters criticizes this proposal since it leaves hardly any room for political debate on the national level. She argues that democratic accountability would be served if national governments and parliaments were to decide about the distribution of the efforts and the tools. This would mean that the EU level would still focus on commitments among states, while the states themselves would focus on the distribution of the commitments among the sectors. However, given the growing tendency of always allocating more powers to the European level it can be doubted whether this criticism will convince the European Commission. However, also with respect to auctioning, some appreciation goes to a decentralized approach, as is indicated by Weishaar in chapter 13: in contrast to a strong degree of harmonization, the decentralized approach allows for more appropriate auction mechanism designs and the realization of the full potential of auctions.

It is striking that this different approach towards the subsidiarity issue by lawyers and economists also reflects differing concerns: economists stress the major danger that as a result of lobbying by national industry national member states would over-allocate allowances to their industry and hence have little incentive for setting efficient standards. Lawyers on the other hand stress that the disadvantage of shifting too many powers to the EU level is precisely that powers are shifted away from the level where the polluting effects are caused, being the local or at least the member state level. Thus one of the key questions is obviously whether it is possible to find the 'best of both worlds', on the one hand constraining powers of member states or at least providing them with incentives not to over-allocate emissions, and on the other hand still keeping decision-making as low as possible to the level where effects are felt. This could perhaps be achieved by letting the EU level decide on the commitments among states (thus avoiding that over-allocation would take place), but letting the member states themselves divide the burden between the sectors.

7. IMPORTANCE OF ENFORCEMENT

Many have already stressed that an emission trading scheme also needs a stringent enforcement. One could be tempted to think that a 'market-based' instrument like emissions trading needs less enforcement than traditional command-and-control approaches. Many have, to the contrary, stressed that for an emissions trading system to function efficiently it is of course required that adequate information is available on the amount of actual emissions by the particular enterprises. In case actual emissions exceed the available allowances the legal system must guarantee that enforcement measures against the violators are taken. If violators were to free ride and could get off the hook without enforcement this would, of course, jeopardize the entire emission trading system since incentives to obtain allowances would be gone if those who emit without having the necessary allowances could easily get away with it.[7] Peeters in chapter 2, but also many other contributors, stress that it is essential for any emissions trading scheme that guarantees should exist that no pollution can be caused unless this is to be covered by a tradable permit (or credit).

[7] See for example Baldwin (2008, p. 16 and p. 24), but also more specifically on the importance of enforcement in the EU ETS Peeters (2006a) and Peeters (2006b).

Efficient enforcement is crucial also for many other aspects of an emissions trading scheme. Bazelmans, for example, stresses in chapter 11 that linking various emissions trading schemes is also only possible when a mutual trust can exist with justification that no emissions will take place for which no allowance exists and that this rule will also effectively be enforced.

In some trading systems the need for monitoring and enforcement can be even stronger. Bluemel makes clear in his chapter 9 discussing the US that their main worry is to prevent so-called leakage of emissions between the states. With respect to GHG emissions associated with imported power Bluemel discusses the potential problem of so-called contract shuffling. Basically this means that load-serving entities could decide to purchase electricity from different types of electricity generators and then decide to allocate all the low GHG-emitting electricity in its generator portfolio to particular states with a GHG reduction policy. A possible solution to this problem of contract shuffling addressed by Bluemel is a so-called load-based cap-and-trade scheme. This, however, requires an emissions tracking system which can be problematic in the long run.

8. LEARNING FROM ALTERNATIVES?

In order to provide a modest attempt to look at the future of the EU ETS a few chapters explicitly discussed alternative regimes, also outside Europe. Bluemel devotes the entire chapter 9 to the US where the so-called problem of 'leakage' mainly plays since, in the absence of a federal greenhouse gas trading regime, separate states have installed state-based trading schemes. The problem of leakage addressed by Bluemel plays strongly in the US where industry could decide to migrate to states not regulated by the GHG regime and subsequently export energy or GHG-intensive products back into the regulated regime. Bluemel discusses *inter alia* the case of California which imports 22–32% of its electricity, whereas those imports account for 39–57% of electricity-related CO_2 emissions. He shows that even though to some extent the leakage problem in the US may be overstated (since the economic incentive to avoid environmental regulation is small compared with other financial incentives) many design options have been worked out to reduce the risk of leakage. Paying attention to those legal-technical solutions applied in the US for the leakage problem remains interesting for Europe and certainly for the international level as well. At first sight one could argue that Europe has less of a leakage risk precisely since there is an EU-wide trading scheme. However, given the relatively limited scope of the EU ETS there is still the risk of leakage in Europe as well if particular member states were to strongly regulate non-covered sectors whereas others did not. Here one recognizes the classic race to the bottom problem often discussed with respect to environmental regulation, and advanced as an argument in favour of centralization. Kuik and Oosterhuis also indicate that before the EU ETS entered into force some studies argued that with respect to particular sectors, more particularly aluminium production, a shift in production capacity to non-EU countries (and thus carbon leakage) could take place. They predicted this for the case of cement and steel. However, Kuik and Oosterhuis show in chapter 8 that *ex post* a direct carbon leakage (in the sense of relocation of companies) may not have taken place, but that this may have largely been due to significant over-allocation of allowances to the iron and steel sector.

Bluemel also rightly states that at the international level leakage will remain a potential difficulty as long as some jurisdictions are covered under the regime whereas others are not. Hence, some of the devices worked out in the US to deal with the leakage problem may also be worthwhile to study as potential solutions for leakage at the international level.

Some contributions also discuss other emissions trading schemes. Bazelmans discusses many of those other schemes, among which are the ETSs in Canada, Australia, New Zealand and Switzerland as well as the voluntary ETS in Japan. A further comparison between those alternatives and the EU ETS may well provide valuable scope for learning. That these comparisons provide interesting insights is also shown in chapter 10 by MacDonald and Makuch dealing with the situation in the UK. For example they mention that the national UK ETS has had as important (additional) benefit in that it raised environmental and climate change-related awareness amongst the public both in the UK and elsewhere. In addition the case of the UK remains interesting for the remarkable and well-known suggestion to come for a so-called personal carbon trading, discussed in the concluding remarks of chapter 10.

When discussing alternatives one can obviously also look at alternatives for particular design issues such as, for example, the allocation problem. We already mentioned above that many contributors discuss a so-called credit and trade (or PSR) system. It is also discussed by Peeters in chapter 2 in connection with the previously mentioned problem of carbon leakage. The risk of a relocation (and thus carbon leakage) is far more serious in a cap-and-trade system (where a relocation could take place to countries where such a cap does not exist) than in a PSR system. Peeters thus sees avoiding carbon leakage as yet another argument in favour of the PSR model.

9. EMISSIONS TRADING VERSUS ALTERNATIVES

Many chapters also discuss the particular characteristics of an emissions trading system in general and the EU ETS in particular to reduce CO_2 emissions.

Emissions trading generally has, as is for example stressed in the contribution by Kuik and Oosterhuis in chapter 8, been advocated by economists as a more cost-effective instrument to reduce emissions in comparison with the command-and-control approach. However, there are still many contributors to this book who stress that, more particularly given particular weaknesses in the legal design of the current EU ETS (more specifically free allocation and over-allocation by member states) one could also have examined whether the alternative of taxation of CO_2 emissions could achieve similar or even better results than emissions trading. Pigovian taxes are indeed traditionally advanced by economists as the instrument to internalize environmental externalities. Peeters for example mentions the possibility of a tax to be applied on the global level in chapter 2 and MacDonald and Makuch discuss environmental taxation in the UK in chapter 10. The UK has indeed introduced a so-called climate change levy of which the revenues are recycled back to business through a 0.3% cut in employers' national insurance contributions. However, not surprisingly, MacDonald and Makuch argue that concerns have arisen in relation to the competitiveness of UK industry relative to other countries that do not impose such a levy. It has of course also been the reason why Europe has been hesitant to impose a carbon tax at EU level. MacDonald and Makuch quote many representatives of industry in the UK who argue that the climate change levy is simply a tax which makes industry uncompetitive. They therefore propose the alternative of a global energy tax which would represent a global and fair burden-sharing system.

Particularly attractive in the UK system is that the tax as such does not increase total costs for business since the revenues are recycled back to industry. An interesting combination of ETS and taxation existing in Switzerland is discussed by Bazelmans in chapter 11 where companies which participate in an ETS are exempted from the domestic CO_2 tax.

However, one should of course also be careful with introducing taxation as the golden solution to reduce CO_2 emissions (if it were at all politically feasible to introduce those). Baldwin for example stresses, as other economists did, that a cap on emissions in a cap-and-trade ETS has at least the advantage that it provides more predictable outcomes since the overall level of emissions is fixed. In a taxation system, to the contrary, emission levels will be contingent on individual firms' cumulative responses to incentives, which makes the outcome (in terms of reducing CO_2 emissions) less predictable.[8]

Another briefly presented alternative instrument to be used, not instead of but probably in addition to emissions trading is the so-called carbon label presented by Peeters in chapter 2. Such a label could unveil the carbon load of

8 Baldwin (2008, p. 6).

a product and service and could be a useful complementary instrument. Particularly interesting also is the discussion in chapter 10 by MacDonald and Makuch of the role of so-called climate change agreements (CCAs) for energy-incentives sectors. These agreements have, so they argue, in absolute terms over-achieved in each target period by nearly double the absolute target compared to initial targets. They also discuss an interesting but complex interaction between the EU ETS and these CCAs (whereby installations were given the opportunity to opt out of the EU ETS until the end of phase I) as well as the interesting relationship between the CCAs and the above-mentioned climate change levy (a tax). Participation in a CCA leads to a 80% discount from the climate change levy for those sectors that agreed to meet certain targets for improving the energy efficiency or reducing carbon emissions.

Notwithstanding the fact that on paper it is hence possible to present alternatives or complements to an ETS, it is not at all certain that these theoretical predictions will be followed. Many contributors stressed that both the choice for an ETS rather than for a tax system as well as the particular legal design chosen in the ETS (grandfathering and allocation by the member states) was to a large extent the result of effective lobbying by industry. Industry has (as also shown in chapter 10 by MacDonald and Makuch) typically always a strong anti-taxation agenda. Hence, one should not be surprised that, also with a view to the future, industry may continue its lobbying efforts to reduce its financial burdens. It therefore still remains to be seen whether, as Peeters argues in chapter 2, the proposal to change to auctioning will survive in the legislative process since this would of course substantially raise costs for industry. Moreover, even though many contributors to this book proposed the theoretically ideal solution of a carbon tax one does not have to be a great public choice scholar to predict that lobbying efforts by industry are very likely to prevent this from happening. Given these political realities one could even (perhaps somewhat cynically) argue that emissions trading is still the best feasible instrument to reduce CO_2 emissions.

10. THE WAY FORWARD

The various contributions to this book have both identified theoretically different ways of dealing with reduction of greenhouse gas emissions or effectively designing an ETS. In addition the contributions have equally made clear what specific plans and proposals are to reform the EU ETS as presented in various ways by the European Commission. Even though at the time of finishing the contributions (April 2008) it remains difficult to predict, for example, whether specific legislative proposals will survive the European legislative process or how the implementing decisions, like the regulation on the timing, administration and other aspects of auctioning will look, a few indications can be given on the basis of the contributions. Peeters summarizes most of the proposals for revision contained in the Commission's proposal to revise the EU emissions trading system which was presented on 23 January 2008. Some of these proposals have already been discussed earlier and refer, for example, to a shift of decision-making to the EU level whereby an EU-wide cap for the installations covered by the EU ETS will be installed, thus fundamentally removing member states' discretion. A consequence of this revision (when it takes place) is undoubtedly that again more powers will be allocated to the European level, thus (as we mentioned above) raising more questions about democratic accountability and possibilities for public participation. Another fundamental change (if it survives the legislative process) is the radical change in the allocation process being that auctioning rather than grandfathering would become the main allocation mechanism. Only for internationally operating industries who strongly suffer from competitive pressures would free allocation still be possible. The latter option is of course again inspired by the desire to avoid carbon leakage. Many of the proposals for revision are also critically discussed and evaluated from the perspective of legal principles by De Cendra De Larragán in chapter 3. Furthermore, the contribution of Weishaar in chapter 13 instructs us about the complexities involved in designing a large-scale auction scheme.

Two chapters deal with specific proposals for revision, being linking the ETS to other ETSs and including emissions caused by aviation. Bazelmans also discusses the importance of linking in the case of the current revision of the EU ETS in chapter 11, arguing that harmonization of the design issues of the various systems remains important to guarantee an optimal use of interdependencies between those trading systems.

Kaminskaite-Salters discusses in chapter 12 the expansion of the EU ETS to the aviation sector. She pays attention both to the 2006 EU proposal to include international aviation in the EU ETS and the legal consequences of this as well as to the interesting question of the interference with international conventions concerning international civil aviation.

As we already mentioned, there is still a lot of uncertainty concerning the way in which the Commission will take the way forward with the EU ETS. Some of the proposals which have been presented are supported by the contributors to this book, like the option to move to auctioning as main allocation mechanisms, but the difficulties with designing such a scheme are equally addressed. On other issues, like the shift of more powers to the EU level, the contributors are somewhat divided. Moreover, proposals have been formulated in the chapters for alternatives whereby, on the one hand, the cap could be set at the EU level (thus avoiding over-allocation to national industry by member states, but now taking away the incentive for member states to adopt further-going measures for the sectors covered by the EU ETs, as was addressed by Peeters in chapter 2). A difficulty with the revision process is, moreover, as was mentioned above, that in theory one would wish the policy maker to take a long-term perspective. There is an emerging trend to refer to the year 2050, whereby the question can of course be asked to what extent it is wise, fair and acceptable to take decisions now that may bind future generations as well.

There are these and many other challenges that need to be addressed when revising the EU ETS. We hope to have contributed, with the many insights included in this book, to the interesting debate on the optimal design of an emissions trading scheme, specifically where it concerns greenhouse gases. We at least hope to have convinced the reader that an integrated legal and economic approach, also paying specific attention to legal design issues, is of utmost importance in guaranteeing the effectiveness of a greenhouse gas emissions trading scheme, and any other emissions trading scheme too.

REFERENCES

Baldwin, R. (2008), 'Regulation lite: the rise of emissions trading', Law Society Economy Working Papers 3/2008 (www.lse.ac.uk/collection/law/wps/wps.atm).

Endres, A. and C. Ohl (2005), 'Kyoto, Europe? An Economic Evaluation of the European Emissions Trading Directive', *European Journal of Law and Economics*, 19, 17–39.

Faure, M., M. Peeters and A. Wibisana (2006), 'Economic Instruments: Suited to Developing Countries?', in Michael Faure and Nicole Niessen (eds.), *Environmental Law in Development. Lessons from the Indonesian Experience*, Cheltenham, Edward Elgar, pp. 218–62.

Mehling, M.A. (2005), 'Emissions Trading and National Allocation in the Member States: an Achilles Heel of European Climate Policy', in Thijs F.M and Etty, Han Somsen (eds.), *The Yearbook of European Environmental Law*, volume 5, Oxford, Oxford University Press, pp. 113–56.

Nash, J.R. (2000), "Too Much Market? Conflict between Tradable Pollution Allowances and the "Polluter Pays Principle"', *Harvard Environmental Law Review*, 24, 465–535.

Peeters, M. (2006a), 'Inspection and Market-Based Regulation Through Emissions Trading: The Striking Reliance on Self-Monitoring, Self-Reporting and Verification', *Utrecht Law Review*, 2(1), June.

Peeters, M. (2006b), 'Enforcement of the EU Greenhouse Gas Emissions Trading Scheme', in Kurt Deketelaere and Marjan Peeters (eds.), *EU Climate Change Policy: The Challenge of New Regulatory Initiatives*, Cheltenham, Edward Elgar, pp. 169–87.

Peeters, M. and K. Deketelaere (eds.) (2006), *EU Climate Change Policy. The Challenge of New Regulatory Initiatives*, Cheltenham, Edward Elgar.

Reitze, A.W. (2007), 'Controlling Greenhouse Gas Emissions from Mobile Sources: Massachusetts v EPA', *Environmental Law Reporter*, 37, 10535–39.

Rodi, M. (2005), 'Legal aspects of the European emissions trading scheme', in Bernd Hansjürgens (ed.), *Emissions Trading for Climate Policy, US and European Perspectives*, Cambridge, Cambridge University Press, pp. 199–221.
Smith, J.A. (2007), 'Massachusetts v EPA: The Way Forward on Climate Change Regulation in the US', *Environmental Liability*, 3, 127–32.
Tietenberg, Th.H. (2006), *Emissions Trading, Principles and Practice*, second edition, Washington DC, Resources for the future.
Victor, D. (2001), *The Collapse of the Kyoto Protocol*, Princeton, New Jersey, Princeton University Press.

Index